Plasma Dynamics

Applications of Mathematics Series

Editor: Alan Jeffrey
Professor of Engineering Mathematics
University of Newcastle-upon-Tyne

WILLIAM F. AMES
Numerical Methods for Partial Differential Equations

T. J. M. BOYD and J. J. SANDERSON
Plasma Dynamics

C. D. GREEN
Integral Equation Methods

I. H. HALL
Deformation of Solids

JEREMY HIRSCHHORN
Dynamics of Machinery

ALAN JEFFREY
Mathematics for Engineers and Scientists

BRIAN PORTER
Synthesis of Dynamical Systems

Plasma Dynamics

T. J. M. BOYD
University College of North Wales

J. J. SANDERSON
University of St. Andrews

Barnes & Noble, Inc.
New York
Publishers ● Booksellers ● Since 1873

First published in Great Britain 1969 by
THOMAS NELSON AND SONS LTD

First published in the United States of America 1969 by
BARNES & NOBLE, INC.

Copyright © T. J. M. Boyd and J. J. Sanderson 1969

155797

Printed in Great Britain

Contents

Preface

This book is intended as an introduction to plasma dynamics for students at advanced undergraduate or graduate level. The approach to plasma dynamics begins with a description of the orbits of charged particles in electromagnetic fields and is followed by a treatment of hydromagnetics which includes discussions of flows, shocks and wave motion. Chapters on radiation processes and plasma kinetic theory complete the text. A feature of the presentation is the comparison of experimental results with predictions from theory.

A background knowledge of electrodynamics and hydrodynamics is a necessary prerequisite. The mathematics needed is elementary and is for the most part in the normal repertoire of undergraduate applied mathematicians, physicists, astronomers and electrical engineers; for completeness, some results on tensors and Bessel functions are included in appendices. We have adopted Gaussian c.g.s. units throughout the book since these are used widely in the literature; there is a short appendix on m.k.s. units.

Since this book is intended as an introductory text no attempt has been made to provide an exhaustive list of references; in particular we are well aware that the many Russian contributions to the plasma physics literature are not reflected in our list. Moreover, in our references to experimental work we have tended to discuss that which is best known to us.

Most of the material in this book has been presented in lecture courses on electrodynamics and hydromagnetics to students at the University of St. Andrews over the past few years and we are indebted to many of them for their questions and comments which have helped to clarify a number of obscurities. We are especially grateful to S. P. Gary for reading and criticizing several chapters of the book and to J. G. Turner for checking many of the examples. Various colleagues have helped us by reading parts of the manuscript and we would like to record our gratitude to A. D. D. Craik, T. R. Carson, I. Fidone, D. E. Evans, A. Hare, S. Abas, D. C. Montgomery and G. Rowlands. Professor Alan Jeffrey provided not only the stimulus to embark on this task but has given us constant encouragement in carrying it through. We are indebted to him, and to our publishers, for their patience. We are grateful also to many people at the University of St. Andrews and the University College of North Wales for their assistance in the preparation of the manuscript, particularly to Miss K. P. Dunne of St. Andrews.

T. J. M. B.

J. J. S.

1

Introduction

In its most general sense a *plasma* is any state of matter which contains enough free, charged particles for its dynamical behaviour to be dominated by electromagnetic forces. Plasma physics therefore encompasses the solid state, since electrons in metals and semiconductors fall into this category; however, the subject is largely concerned with ionized gases. Very low degrees of ionization are sufficient for a gas to exhibit electromagnetic properties; at about 0·1 per cent ionization a gas achieves an electrical conductivity of the order of half its possible maximum and at 1 per cent ionization the conductivity is nearly that of the fully ionized gas.

The sun and stars are hot enough to be almost completely ionized and interstellar gas is also ionized, due to the action of stellar radiation. This means, in fact, that almost all matter in the universe can be described as plasma.[1] In this respect the earth is exceptional; the natural occurrence of plasma on the earth's surface is non-existent (except for transient plasmas caused by lightning). However, the upper atmosphere of the earth is ionized, this region being known as the *ionosphere*; still farther out, the earth's radiation belts (the *Van Allen belts*) consist of electrons and protons trapped by the earth's magnetic field.

It was in terms of the plasma properties of the ionosphere that the transmission of radio waves across the Atlantic was explained. This led eventually to the magneto-ionic theory of Appleton and Hartree[2] which was amongst the earliest work on plasma wave propagation. About the same time as the development of magneto-ionic theory came the first observations of plasma oscillations. This work was carried out principally by Tonks and Langmuir;[3] it was Langmuir who first coined the name 'plasma' in 1928. Subsequently Vlasov and Landau contributed the kinetic description of plasma oscillations; this marked the beginning of plasma kinetic theory[4] as opposed to the earlier adaptations of Boltzmann's molecular gas theory.

More recently, research in plasma physics has been concerned with possible technological applications, particularly controlled thermonuclear fusion,[5, 6] but also the direct conversion of heat energy into electricity,[7] ionic propulsion for space flight,[8] and other possibilities. Some of the problems encountered in these applications (such as magnetic containment of plasma) are difficult enough, and the goals (such as more plentiful fuel) desirable enough, for research to broaden out into attempts at general and fundamental understanding of plasmas. The subject has thus remained a fertile field of research for physicists, applied mathematicians, and engineers.

One of the attractions of plasma physics is its common ground with so much of classical physics, ranging from particle dynamics through fluid mechanics to statistical mechanics. From a theoretical point of view the basic description of a plasma lies in the kinetic theory of matter. One defines a function of position, velocity, and time, $f(\mathbf{r}, \mathbf{v}, t)$, such that $f\, d\mathbf{r}\, d\mathbf{v}$ is the probability of finding particles within the 6-dimensional volume element $d\mathbf{r}\, d\mathbf{v}$, centred at the point $(\mathbf{r}, \mathbf{v})$ in coordinate and velocity space. Observable properties of the plasma can then be obtained from this function, known as the *distribution function*, by taking various velocity *moments* of f. For example, the number density of particles in the spatial volume element $d\mathbf{r}$ is

$$n(\mathbf{r}, t) = \int f(\mathbf{r}, \mathbf{v}, t)\, d\mathbf{v} \tag{1-1}$$

where the integration is taken over the whole of velocity space. Similarly the mean velocity of particles in $d\mathbf{r}$ is

$$\mathbf{u}(\mathbf{r}, t) = \frac{\displaystyle\int \mathbf{v}\, f(\mathbf{r}, \mathbf{v}, t)\, d\mathbf{v}}{\displaystyle\int f(\mathbf{r}, \mathbf{v}, t)\, d\mathbf{v}} = \frac{1}{n}\int \mathbf{v} f\, d\mathbf{v} \tag{1-2}$$

The density corresponds to a zero-order moment (f is multiplied by $\mathbf{v}^0$) while the mean velocity is a first-order moment (f is multiplied by $\mathbf{v}^1$); pressure and temperature are second-order moments and the heat flux vector is an example of a third-order moment.

The equation determining the distribution function is called the kinetic equation. The usual heuristic argument[9, 10] used to obtain this equation treats the macroscopic forces (such as the applied fields) separately from the microscopic, collisional forces. Suppose first of all that particle interactions may be neglected. Then considering some given volume in $(\mathbf{r}, \mathbf{v})$ space one argues that the number of particles in this volume will change with time and that the rate of change will be given by the flux of particles through the surface of the given volume. In coordinate ($\mathbf{r}$) space this flux is $\int\!\!\int \mathbf{v}.\hat{\mathbf{n}}_r\, dS_r$, where $\hat{\mathbf{n}}_r$ is the unit vector perpendicular to the element of surface dS_r and the integral is taken over the whole surface in $\mathbf{r}$ space; if $\hat{\mathbf{n}}_r$ is directed outwards from the volume the integral represents a net outflow of particles. Similarly, the net flux of particles out of the volume in velocity ($\mathbf{v}$) space is $\int\!\!\int \mathbf{a}.\hat{\mathbf{n}}_v\, dS_v$, where $\mathbf{a}$ is the acceleration experienced by particles in the velocity surface element dS_v with unit normal vector $\hat{\mathbf{n}}_v$, the integral being over the whole surface in $\mathbf{v}$ space. Hence

$$\frac{\partial}{\partial t}\int f\, d\mathbf{r}\, d\mathbf{v} = -\int f\mathbf{v}.\hat{\mathbf{n}}_r\, dS_r\, d\mathbf{v} - \int f\mathbf{a}.\hat{\mathbf{n}}_v\, dS_v\, d\mathbf{r} \tag{1-3}$$

Taking the time derivative inside the integral and using Gauss' theorem to convert the surface integrals into volume integrals, (1–3) may be written

$$\int d\mathbf{r}\, d\mathbf{v} \left[\frac{\partial f}{\partial t} + \frac{\partial}{\partial \mathbf{r}} \cdot (\mathbf{v}f) + \frac{\partial}{\partial \mathbf{v}} \cdot (\mathbf{a}f) \right] = 0 \qquad (1\text{–}4)$$

Since the volume in $(\mathbf{r}, \mathbf{v})$ space is arbitrary, (1–4) implies that

$$\frac{\partial f}{\partial t} + \frac{\partial}{\partial \mathbf{r}} \cdot (\mathbf{v}f) + \frac{\partial}{\partial \mathbf{v}} \cdot (\mathbf{a}f) = 0$$

and on observing that $\mathbf{r}$ and $\mathbf{v}$ are independent variables this may be written

$$\frac{\partial f}{\partial t} + \mathbf{v} \cdot \frac{\partial f}{\partial \mathbf{r}} + \frac{\mathbf{F}}{m} \cdot \frac{\partial f}{\partial \mathbf{v}} = 0 \qquad (1\text{–}5)$$

where it has been assumed that the only velocity dependent forces, $\mathbf{F}$, are those due to magnetic fields, that is, of the form $e(\mathbf{v} \times \mathbf{B})/c$.

Now suppose one allows for particle interactions. In principle it is easy to deal with the average field due to the charged particles since this is given by Poisson's equation

$$\nabla \cdot \mathbf{E}(\mathbf{r}, t) = 4\pi q(\mathbf{r}, t) = 4\pi \sum_i e_i \int f_i(\mathbf{r}, \mathbf{v}, t)\, d\mathbf{v} \qquad (1\text{–}6)$$

where the index has been introduced to denote particle type and e_i is the charge on particles of type i. The field $\mathbf{E}$ may thus be included with the macroscopic forces $\mathbf{F}$. However, the rapidly fluctuating microscopic interactions which a particle suffers cannot be so simply represented and this is why 'collisions' are treated separately. One writes

$$\frac{\partial}{\partial t} \int f\, d\mathbf{r}\, d\mathbf{v} = \int \frac{\partial f}{\partial t}\, d\mathbf{r}\, d\mathbf{v} = \int \left[\left(\frac{\partial f}{\partial t} \right)_F + \left(\frac{\partial f}{\partial t} \right)_c \right] d\mathbf{r}\, d\mathbf{v}$$

where the first term in the square bracket represents the time rate of change due to the macroscopic forces $\mathbf{F}$ and the second term is the rate of change due to collisions. Then treating the first term as before, while leaving the second alone, leads to the equation

$$\frac{\partial f}{\partial t} + \mathbf{v} \cdot \frac{\partial f}{\partial \mathbf{r}} + \frac{\mathbf{F}}{m} \cdot \frac{\partial f}{\partial \mathbf{v}} = \left(\frac{\partial f}{\partial t} \right)_c \qquad (1\text{–}7)$$

This is the (collisional) *kinetic equation*. Of course, one still has to write down some mathematical expression for $(\partial f/\partial t)_c$ but, assuming this can be done in terms of f and the other variables occurring on the left-hand side of

(1–7), the kinetic equation determines the distribution function. The field $\mathbf{E}$ is often referred to as the *self-consistent field* since it depends on f and its evaluation therefore requires a self-consistent solution of the kinetic equation (for each species of particles) together with the *Maxwell equations* for the electric and magnetic fields; with (1–6) these are†

$$
\left.
\begin{aligned}
\nabla \times \mathbf{E} &= -\frac{1}{c}\frac{\partial \mathbf{B}}{\partial t} \\[2mm]
\nabla \times \mathbf{B} &= \frac{1}{c}\frac{\partial \mathbf{E}}{\partial t} + \frac{4\pi}{c}\mathbf{j} \\[2mm]
\nabla \cdot \mathbf{B} &= 0
\end{aligned}
\right\}
\tag{1–8}
$$

where the current density

$$
\mathbf{j}(\mathbf{r}, t) = \sum_i e_i \int \mathbf{v} f_i(\mathbf{r}, \mathbf{v}, t)\, d\mathbf{v}
\tag{1–9}
$$

The form of the right-hand side of (1–7) depends upon the assumptions one makes about collisions. The best known expression is the *Boltzmann collision integral* which is based on the dynamics of a binary collision and, for interactions between particles of types i and j, is given by

$$
\left(\frac{\partial f_i(\mathbf{r}, \mathbf{v}, t)}{\partial t}\right)_c = \int d\mathbf{v}' \int d\Omega\, \sigma(|\mathbf{v} - \mathbf{v}'|, \theta)\,|\mathbf{v} - \mathbf{v}'|
$$
$$
[f_i(\mathbf{r}, \bar{\mathbf{v}}, t)\, f_j(\mathbf{r}, \bar{\mathbf{v}}', t) - f_i(\mathbf{r}, \mathbf{v}, t)\, f_j(\mathbf{r}, \mathbf{v}', t)]
\tag{1–10}
$$

In this $\mathbf{v}, \mathbf{v}'$ are the initial velocities, $\bar{\mathbf{v}}, \bar{\mathbf{v}}'$ the final velocities, σ is the differential scattering cross-section, θ the scattering angle, and $d\Omega$ the element of solid angle $\sin\theta\, d\theta\, d\phi$. It is not appropriate to derive this expression here (this will be done in Section 10–7) but it may be made plausible by the following observations. A particle i approaches a particle j with a speed $|\mathbf{v} - \mathbf{v}'|$ and those entering the cross-sectional area σ scatter through the angle θ. Thus multiplying by the probability $(f_i f_j)$ of finding these particles and integrating over all angles and all $\mathbf{v}'$ gives the rate at which particles of type i are lost from the volume element at the point $(\mathbf{r}, \mathbf{v})$; this is represented by the term in $f_i(\mathbf{v}) f_j(\mathbf{v}')$. The term in $f_i(\bar{\mathbf{v}}) f_j(\bar{\mathbf{v}}')$ represents the gain in particles due to collisions and is arrived at in a similar manner, but requires the use of certain properties of collisions. With $(\partial f/\partial t)_c = \sum_j (\partial f/\partial t)_{c_j}$ the kinetic equation is known as the *Boltzmann equation*.[9–12]

---

† In plasma physics it is convenient to treat all current and charge densities explicitly; the Maxwell equations then appear in terms of $\mathbf{E}$ and $\mathbf{B}$ only and do not involve $\mathbf{D}$ and $\mathbf{H}$ (cf. Appendix 4).

One can see from this example that the collision term is in general a complicated non-linear function of f and it is the source of most of the mathematical difficulty in kinetic theory. Fortunately one may often neglect collisions completely. The mean time it takes for a particle to suffer an appreciable deflection due to collisions is called the *collision time* (τ_c) and its inverse is the *collision frequency* (ν_c). Thus a necessary condition for the neglect of collisions is $\nu_c \ll \omega$, where ω is a frequency which characterizes the time rate of change of the macroscopic forces. This condition could also be written in terms of the *mean free path* (the mean distance a particle travels between appreciable deflections); if λ_c is the mean free path the condition is $\lambda_c \ll L$, where the length L characterizes the distance over which the macroscopic fields vary. With the neglect of collisions (1–7) reduces to (1–5), which is known as the *collisionless Boltzmann equation* or *Vlasov equation*. However, even with this simplification the simultaneous solution of equations (1–5) to (1–9) is seldom less than formidable. From a pedagogical point of view it is desirable to start a description of plasma dynamics from imspler theories.

In Chapter 2 *particle orbit theory* is presented. Basically the theory is equivalent to the collisionless kinetic theory (cf. Section 10–4) but further simplifying assumptions are introduced. In essence one describes the plasma by calculating the orbits of single particles in given electromagnetic fields. It is fairly clear that orbit theory will not be applicable whenever distribution of charge or current plays a leading role in the dynamics; two further conditions are clearly that the induced fields must be small compared with the applied fields and the plasma must be collisionless. In spite of these restrictions, orbit theory gives a useful picture of plasma motion, particularly in strong magnetic fields where the basic particle motion (cf. Section 2–2) consists of rotation about the magnetic field lines with frequency $\Omega = eB/mc$ and radius $r_L = v_\perp/\Omega$ ($v_\perp$ is the component of the particle velocity perpendicular to the magnetic field **B**); the quantities Ω and r_L are known as the *Larmor frequency* and *Larmor radius*.

Like kinetic theory, orbit theory is a microscopic description; it is simpler by virtue of the assumptions made. However, an alternative method of gaining simplification is to derive a *macroscopic* theory. This is done in Chapter 3, in which a set of macroscopic equations is obtained from the kinetic equation in the same way that the macroscopic variables n and **u** are obtained from f, that is, by taking velocity moments. For example, it is not difficult to see, using (1–1) and (1–2), that the zero-order moment of (1–7) gives rise to the density equation

$$\frac{\partial n}{\partial t} + \mathbf{\nabla}.(n\mathbf{u}) = 0 \qquad (1\text{–}11)$$

The force term in (1–7) gives no contribution since $f \to 0$ as $|v| \to \infty$ and provided collisions do not change the number of particles (no ionization, recombination, etc.) $\int d\mathbf{v}(\partial f/\partial t)_c = 0$. In a similar way one obtains momentum and energy equations giving a *fluid* or *hydromagnetic* description (when collisions are dominant) and a wave description (if the fields are dominant).

Finally, by way of introduction one must mention some characteristics which are fundamental to the nature of a plasma and which therefore appear again and again in discussions of plasma physics. The first point to note is that a plasma maintains approximate charge neutrality. The reason for this is simply that any significant imbalance of positive and negative charge could only be maintained by huge electric fields. For example, a 1 per cent deviation from neutrality in even a diffuse plasma, say $n \sim 10^{11}/\text{cm}^3$, would give rise to a field

$$|\mathbf{E}| = \frac{4}{3}\pi r^3 \cdot \frac{n}{100} \cdot \frac{e}{r^2} \approx 600r \text{ volt/cm}$$

Taking $r \sim 1$ cm such a field would give an electron acceleration $\sim 10^{18}$ cm/s² so the imbalance would be very quickly neutralized by transfer of electrons.

This brings us to a second characteristic of plasmas. The movement of electrons to neutralize a charge inhomogeneity would be followed by oscillatory motion and it is of interest to calculate this natural frequency of the plasma. Suppose the ion and electron densities are given by

$$n^+ = n_0 \qquad n^-(\mathbf{r}, t) = n_0 + n_1(\mathbf{r}, t) \tag{1–12}$$

where $n_1(\mathbf{r}, t)$ is zero everywhere except in some small region and $|n_1| \ll n_0$. The electric field $\mathbf{E}(\mathbf{r}, t)$ which arises is directly proportional to n_1; this field then superimposes on the random thermal motion of the electrons a small flow velocity $\mathbf{u}^-(\mathbf{r}, t)$ which is also proportional to n_1. Thus the electron density equation (1–11) may be linearized to give

$$\frac{\partial n_1}{\partial t} + n_0 \nabla . \mathbf{u}^- = 0 \tag{1–13}$$

Similarly the momentum equation is

$$m^- \frac{\partial \mathbf{u}^-}{\partial t} = -e\mathbf{E} \tag{1–14}$$

where $\mathbf{E}$ is given by Poisson's equation

$$\nabla . \mathbf{E} = -4\pi n_1 e \tag{1–15}$$

Taking the partial time derivative of (1–13) and the divergence of (1–14) one finds on using (1–15)

$$\frac{\partial^2 n_1}{\partial t^2} + \frac{4\pi n_0 e^2}{m^-} n_1 = 0$$

Thus the density disturbance oscillates with a frequency

$$\omega_p = \left(\frac{4\pi n_0 e^2}{m^-}\right)^{1/2}$$

called the *plasma frequency*. Note that any applied fields with frequencies less than ω_p cannot penetrate the plasma since the more rapid plasma oscillations neutralize the applied field; thus a plasma is opaque to electromagnetic radiation having a frequency $\omega < \omega_p$.

Of course one cannot talk of plasma oscillations unless a large number of particles are involved in the motion; it is of interest therefore to examine the spatial range of the field set up by the charge inequality n_1. At first sight this may seem pointless since the Coulomb field varies as the inverse square of the distance from the region of charge imbalance. However, as we have seen already the situation is not a static one and the flow of electrons to counteract the charge inequality has the effect of 'shielding' the field and thus limiting its range. By a dimensional analysis of (1–14) and (1–15) the field extends over a distance

$$l \sim \frac{E}{4\pi n_1 e} \sim \frac{m^- \omega_p u^-}{4\pi n_1 e^2} = \frac{n_0 u^-}{n_1 \omega_p} \sim \frac{v_{th}}{\omega_p}$$

where v_{th} is the electron mean thermal velocity. Since $v_{th} \sim (\kappa T/m^-)^{1/2}$, where κ is Boltzmann's constant and T is the temperature,

$$l \sim \left(\frac{\kappa T}{4\pi n_0 e^2}\right)^{1/2} = \lambda_D$$

where λ_D is known as the *Debye shielding distance*, being first derived by Debye in electrolyte theory. The same result can be obtained from equilibrium statistical mechanics. In equilibrium the spatial distribution of the charge is given by the Maxwell–Boltzmann distribution[9] and so $n_0^- = n_0 \exp(e\phi/\kappa T)$, where ϕ is the potential due to the charge imbalance. Ignoring proton motion and taking $n_1 = n_1 \delta(\mathbf{r})$, where $\delta(\mathbf{r})$ is the Dirac delta function, Poisson's equation is

$$\nabla^2 \phi = -4\pi[en_0 - en_0 \exp(e\phi/\kappa T) - en_1 \delta(\mathbf{r})]$$
$$\simeq \frac{4\pi n_0 e^2}{\kappa T} + 4\pi en_1 \delta(\mathbf{r})$$

at distances such that $e\phi(r)/\kappa T \ll 1$. The solution is

$$\phi(r) = -\frac{en_1}{r} \exp(-r/\lambda_D)$$

that is, a *shielded Coulomb potential*.

It is clear that λ_D also gives the magnitude of the sheath thickness—the distance over which charge neutrality may not be valid near a boundary of the plasma. More important still, the Debye length is an effective range for collisions, the potential between charged particles being the shielded Coulomb potential rather than the Coulomb potential which would apply in a vacuum.

The number of particles involved in the plasma oscillations is clearly $\sim n_0 \lambda_D^3$, so the description of plasma characteristics we have given only makes sense if

$$n_0 \lambda_D^3 \gg 1 \qquad \lambda_D \ll L$$

where L is a length characterizing the dimensions of the plasma; these conditions, together with that of quasineutrality, are usually incorporated into the definition of a plasma.

We have shown that the effective range of a field, and hence of a collision, is λ_D. Thus any particle interacts at a given moment with the large number of neighbouring particles at distances $\leqslant \lambda_D$. Plasma collisions are therefore many-body interactions and since

$$\frac{e^2/\lambda_D}{mv_{th}^2} \sim \frac{e^2}{\kappa T \lambda_D} \sim \frac{1}{4\pi n_0 \lambda_D^3} \ll 1$$

collisions are predominantly weak. This provides a sharp contrast between a plasma and a neutral gas in which collisions are binary and strong. In the latter case one defines the collision frequency τ_c by

$$\tau_c = \frac{1}{nv_{th}\sigma(\pi/2)} \tag{1-16}$$

where $\sigma(\pi/2)$ is the total cross-section for a $\pi/2$ deflection. Such a deflection in a plasma would occur for particles interacting at a distance b_0 such that $e^2/b_0 \sim \kappa T$. Using formula (1–16) one would thus arrive at a collision time $\tau_c = (nv_{th}\pi b_0^2)^{-1}$. However, the cumulative effect of the much more frequent weak interactions is to reduce this by a factor $\sim 8\ln(\lambda_D/b_0)$; the plasma collision frequency is thus given by (cf. Section 10–9)

$$\nu_c \sim 8nv_{th}\pi b_0^2 \ln(\lambda_D/b_0) = \frac{\omega_p \ln(4\pi n \lambda_D^3)}{2\pi n \lambda_D^3} \equiv \frac{\omega_p \ln \Lambda}{2\pi \, n \lambda_D^3}$$

Table 1–1 gives a list of characteristics for various plasmas.

In this book Chapter 2 is self-contained but is a prerequisite for Chapter 9. Chapter 3 sets up the equations for the following five chapters but these fall into the separate categories of hydromagnetics (Chapters 4 to 6) and plasma waves (Chapters 7 and 8). Chapter 10 is also self-contained. Possible reading schemes may be obtained by selecting any number of the following groups of chapters: (i) 2; (ii) 3, 4, 5, 6; (iii) 3, 7, 8; (iv) 2, 9; and (v) 10. In group (ii) Chapters 5 or 6 could be omitted and in (iii) Chapter 8 could be omitted.

Table 1-1 Approximate magnitudes of the parameters characterizing some typical plasmas

Plasma	n (cm^{-3})	T (°K)	B (G)	ω_p (s^{-1})	λ_D (cm)	$n\lambda_D^3$	$\ln \Lambda$	ν_c (s^{-1})	Ω_e (s^{-1})	Ω_i (s^{-1})	r_L^- (cm)	r_L^+ (cm)
Interstellar gas	1	10^4	10^{-6}	6.10^4	7.10^2	3.10^8	22	6.10^{-4}	20	10^{-2}	2.10^6	10^8
Solar corona	10^6	10^6	10^{-4}	6.10^7	7	3.10^8	22	0·6	2.10^3	1	2.10^5	10^7
Ionosphere	10^6	10^3	10^{-1}	6.10^7	0·2	10^4	12	10^4	2.10^6	10^3	7	3.10^2
Q-plasma	10^8	10^5	10^2	6.10^8	0·2	10^6	16	10^3	2.10^9	10^6	7.10^{-2}	3
Dense Q-plasma	10^{11}	10^4	10^3	2.10^{10}	2.10^{-3}	10^3	10	3.10^7	2.10^{10}	10^7	2.10^{-3}	0·1
Tenuous, hot plasma	10^{12}	10^6	10^3	6.10^{10}	7.10^{-3}	3.10^5	15	4.10^5	2.10^{10}	10^7	2.10^{-2}	1
Arc discharge	10^{14}	10^4	10^3	6.10^{11}	7.10^{-5}	30	6	2.10^{10}	2.10^{10}	10^7	2.10^{-3}	0·1
Dense, hot plasma	10^{16}	10^6	10^4	6.10^{12}	7.10^{-5}	3.10^3	11	3.10^9	2.10^{11}	10^8	2.10^{-3}	0·1
Thermonuclear plasma	10^{16}	10^8	10^5	6.10^{12}	7.10^{-4}	3.10^6	18	5.10^6	2.10^{12}	10^9	2.10^{-3}	0·1

2

Particle orbit theory

2–1 Introduction

We shall begin the development of plasma dynamics by discussing the trajectories of individual charged particles, since this involves familiar techniques used in classical mechanics and electrodynamics. In principle, one solves the equation of motion for the particle moving in an electric field, **E**, and a magnetic field, **B**, under the condition that the fields obey Maxwell's equations. Provided that radiative effects may be neglected, the equation of motion is the Lorentz equation which, for non-relativistic velocities, is

$$m\ddot{\mathbf{r}} = e\left[\mathbf{E}(\mathbf{r}, t) + \frac{\dot{\mathbf{r}} \times \mathbf{B}(\mathbf{r}, t)}{c}\right] \tag{2–1}$$

where m, e, and **r** are the mass, charge, and position of the particle, t is the time, and c the velocity of light. Maxwell's equations are

$$\nabla \times \mathbf{E} = -\frac{1}{c}\frac{\partial \mathbf{B}}{\partial t} \tag{2–2}$$

$$\nabla \times \mathbf{B} = \frac{1}{c}\frac{\partial \mathbf{E}}{\partial t} + \frac{4\pi}{c}\mathbf{j} \tag{2–3}$$

$$\nabla.\mathbf{E} = 4\pi q \tag{2–4}$$

$$\nabla.\mathbf{B} = 0 \tag{2–5}$$

where $\mathbf{j}(\mathbf{r}, t)$ and $q(\mathbf{r}, t)$ are the current and charge densities. For a plasma containing N particles there are, of course, N equations like (2–1); the motion of the plasma is determined by these equations together with the field equations (2–2) to (2–5) and the definitions of **j** and q,

$$\mathbf{j}(\mathbf{r}, t) = \sum_{i=1}^{N} e_i\dot{\mathbf{r}}_i(t)\,\delta(\mathbf{r} - \mathbf{r}_i(t)) \tag{2–6}$$

$$q(\mathbf{r}, t) = \sum_{i=1}^{N} e_i\,\delta(\mathbf{r} - \mathbf{r}_i(t)) \tag{2–7}$$

where δ is the Dirac delta function.†

† The set of Eqs (2–1) to (2–7) contain two more equations than there are unknowns. However, (2–4) and (2–5) may be regarded as initial conditions for **E** and **B** (cf. Problem 2–1).

In general, the solution of (2–1) to (2–7) is impossible. Not only are the equations non-linear, but one never knows sufficient boundary conditions. Because of this one must resort to the techniques of statistical mechanics. However, under certain conditions the plasma motion may be adequately described by determining the orbit characteristic of each kind of particle. The procedure is to solve (2–1) for each type of particle assuming that **E** and **B** are given quantities; the solution is then substituted in (2–2) to (2–7) to check for self-consistency. Provided **E** and **B** are simple functions or slowly varying, the solution of (2–1) is straightforward. The check for self-consistency, though not often stressed, is important since it may impose limits on the use of orbit theory and, in some cases, necessary conditions on the plasma (such as non-uniform density) which would not otherwise be obvious. In the following applications of orbit theory we discuss the question of self-consistency only when it gives rise to such limitations. Particle collisions are ignored. Whenever charge distributions or current densities are significant, statistical or fluid descriptions are preferable and orbit theory is usually inadequate.

2–2 Constant uniform magnetic field

The simplest problem in orbit theory is that of the motion of a charged particle in a constant, spatially uniform magnetic field. In addition, most of the more general cases may be solved as perturbations of this basic motion. Taking the direction of **B** to define the z-axis, that is $\mathbf{B} = B\hat{\mathbf{k}}$, the scalar product of (2–1) with $\hat{\mathbf{k}}$ gives, since $\mathbf{E} = 0$,

$$\ddot{z} = 0 \qquad (2\text{–}8)$$

i.e.,
$$\dot{z} = v_{\parallel} = \text{const.}$$

Also from (2–1),

$$m\ddot{\mathbf{r}} \cdot \dot{\mathbf{r}} = 0$$

so that
$$\tfrac{1}{2}m\dot{r}^2 = W = \text{const.}$$

Hence, the velocity components perpendicular $(v_{\perp})$ and parallel $(v_{\parallel})$ to **B** are constant and the kinetic energy

$$W = W_{\perp} + W_{\parallel} = \tfrac{1}{2}m(v_{\perp}^2 + v_{\parallel}^2)$$

It is not surprising, of course, that kinetic energy is conserved since the force is at all times perpendicular to the velocity of the particle and, in consequence, does no work on it. Clearly, this conservation of kinetic energy is not restricted to uniform magnetic fields.

The particle trajectory is determined by (2–8) together with the x and y

components of (2–1):

$$\ddot{x} = \Omega\dot{y} \qquad \ddot{y} = -\Omega\dot{x}$$

where $\Omega = eB/mc$. These equations may be decoupled by differentiating with respect to time:

$$\dddot{x} + \Omega^2\dot{x} = 0 \qquad \dddot{y} + \Omega^2\dot{y} = 0$$

which have solutions

$$\dot{x} = v_\perp \cos(\Omega t + \alpha) \qquad \dot{y} = -v_\perp \sin(\Omega t + \alpha) \tag{2–9}$$

i.e.,

$$\left.\begin{aligned} x &= \left(\frac{v_\perp}{\Omega}\right)\sin(\Omega t + \alpha) + x_0 \\[2mm] y &= \left(\frac{v_\perp}{\Omega}\right)\cos(\Omega t + \alpha) + y_0 \end{aligned}\right\} \tag{2–10}$$

and

$$z = v_{||}t + z_0 \tag{2–11}$$

where α, x_0, y_0, and z_0 are determined by the initial conditions. Thus, the particle moves on a helix whose axis is parallel to **B**. Referred to the moving plane $z = v_{||}t + z_0$ the orbit is a circle with centre (x_0, y_0). The point $\mathbf{r}_g = (x_0, y_0, v_{||}t + z_0)$, which describes the locus of the centre of the circle, is called the *guiding centre*. The radius of the circle, $r_L = (v_\perp/\Omega)$, is known as

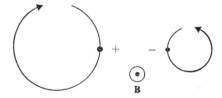

Fig. 2–1 Gyration of charged particles in a constant and uniform magnetic field. The field points upwards perpendicular to the plane of the paper

the *Larmor radius* and the frequency of rotation, Ω, as the *Larmor frequency*, *cyclotron frequency*, or *gyro-frequency*. Viewed from $z = +\infty$, positive and negative particles rotate in clockwise and anticlockwise directions respectively (Fig. 2–1). The net effect of many gyrating particles, therefore, is to produce a circular current which, in turn, induces a magnetic field. However, for a sufficiently strong applied magnetic field the induced field is negligible. Its magnitude relative to the applied field is

$$\frac{B_{ind}}{B} \sim \frac{jr_L}{cB} \sim \frac{nev_\perp r_L}{cB} \sim \frac{nW_\perp}{B^2}$$

where n is the particle density. Since this induced field opposes the applied field it is clear that a plasma is diamagnetic.

2–3 Constant uniform electric and magnetic fields

We now introduce a constant, uniform electric field which may be resolved into components $E_{||}$, in the direction of B, and $E_\perp$ which is taken to define the y-axis. Thus the fields are $B = (0, 0, B)$ and $E = (0, E_\perp, E_{||})$, and the components of (2–1) are

$$\ddot{x} = \Omega \dot{y} \tag{2-12}$$

$$\ddot{y} = \frac{eE_\perp}{m} - \Omega \dot{x} \tag{2-13}$$

$$\ddot{z} = \frac{eE_{||}}{m} \tag{2-14}$$

Integrating (2–14),

$$\dot{z} = v_{||} + \frac{eE_{||}t}{m}$$

from which it is clear that for sufficiently long times the non-relativistic approximation breaks down unless $E_{||} = 0$. Further, since charges of opposite sign are accelerated in opposite directions, a non-zero $E_{||}$ gives rise to arbitrarily large currents and charge separation. Therefore, from (2–3) and (2–4) significant fluctuating fields are induced and this is inconsistent with the assumption of constant fields. Thus it is necessary in orbit theory to assume that $E_{||} = 0$.

Equations (2–12) and (2–13) are solved as in Section 2–2. Again,

$$\ddot{y} + \Omega^2 \dot{y} = 0$$

while
$$\ddot{x} + \Omega^2 \dot{x} = \frac{eE\Omega}{m} \tag{2-15}$$

where $E_\perp$ has been replaced by E. Hence,†

$$\dot{y} = -u \sin (\Omega t + \alpha)$$

$$\dot{x} = u \cos (\Omega t + \alpha) + v_E$$

---

† The solution of (2–15) is given by the complementary function plus a particular integral. The latter is obtained by the standard technique; writing d^2/dt^2 as D^2

$$\text{P.I.} = (D^2 + \Omega^2)^{-1} eE\Omega/m = eE/m\Omega$$

For future reference we note that
$$(D^2 + \Omega^2)^{-1} \exp (ip\Omega t) = \exp (ip\Omega t)/\Omega^2(1 - p^2)$$

where u is a constant and

$$v_E = \frac{eE}{m\Omega} = \frac{cE}{B} \qquad (2\text{–}16)$$

The velocity of the guiding centre is now

$$\mathbf{v}_g = (v_E, 0, v_{||})$$

Thus the effect of an electric field perpendicular to the magnetic field is to produce a drift orthogonal to both. The drift velocity, which may be written

$$\mathbf{v}_E = c(\mathbf{E} \times \mathbf{B})/B^2$$

depends only on the fields; it is the same for all particles and does not give rise to a current. The non-relativistic approximation implies a further restriction on the electric field; by (2–16), $E \ll B$. The trajectories of positive and negative particles in the plane (2–11) are shown in Fig. 2–2.

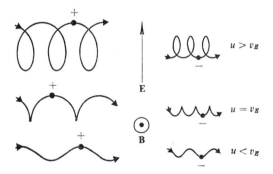

Fig. 2–2   Drift produced by a constant uniform electric field

2–4   Inhomogeneous magnetic field

The general solution of (2–1) for inhomogeneous fields is no simple matter. However, if the inhomogeneity is small—that is, the field experienced by the particle in traversing a Larmor orbit is almost constant—we may determine the trajectory as a perturbation of the basic motion found in Section 2–2. The requirement is that $\delta\mathbf{B}$, the change in $\mathbf{B}$ over a distance r_L, be such that

$$|\delta\mathbf{B}| \ll |\mathbf{B}|$$

i.e.,
$$r_L \ll L$$

where L is a distance over which the field changes significantly.

With this assumption, suppose $\mathbf{B} = (0, 0, B(y))$ so that (2–1) gives

$$\ddot{x} = \Omega(y)\dot{y} \qquad (2\text{–}17)$$

$$\ddot{y} = -\Omega(y)\dot{x} \qquad (2\text{--}18)$$

$$\ddot{z} = 0 \qquad (2\text{--}19)$$

Differentiating (2–17) and (2–18) as before,

$$\dddot{x} = -\Omega^2(y)\dot{x} + \dot{y}^2\frac{d\Omega}{dy}$$

$$\dddot{y} = -\Omega^2(y)\dot{y} - \dot{x}\dot{y}\frac{d\Omega}{dy}$$

Since **B** is almost constant during one orbit Ω may be expanded about the guiding centre,

$$\Omega(y) = \Omega(y_0) + (y - y_0)\frac{d\Omega}{dy}\bigg|_{y_0} + \cdots$$

$$= \Omega_0 + (y - y_0)\Omega_0' + O(\Omega_0(r_L/L)^2)$$

Hence

$$\dddot{x} + \Omega_0^2\dot{x} = -2\Omega_0\Omega_0'(y - y_0)\dot{x} + \Omega_0'\dot{y}^2$$

$$\dddot{y} + \Omega_0^2\dot{y} = -2\Omega_0\Omega_0'(y - y_0)\dot{y} - \Omega_0'\dot{x}\dot{y}$$

where all the terms on the right-hand sides of these equations are small quantities. Therefore, in these terms, y, $\dot{x}$, and $\dot{y}$ may be replaced by their zero-order (that is, uniform **B**) values (2–9) and (2–10),

$$\dddot{x} + \Omega_0^2\dot{x} = -\tfrac{1}{2}v_\perp^2\Omega_0'[1 + 3\cos 2(\Omega_0 t + \alpha)]$$

$$\dddot{y} + \Omega_0^2\dot{y} = \tfrac{3}{2}v_\perp^2\Omega_0'\sin 2(\Omega_0 t + \alpha)$$

from which (cf. footnote on p. 13)

$$\dot{x} = v_\perp\cos(\Omega_0 t + \alpha) + \tfrac{1}{2}v_\perp^2(\Omega_0'/\Omega_0^2)\cos 2(\Omega_0 t + \alpha) - v_\perp^2\Omega_0'/2\Omega_0^2 \quad (2\text{--}20)$$

$$\dot{y} = -v_\perp\sin(\Omega_0 t + \alpha) - \tfrac{1}{2}v_\perp^2(\Omega_0'/\Omega_0^2)\sin 2(\Omega_0 t + \alpha) \qquad (2\text{--}21)$$

Also, from (2–19)

$$\dot{z} = v_{\|}$$

From (2–20) and (2–21), we see that oscillations occur at $2\Omega_0$ in addition to those at Ω_0. In the present context, however, the term of interest is the non-oscillatory one in (2–20). The average of the velocity over one period

$(T = 2\pi/\Omega_0)$ is

$$\langle \mathbf{v} \rangle = (-v_\perp^2 \Omega_0'/2\Omega_0^2, 0, v_{||})$$

Thus a magnetic field in the z direction with a gradient in the y direction gives rise to a drift in the x direction. This drift velocity may be written

$$\mathbf{v}_G = [cW_\perp(\mathbf{B} \times \nabla)B]/eB^3 \qquad (2\text{--}22)$$

In this case, the drift depends on properties of the particle and, in particular, occurs in opposite directions for positive and negative charges (cf. Fig. 2–3).

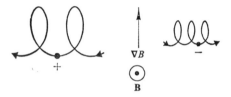

Fig. 2–3   Drift produced by an inhomogeneous magnetic field

However, this alone does not imply a flow of current. For, suppose that were the case, the x component of (2–3) gives (cf. Problem 2–4)

$$\dot{E}_x = c\frac{dB}{dy} - 4\pi j_x$$

which is non-zero since positive dB/dy would give a negative j_x. A consistent treatment, therefore, should include an electric field. If this is done assuming† $\mathbf{j} = ne\mathbf{v}$, differentiating (2–1) with respect to time and using (2–3) gives

$$\left. \begin{array}{l} \ddot{x} = -\omega_p^2 \dot{x} + \Omega\ddot{y} + \dot{y}^2\Omega' + c^2\Omega' \\ \ddot{y} = -\omega_p^2 \dot{y} - \Omega\ddot{x} - \dot{x}\dot{y}\Omega' \end{array} \right\} \qquad (2\text{--}23)$$

where $\omega_p = (4\pi ne^2/m)^{1/2}$ is the plasma frequency. If $\omega_p > \Omega$, which is often the case, the zero-order solution is an oscillation at the plasma frequency and the main drift velocity comes from the $c^2\Omega'$ term.

However, the use of orbit theory, which is essentially a single-particle model, is of doubtful value in situations where currents or charge separation may be significant. Such problems require a knowledge of the particle distribution function or its moments. For example, in the present problem, a gradient in $\mathbf{B}$ is usually accompanied by an opposite gradient in the particle pressure so that either n or the average kinetic energy is inhomogeneous and the computation of $\mathbf{j}$ is non-trivial.

In general, if $\dot{\mathbf{E}} \neq 0$, equations similar to (2–23) may be derived and plasma oscillations are again dominant. It is necessary, therefore, in calculations

† Strictly this is true only for a set of particles with the same velocity.

based on Larmor orbits, to assert that $\dot{\mathbf{E}} = 0$ and hence $\mathbf{j} = (c/4\pi)\nabla \times \mathbf{B}$. This may imply, as it does above, that the plasma distribution function is spatially non-uniform.

A second type of inhomogeneity in $\mathbf{B}$ giving rise to a drift velocity is that in which the field lines are slightly curved. For example, if $\mathbf{B} = (0, B_y(z), B)$, where B_y and dB_y/dz are small quantities,† one has

$$\ddot{x} = \Omega\dot{y} - \Omega_y\dot{z}$$
$$\ddot{y} = -\Omega\dot{x}$$
$$\ddot{z} = \Omega_y\dot{x}$$

where $\Omega_y = (eB_y/mc)$. Hence,

$$\dddot{x} + \Omega^2\dot{x} = -\Omega'_y v^2_{||}$$
$$\dddot{y} + \Omega^2\dot{y} = \Omega\Omega_y v_{||}$$

neglecting squares of small quantities. There is, therefore, a drift in the x direction given by

$$v_C = -v^2_{||}\Omega'_y/\Omega^2 = -mv^2_{||}c(dB_y/dz)/eB^2$$

This *curvature drift* may also be written

$$\mathbf{v}_C = 2cW_{||}(\mathbf{B} \times (\mathbf{B}.\nabla)\mathbf{B})/eB^4$$

There is also a drift in the y direction given by $(v_{||}\Omega_y/\Omega)$. This merely keeps the guiding centre moving parallel to $\mathbf{B}$ in the Oyz plane since, on average,

$$\frac{\langle\dot{y}\rangle}{\langle\dot{z}\rangle} = \frac{v_{||}\Omega_y/\Omega}{v_{||}} = \frac{B_y}{B}$$

If both types of inhomogeneity are present and there are no currents, $\nabla \times \mathbf{B} = 0$; that is, $\partial B_y/\partial z = \partial B/\partial y$, so that the total drift velocity is given by

$$\mathbf{v}_B = [c(W_\perp + 2W_{||})(\mathbf{B} \times \nabla)B]/eB^3$$

One may consider other inhomogeneities in the magnetic field; in particular, $\mathbf{B} \approx B(z)\hat{\mathbf{k}}$ is considered in Section 2–7. None of these gives rise to a drift motion (cf. Problem 2–5).

2–5   Constant, non-electromagnetic forces

It may happen that the particles are subject to non-electromagnetic forces. If such a force (per unit mass), $\mathbf{F}$, is constant it is equivalent to an electric

---

† It is convenient to keep B for the z component of $\mathbf{B}$. Since $|\mathbf{B}|$ does not appear in the remainder of this section no confusion arises.

field $\mathbf{E}' = m\mathbf{F}/e$ and is subject to the same restrictions as found for $\mathbf{E}$, i.e., its component parallel to $\mathbf{B}$ must be negligible and $mF \ll eB$. The drift velocity is clearly

$$\mathbf{v}_F = cm(\mathbf{F} \times \mathbf{B})/eB^2$$

Gravity is an obvious example of such a force. Gravitational drifts, however, are usually negligible.

2–6 Time-varying magnetic field

In Section 2–4 we introduced certain inhomogeneous magnetic fields. In practice, one is often faced with situations in which the magnetic field may be not only inhomogeneous but also time-dependent. In keeping with Section 2–4, which was restricted to weakly inhomogeneous fields, we shall only consider magnetic fields varying slowly in time ($\dot{B}/B \ll \Omega$). For simplicity, consider a magnetic field which varies in time but not in space. Are there any invariants here analogous to $W_\perp$, $W_{||}$ found in Section 2–2 for motion in a constant and uniform magnetic field? In fact, one may demonstrate that such invariants do exist in particular situations.

An axial magnetic field which varies in time induces an azimuthal electric field (cf. (2–2)); hence, the scalar product of (2–1) with $\mathbf{v}_\perp$ (no longer constant as in Section 2–2) is

$$m\mathbf{v}_\perp \cdot \ddot{\mathbf{r}} = e\mathbf{v}_\perp \cdot \left[\mathbf{E} + \frac{\dot{\mathbf{r}} \times \mathbf{B}}{c} \right] \tag{2–24}$$

Since $\mathbf{E} \perp \mathbf{B}$, $\mathbf{v}_{||}$ is constant and (2–24) becomes

$$\frac{d}{dt}(\tfrac{1}{2}mv_\perp^2) = e\mathbf{E} \cdot \mathbf{v}_\perp$$

Thus in traversing a Larmor orbit the particle energy changes by

$$\delta(\tfrac{1}{2}mv_\perp^2) = e \oint \mathbf{E} \cdot d\mathbf{r}_\perp = e \int (\nabla \times \mathbf{E}) \cdot d\mathbf{S}$$

where $d\mathbf{r}_\perp = \mathbf{v}_\perp \, dt$ and $d\mathbf{S}$ is an element of the surface enclosed by the path. Hence, from (2–2)

$$\delta(\tfrac{1}{2}mv_\perp^2) = -\frac{e}{c} \int \frac{\partial \mathbf{B}}{\partial t} \cdot d\mathbf{S}$$

Since the field changes slowly†

$$\delta(\tfrac{1}{2}mv_\perp^2) \simeq \pi r_L^2 \, | \, e \, | \, \dot{B}/c = \frac{mv_\perp^2}{2} \frac{2\pi}{|\Omega|} \frac{\dot{B}}{B}.$$

† The negative sign disappears since for positive (negative) charges $e > 0$ (<0) and $\mathbf{B} \cdot d\mathbf{S} < 0$ (>0).

i.e.,
$$\delta W_\perp = W_\perp \frac{\delta B}{B}$$

δB being the change in magnitude of the magnetic field during one orbit.

Thus,
$$\delta(W_\perp / B) = 0 \qquad (2\text{--}25)$$

and the quantity $W_\perp / B$ is an approximate constant of the motion; it may be identified with the *magnetic moment* of the particle, μ (cf. Problem 2–6). Physical quantities which are approximately constant for slow (adiabatic) changes of the magnetic field in time or in space are called *adiabatic invariants*.

Equation (2–25) also implies the invariance of $\pi r_L^2 B$, the magnetic flux through a Larmor orbit.

2–7 Invariance of μ in an inhomogeneous field

The magnetic moment, μ, is an adiabatic invariant for motion in a spatially inhomogeneous magnetic field as well as for a time-dependent field. To

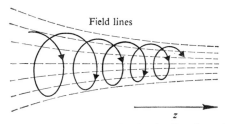

Fig. 2-4 Magnetic field increasing in the direction of the field

demonstrate this, consider an axially symmetric magnetic field which increases slowly with z (Fig. 2–4). From (2–5), in cylindrical polar coordinates

$$\frac{1}{r}\frac{\partial}{\partial r}(rB_r) + \frac{\partial B_z}{\partial z} = 0$$

i.e.,
$$rB_r = -\int r \frac{\partial B_z}{\partial z}dr$$

Since the field is approximately constant over one Larmor orbit

$$B_r \simeq -\frac{r_L}{2}\frac{\partial B_z}{\partial z} \simeq -\frac{r_L}{2}\frac{\partial B}{\partial z}$$

since B_r is small. To the same approximation (first-order in small quantities), the z component of (2–1) gives

$$m\frac{dv_{||}}{dt} = \frac{|e|}{c}v_\perp B_r = -\frac{|e|}{c}\frac{r_L}{2}v_\perp\frac{\partial B}{\partial z}$$

i.e.,
$$m\frac{dv_{||}}{dt} = -\mu\frac{\partial B}{\partial z} \tag{2-26}$$

Thus,
$$\frac{d}{dt}(\tfrac{1}{2}mv_{||}^2) = -\mu v_{||}\frac{\partial B}{\partial z} = -\mu\frac{dB}{dt} \tag{2-27}$$

From the definition of μ

$$\frac{d}{dt}(\tfrac{1}{2}mv_{\perp}^2) \equiv \frac{d}{dt}(\mu B) \tag{2-28}$$

and hence, adding (2–27) and (2–28) and using energy conservation

$$\frac{d}{dt}(\mu B) - \mu\frac{dB}{dt} = 0$$

i.e.,
$$\frac{d\mu}{dt} = 0$$

The invariance of μ in such fields has important implications which are considered next.

2–8 Magnetic mirrors

Consider a particle moving in the inhomogeneous field introduced in Section 2–7 towards the region of increasing B. It follows from the invariance of $W_{\perp}/B$ that $W_{\perp}$ must increase. Since energy is conserved this increase

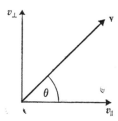

Fig. 2–5 Pitch angle

must be at the expense of $W_{||}$. Thus it may happen that for some value of B (B_R say) $W_{||} = 0$, in which case the particle cannot penetrate further into the magnetic field and suffers reflection at this point (provided $(\mathbf{v} \times \mathbf{B}).\hat{\mathbf{k}} \neq 0$). Such a field configuration is known as a *magnetic mirror*.

It is convenient to define the *pitch angle*, θ, of the particle by (Fig. 2–5)

$$\tan\theta = \frac{v_{\perp}}{v_{||}} \tag{2-29}$$

Then,
$$\frac{W_{\perp}}{B} = \frac{mv^2\sin^2\theta}{2B}$$

i.e., $$\sin^2 \theta / B = \text{const.} \tag{2-30}$$

or $$\sin \theta = (B/B_R)^{1/2}$$

For a particle which penetrates to the point M where B is maximum, B_M, before being reflected, $B_R = B_M$. Hence particles with pitch angles such that $\sin \theta > (B/B_M)^{1/2}$ suffer reflection before reaching the region of maximum field; those having $\sin \theta < (B/B_M)^{1/2}$ are not reflected.

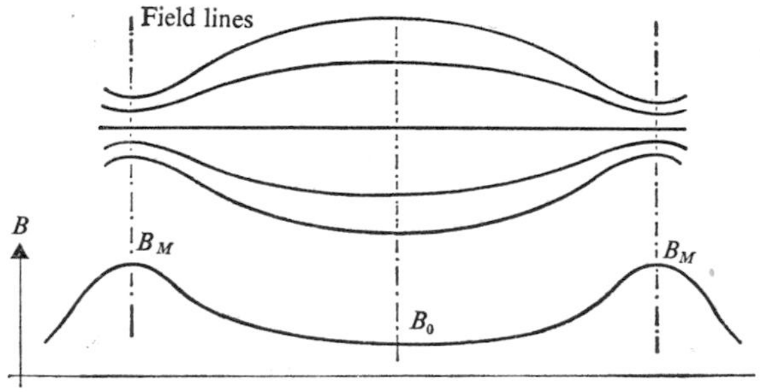

Fig. 2–6   Arrangement of magnetic field in a magnetic bottle

If one arranges two mirror fields as shown in Fig. 2–6 then particles with $\sin \theta \geqslant (B/B_M)^{1/2}$ will be reflected to and fro. This configuration is known as a *magnetic bottle* or *adiabatic mirror trap*. Taking B_0 to be the value of the magnetic field in the mid-plane of the bottle, the *mirror ratio* is defined by

$$R = \frac{B_M}{B_0}$$

and particles will be reflected if

$$\sin \theta_0 \geqslant R^{-1/2}$$

The particles which are lost from the magnetic bottle are those within the

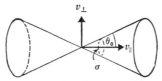

Fig. 2–7   Loss cone for a magnetic bottle. Those particles with resultant velocities lying within the cone will penetrate the magnetic mirror and be lost from the bottle

solid angle σ (Fig. 2–7; note that this figure refers to velocity space and not to configuration space). This solid angle σ defines the *loss cone*. The probability, P, of loss from the bottle is

$$P = \frac{\sigma}{2\pi} = \int_0^{\theta_0} \sin \theta \, d\theta$$

i.e.,
$$P = 1 - \sqrt{\left(\frac{R-1}{R}\right)}$$

$$\simeq \frac{1}{2R} \quad \text{if } R \gg 1$$

This means that the higher the mirror ratio the less likely it is that particles will escape.

In laboratory research into controlled thermonuclear reactions, adiabatic mirror trapping has been an important way of containing plasma.[13] One such containment system is the PHOENIX experiment at the Culham Laboratory of the United Kingdom Atomic Energy Authority.[14] In this device, particles accelerated to tens of kilovolts are injected as neutral hydrogen atoms across the magnetic field ($B \sim 50,000$ G) of the mirror trap. Those atoms in highly excited states suffer ionization by the $e(\mathbf{v} \times \mathbf{B})/c$ Lorentz force, and the electrons and protons are then trapped by the mirror field.

2–9   Earth's radiation belts

Adiabatic mirror traps are not only important in containing laboratory plasmas; in this section, we shall briefly describe a naturally occurring magnetic mirror trap.

Perhaps the most notable discovery in the exploration of the earth's environment by satellite was the identification of its radiation belts by Van Allen and his co-workers in 1958.[15] These workers found that cosmic radiation detectors on board the satellite, *Explorer* I, stopped functioning at distances about 700 km from the earth and beyond; however, when the orbit of the satellite brought it within this distance, the counters behaved normally again. They concluded that this behaviour was consistent with the theory that above about 700 km the counters were being choked by an anomalously high radiation intensity. Laboratory tests showed that this would happen if the intensity of radiation exceeded a value about 10^4 times greater than that due to cosmic radiation. The hypothesis that the satellite passed through regions of high radiation intensity was tested by further experiments on board other satellites in the Explorer series. By putting counters on board lunar probes, regions of high radiation intensity were detected at great distances from the earth. This, and later work, helped to establish the now familiar morphology of the radiation belts shown in Fig. 2–8. Two zones of high-intensity particle radiation surround the earth, although in reality there is no sharp boundary dividing them. The inner belt is a few thousand kilometres from the earth's surface on average while the outer belt is approximately 20,000 km away. The shape of the belts—particularly the outer—is reminiscent of the geomagnetic field lines.

The particles in the belts are mostly electrons and protons. The protons in the inner belt have energies 40 MeV and above;[15] deuterons and tritons are also present, but typically only to the extent of 1 per cent of the number of protons. The energies of protons in the outer belt are much lower (0·1 MeV to a few MeV): in fact, the picture is not quite so simple as implied by Fig. 2–8 since we now know that there are several low-energy proton belts around the earth and even these overlap. The outer belt of electrons is peaked

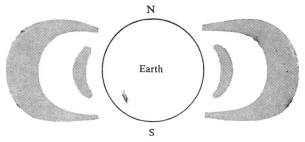

Fig. 2–8 Morphology of the earth's radiation belts

at about 30,000 km, and so it is now customary to consider proton and electron belts separately. Electrons in this outer belt have energies typically greater than 40 keV.

The origin of these radiation belts will not be discussed here; indeed, there are still many unsolved problems concerning them. However, since the geomagnetic field may be pictured as a dipole (at any rate out to distances of about 10^5 km) with the magnetic lines of force bunching at the north and south poles, it is clear that this field may act as a magnetic trap for charged

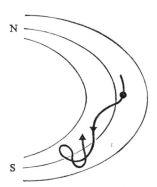

Fig. 2–9 Reflection of a charged particle as it approaches a magnetic pole

particles. Energetic particles injected into the earth's magnetic field will describe helical trajectories as in Fig. 2–9 and undergo reflection in the stronger field regions around the magnetic poles. Transit times from one mirror point to the other are of the order of a second. In addition, the results of Section 2–4 imply that the particles will drift azimuthally since the field lines are curved and there exists a magnetic field gradient normal to the direction of **B**. Electrons drift from west to east and protons vice versa; the guiding centre of a particle traces out a surface of rotation known as the longitudinal invariant surface.† For an electron with energy 40 keV, the time taken for the guiding centre to make one circuit around the longitudinal invariant surface is of the order of an hour.

The trapping of charged particles in the geomagnetic field has actually been demonstrated experimentally. In 1958, American scientists carried out a series of experiments (Project Argus) in which atomic bombs were carried by rockets to a height of about 650 km and exploded. Electrons were injected into the geomagnetic field and an artificial radiation belt established on each occasion which persisted for times varying from a few days to a few weeks. Counters on board the satellite *Explorer* IV were used to observe these artificial belts; their thickness was estimated at about 40 km and the azimuthal drift of the electron shell was measured.

2–10   Another adiabatic invariant

There is another adiabatic invariant of importance in mirror traps; this is the *longitudinal invariant, J*, defined by

$$J = \oint v_{||}\, ds \qquad (2\text{--}31)$$

where ds is an element of the guiding centre path and the integral is evaluated over one complete traverse of the guiding centre (from s_1 to s_2 in Fig. 2–10).

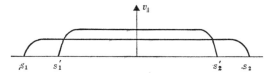

Fig. 2–10   Variation of $v_{||}$ between mirror points s_1 and s_2. If the field maxima approach one another, so do the mirror points and at some later time the phase space trajectory is given by the curve from s_1' to s_2'. The area under the trajectory is an adiabatic invariant

The invariance of J is useful in situations in which the mirror points are no longer stationary; therefore B is taken to be slowly varying in both space and time. Since J is a function of $\dot{s} \equiv v_{||}$, s, t, using

$$W = \tfrac{1}{2}mv_{||}^2 + \mu B$$

one may write

$$J(W, s, t) = \int_{s_1}^{s} \left[\frac{2}{m}(W - \mu B) \right]^{1/2} ds \qquad (2\text{--}32)$$

† The magnetic flux enclosed by this surface has been shown to be an adiabatic invariant. (Cf. T. G. Northrop, *The Adiabatic Motion of Charged Particles*, p. 61, John Wiley (Interscience), New York, 1963.)

Then
$$\frac{\mathrm{d}J}{\mathrm{d}t} = \left(\frac{\partial J}{\partial t}\right)_{W,s} + \left(\frac{\partial J}{\partial W}\right)_{s,t}\frac{\mathrm{d}W}{\mathrm{d}t} + \left(\frac{\partial J}{\partial s}\right)_{W,t}\frac{\mathrm{d}s}{\mathrm{d}t}$$

$$= -\int_{s_1}^{s}\left[\frac{2}{m}(W-\mu B)\right]^{-1/2}\left(\frac{\mu}{m}\frac{\partial B}{\partial t}\right)\mathrm{d}s$$

$$+ \left\{v_{\|}\dot{v}_{\|} + \frac{\mu}{m}\frac{\partial B}{\partial t} + \frac{\mu}{m}v_{\|}\frac{\partial B}{\partial s}\right\}\int_{s_1}^{s}\left[\frac{2}{m}(W-\mu B)\right]^{-1/2}\mathrm{d}s$$

$$+ \left[\frac{2}{m}(W-\mu B)\right]^{1/2}v_{\|} - v_{\|}\int_{s_1}^{s}\left[\frac{2}{m}(W-\mu B)\right]^{-1/2}\left(\frac{\mu}{m}\frac{\partial B}{\partial s}\right)\mathrm{d}s$$

Now, at the turning point $s = s_2$, $v_{\|} = 0$; i.e.,

$$\frac{\mathrm{d}J(W,s_2,t)}{\mathrm{d}t} = -\int_{s_1}^{s_2}\left[\frac{2}{m}(W-\mu B)\right]^{-1/2}\left(\frac{\mu}{m}\frac{\partial B}{\partial t}\right)\mathrm{d}s$$

$$+ \frac{\mu}{m}\frac{\partial B}{\partial t}\int_{s_1}^{s_2}\left[\frac{2}{m}(W-\mu B)\right]^{-1/2}\mathrm{d}s \qquad (2\text{--}33)$$

Hence, provided the variation of B with time over an interval of the order of the transit time from s_1 to s_2 is negligible, $\partial B/\partial t$ may be taken outside the integral in the first term in (2–33) and

$$\frac{\mathrm{d}J(W,s_2,t)}{\mathrm{d}t} = 0$$

Thus J is an adiabatic invariant. Since the transit time between mirror points is many Larmor periods, the condition on $\partial B/\partial t$ in order that J should be invariant is clearly more stringent than that required for the invariance of μ.

2–11. Some consequences of J as an invariant

The motion of a charged particle in a magnetic mirror was considered in Section 2–8. What happens if the mirror points are no longer stationary but move towards one another? The existence of J as an adiabatic invariant gives us the answer provided the mirrors approach one another sufficiently slowly. Suppose that at $t = 0$ the coordinates of the mirror points in the phase space diagram (Fig. 2–10) are s_1 and s_2, while at some later time, t', they shift to s_1' and s_2', respectively. The invariance of J means that the area enclosed by the phase space orbit is constant; denoting $s_1 - s_2$ by s_0, $s_1' - s_2'$ by s_0' gives

$$v_{\|}' \simeq v_{\|}\frac{s_0}{s_0'}$$

where $v_{\|}$, $v_{\|}'$ are the parallel components of velocity at the mid-plane at $t = 0$, $t = t'$, respectively.

i.e.,
$$W' = \frac{m}{2}\left[v_{\perp}'^2 + v_{\parallel}'^2\right] = \frac{m}{2}\left[v_{\perp}^2 + \left(\frac{s_0}{s_0'}\right)^2 v_{\parallel}^2\right]$$

by making use of the invariance of μ (assuming the field is constant at the mid-plane). Thus it appears that the energy of a particle trapped between slowly approaching magnetic mirrors increases.

This gain in energy by a charged particle in such fields is the basis of a mechanism which has been proposed to account for the very energetic particles in cosmic radiation (as high as 10^{18} eV). How such enormous energies are attained—they are still far in excess of those available from laboratory particle accelerators—is obviously a key question in any cosmic ray theory. The limitations on the magnitude of electrostatic fields in interstellar space rule these out of consideration. However, since magnetic fields exist in space, and since it is plausible to suppose that there are regions of space in which 'clumps' of magnetic field of higher intensity occur, charged particles may become trapped between these regions. Moreover, the magnetic clumps will not be static and trapped particles could therefore be accelerated if such regions were approaching one another; by the same token, of course, particles would lose energy if these mirror regions were separating. This mechanism involving the interchange of energy between charged particles and moving magnetic clouds is the basis of a theory proposed by Fermi in 1949 to account for the origin of cosmic rays. Fermi showed that there is a slightly higher probability of head-on collisions than of overtaking ones; their relative frequencies, in fact, are proportional to $(v_{\parallel} + v_B)/(v_{\parallel} - v_B)$ where v_B is the velocity of the magnetic cloud.

As proposed originally, the mechanism suffers from a rather severe limitation. While $v_{\parallel}$ increases, the pitch angle (2–29) decreases so that ultimately the particle undergoing acceleration falls into the loss cone and escapes from the mirror trap. (Since μ is adiabatically invariant, $v_{\perp}$ cannot increase while the magnetic field between the clouds does not change.) Thus, for the Fermi mechanism to be viable, some additional feature (such as inter-particle collisions) is needed to restore the isotropic nature of the velocity distribution.

2–12 Motion in a monochromatic, plane wave

Another problem susceptible to orbit theory treatment is that of motion of an electron in a monochromatic, plane wave. This differs from other problems in this chapter in that the basic frequency is the wave frequency rather than the Larmor frequency, and also the motion is now dominated by the electric field in the non-relativistic approximation. Indeed, since $|\mathbf{E}| = |\mathbf{B}|$ for plane waves, the equation of motion to lowest order in (v/c) is

$$m\ddot{\mathbf{r}} = e\mathbf{E} \tag{2–34}$$

and the electron oscillates with the electric field. This is the basic motion in the theory of Thomson scattering (the scattering of light by free electrons).

Since the velocity of the electron is proportional to the amplitude of the wave, relativistic velocities may be attained in the scattering of sufficiently intense light (as from the most powerful lasers). Relativistic calculation of the trajectory reveals a drift in the direction of propagation of the wave.

For a plane polarized wave propagating in the x direction, $\mathbf{E} = (0, E, 0)$, $\mathbf{B} = (0, 0, B)$. We can conveniently describe the fields by their vector potential, $\mathbf{A} = (0, a(\tau) \cos \omega \tau, 0)$, where a is the amplitude of the wave,† ω the frequency, and $\tau = t - x/c$. $\mathbf{E}$ and $\mathbf{B}$ are given by the usual equations

$$\mathbf{E} = -\frac{1}{c}\frac{\partial \mathbf{A}}{\partial t} \qquad \mathbf{B} = \nabla \times \mathbf{A}$$

from which it follows in the case of a plane polarized wave

$$E = B = -\frac{1}{c}\frac{dA}{d\tau}$$

The relativistic Lorentz equation gives[16]

$$\frac{d}{dt}\left(\frac{\dot{x}}{(1 - \beta^2)^{1/2}}\right) = \frac{eB}{mc}\dot{y} = -\frac{e}{mc^2}\frac{dA}{d\tau}\dot{y} \tag{2-35}$$

$$\frac{d}{dt}\left(\frac{\dot{y}}{(1 - \beta^2)^{1/2}}\right) = \frac{e}{mc}(Ec - B\dot{x}) = -\frac{e}{mc^2}(c - \dot{x})\frac{dA}{d\tau} \tag{2-36}$$

$$\frac{d}{dt}\left(\frac{\dot{z}}{(1 - \beta^2)^{1/2}}\right) = 0 \tag{2-37}$$

$$\frac{d}{dt}\left(\frac{c}{(1 - \beta^2)^{1/2}}\right) = \frac{eE}{mc}\dot{y} = -\frac{e}{mc^2}\frac{dA}{d\tau}\dot{y} \tag{2-38}$$

where $\beta = v/c$. Subtracting (2-35) from (2-38) gives, on integrating,

$$1 - \dot{x}/c = (1 - \beta^2)^{1/2} \tag{2-39}$$

assuming that the electron is initially at rest at the origin. Since

$$\frac{d\tau}{dt} = 1 - \frac{\dot{x}}{c} \tag{2-40}$$

---

† For a truly monochromatic wave a is constant. In practice, however, a can never be constant since the amplitude must grow from zero when the wave is switched on. Moreover, treatments which assume a is constant lead to solutions depending critically on the initial phase and thereby predict electron drifts in arbitrary directions (cf. Problem 2-8). For an almost monochromatic wave, $a(\tau)$ must be a slowly varying function and the only significant effect of including its variation is to ensure that the fields are initially zero; dependence on the initial phase does not then appear.

(2–36) may be integrated directly:

$$\frac{\dot{y}/c}{(1 - \beta^2)^{1/2}} = -\frac{eA}{mc^2} \tag{2-41}$$

Substituting for $\dot{y}$ from (2–41) and using (2–39) and (2–40), (2–35) becomes

$$\frac{d}{dt}\left(\frac{\dot{x}/c}{(1 - \beta^2)^{1/2}}\right) = \left(\frac{e}{mc^2}\right)^2 A\frac{dA}{d\tau}\frac{d\tau}{dt}$$

Hence,

$$\frac{\dot{x}/c}{(1 - \beta^2)^{1/2}} = \frac{1}{2}\left(\frac{eA}{mc^2}\right)^2$$

Finally, using (2–39) and (2–40) again,

$$\dot{x} = \frac{c}{2}\left(\frac{eA}{mc^2}\right)^2 \frac{d\tau}{dt} = \frac{c}{2}\left(\frac{\Omega}{\omega}\right)^2 \frac{d\tau}{dt}\cos^2 \omega\tau \tag{2-42}$$

$$\dot{y} = -c\left(\frac{eA}{mc^2}\right)\frac{d\tau}{dt} = -c\left(\frac{\Omega}{\omega}\right)\frac{d\tau}{dt}\cos \omega\tau \tag{2-43}$$

and from (2–37)

$$\dot{z} = 0$$

where $\Omega = (ea\omega/mc^2)$. Averaging $\dot{x}$ over one period $(T = 2\pi/\omega)$, $a(\tau)$ may be treated as constant,

$$\langle \dot{x} \rangle = \frac{\int \dot{x}\,dt}{\int dt} = \frac{\frac{c}{2}\left(\frac{\Omega}{\omega}\right)^2 \int \cos^2 \omega\tau\,d\tau}{\int (d\tau + dx/c)}$$

$$= \frac{\frac{c}{2}\left(\frac{\Omega}{\omega}\right)^2 \int \cos^2 \omega\tau\,d\tau}{\int d\tau\left(1 + \frac{1}{2}\left(\frac{\Omega}{\omega}\right)^2 \cos^2 \omega\tau\right)} = \frac{c(\Omega/2\omega)^2}{1 + (\Omega/2\omega)^2}$$

Similarly,

$$\langle \dot{y} \rangle = 0$$

Thus the electron drifts in the direction of propagation of the wave. If $\Omega < \omega$, the motion is mainly in the y direction (that is, oscillation due to the **E** field); also the drift velocity is approximately

$$v_P = \frac{e^2}{2m^2c\omega^2}\langle \mathbf{E} \times \mathbf{B} \rangle$$

An interesting effect of this drift velocity is to predict a Doppler shift in the frequency of light scattered by free electrons.[17] This average motion in the direction of the wave should not be confused with that due to radiation pressure. To calculate the latter, we must include the radiation damping term in the equation of motion (cf. Chapter 9).

Spatial inhomogeneities in plane waves also give rise to drift velocities. To discuss this, one needs only the non-relativistic approximation (2–34). Assuming that the amplitude of $\mathbf{E}$ has a small gradient, $\mathbf{E} = (0, E_0(y) \cos \omega t, 0)$, since we know from (2–42) and (2–43) that $(x/c) \sim (v/c)^2 t$ and hence it may be left out of the argument of $\mathbf{E}$.

When the gradient is ignored, the solution of (2–34) is

$$y = -\frac{eE_0}{m\omega^2} \cos \omega t$$

Thus, expanding $E_0(y)$ about the mean position of the electron

$$E_0(y) = E_0(0) + y\frac{\partial E_0(0)}{\partial y} + \ldots = E_0(0) - \frac{eE_0}{m\omega^2}\frac{\partial E_0(0)}{\partial y} \cos \omega t + \ldots$$

To first order in the gradient of E_0, therefore,

$$\ddot{y} = \frac{e}{m} \cos \omega t \left[E_0(0) - \frac{eE_0(0)}{m\omega^2}\frac{\partial E_0(0)}{\partial y} \cos \omega t \right]$$

which gives, when averaged over the field oscillations,

$$\langle \ddot{y} \rangle = -\left(\frac{e}{2m\omega}\right)^2 \frac{\partial E_0^2(0)}{\partial y}$$

This mean acceleration gives rise to a drift velocity carrying the electron away from regions of higher field intensity. More generally, it may be shown (cf. Problem 2–11) that

$$\langle \ddot{\mathbf{r}} \rangle = -\left(\frac{e}{2m\omega}\right)^2 \nabla E_0^2(0) \tag{2–44}$$

This deflection of electrons by light of variable intensity is analogous to the refraction of light by matter of variable density.[18]

Summary

1. *Conditions*

These conditions are necessary if the results given in Sections 2–1 to 2–11 are to be valid:

(i) Radiation negligible.
(ii) Non-relativistic velocities; this implies that $|\mathbf{E}| \ll |\mathbf{B}|$.

(iii) Inter-particle collisions negligible.

(iv) $E_{\parallel} = 0$.

(v) $r_L \ll L$.

(vi) $\dot{E} = 0$.

(vii) $\dot{B}/B \ll \Omega$.

2. Particle drifts

(i) **E** × **B** drift: $\mathbf{v}_E = c(\mathbf{E} \times \mathbf{B})/B^2$.

(ii) Grad B drift: $\mathbf{v}_G = cW_{\perp}[(\mathbf{B} \times \nabla)B]/eB^3$.

(iii) Curvature drift: $\mathbf{v}_C = 2cW_{\parallel}[\mathbf{B} \times (\mathbf{B}.\nabla)\mathbf{B}]/eB^4$.

(iv) Constant non-electromagnetic force drift: $\mathbf{v}_F = cm(\mathbf{F} \times \mathbf{B})/eB^2$.

3. Adiabatic invariants

(i) Magnetic moment: $\mu = W_{\perp}/B$.

(ii) Longitudinal invariant: $J = \oint v_{\parallel} \, ds$.

4. Motion in a monochromatic plane wave

(i) Drift in direction of wave propagation:

$$\mathbf{v}_P = \frac{e^2}{2m^2c\omega^2}\langle \mathbf{E} \times \mathbf{B} \rangle = \frac{c}{2B^2}\left(\frac{\Omega}{\omega}\right)^2\langle \mathbf{E} \times \mathbf{B} \rangle$$

(ii) Acceleration due to wave intensity gradient:

$$\langle \ddot{\mathbf{r}} \rangle = -\left(\frac{e}{2m\omega}\right)^2 \nabla E_0^2$$

Problems

2–1 Show that the current and charge densities as defined by (2–6) and (2–7) satisfy the charge conservation equation

$$\frac{\partial q}{\partial t} + \nabla.\mathbf{j} = 0$$

Hence, using (2–2) and (2–3), show that (2–4) and (2–5) are true for all t provided they are valid at some particular time t_0.

2–2 A particle of mass m and charge e, initially at rest at the origin, is subjected to the constant fields $\mathbf{E} = (0, E, 0)$ and $\mathbf{B} = (0, 0, B)$. Show that it moves on the cycloid

$$x = \frac{cE}{\Omega B}(\Omega t - \sin \Omega t)$$

$$y = \frac{cE}{\Omega B}(1 - \cos \Omega t)$$

in the plane $z = 0$. Sketch the paths for $e > 0$ and $e < 0$. What is the wavelength of the motion?

2–3 Verify (2–22) for a magnetic field $\mathbf{B} = (0, 0, B(\mathbf{r}))$ in which the spatial inhomogeneities are small but otherwise arbitrary.

Hint: use (2–5) to show that B is not a function of z.

2–4 In the discussion following (2–22) use the Maxwell equations to verify $\partial E_x/\partial t = dE_x/dt \equiv \dot{E}_x$.

2–5 Show that there is no mean drift velocity when charged particles move in the field $\mathbf{B} = (0, B_y(x), B)$ where B_y and $\partial B_y/\partial x$ are small quantities.

2–6 The magnetic moment μ of a current flowing in a closed loop is given by $\mu = iA/c$, where i is the current and A is the area enclosed by the loop. Hence verify that the adiabatic invariant $W\perp/B$ is equivalent to the magnetic moment of the gyrating charged particle.

2–7 In Hamiltonian theory it may be shown that $\oint p_i \, dq_i$ is an adiabatic invariant whenever the generalized coordinate q_i and its corresponding generalized momentum p_i are periodic with the same frequency. The integral is taken over one complete orbit of q_i. Given that the Lagrangian for a charged particle in an electromagnetic field is

$$L = \tfrac{1}{2}m\dot{r}^2 - e\phi(\mathbf{r}, t) + \frac{e}{c}\dot{\mathbf{r}} \cdot \mathbf{A}(\mathbf{r}, t)$$

and $$p_i \equiv \partial L/\partial \dot{q}_i$$

where ϕ and $\mathbf{A}$ are the electric and magnetic potentials, i.e.,

$$\mathbf{E} = -\nabla\phi - \frac{1}{c}\frac{\partial \mathbf{A}}{\partial t} \qquad \mathbf{B} = \nabla \times \mathbf{A}$$

show that the adiabatic invariants μ and J may be obtained from

$$\oint \mathbf{p}_\perp \cdot d\mathbf{r}_\perp \quad \text{and} \quad \oint p_{||} \, dr_{||}$$

respectively.

2–8 (i) Calculate the mean velocity of a charged particle in the electric field $\mathbf{E} = \mathbf{E}_0 \cos(\omega t + \theta_0)$, where $\mathbf{E}_0$ and θ_0 are constants, assuming that relativistic effects are negligible.

(ii) Repeat the calculation assuming that $\mathbf{E}_0$ is a slowly varying function (i.e., terms of order $\dfrac{dE_0/dt}{\omega E_0}$ may be neglected) with $\mathbf{E}_0(0) = 0$.

Comment on the different results in (i) and (ii).

2–9 The vector potential for a monochromatic, plane wave of arbitrary polarization propagating in the x direction is given by

$$\mathbf{A} = a(\tau)[0, \, \alpha \cos \omega\tau, \, (1 - \alpha^2)^{1/2} \sin \omega\tau]$$

where $\tau = t - x/c$, $0 < \alpha < 1$ and the amplitude $a(\tau)$ is slowly varying. Integrate the relativistic Lorentz equation, which has components[16]

$$\frac{d}{dt}\left(\frac{\dot{x}}{(1 - \beta^2)^{1/2}}\right) = -\frac{e}{mc^2}\left(\dot{y}\frac{dA_y}{d\tau} + \dot{z}\frac{dA_z}{d\tau}\right)$$

$$\frac{d}{dt}\left(\frac{\dot{y}}{(1 - \beta^2)^{1/2}}\right) = -\frac{e}{mc^2}(c - \dot{x})\frac{dA_y}{d\tau}$$

$$\frac{d}{dt}\left(\frac{\dot{z}}{(1 - \beta^2)^{1/2}}\right) = -\frac{e}{mc^2}(c - \dot{x})\frac{dA_z}{d\tau}$$

$$\frac{d}{dt}\left(\frac{c}{(1 - \beta^2)^{1/2}}\right) = -\frac{e}{mc^2}\left(\dot{y}\frac{dA_y}{d\tau} + \dot{z}\frac{dA_z}{d\tau}\right)$$

where $\beta = v/c$, to obtain the velocity of an electron interacting with this wave, assuming that the electron starts from rest at the origin. Hence show that the drift velocity is the same as that for a plane polarized wave ($\alpha = 1$).

2-10 Integrate (2–42) and (2–43), ignoring terms of order $(da/d\tau)/a\omega$. Sketch the trajectory of the particle.

2-11 Using (2–1) and (2–2), keeping only first-order terms in gradients of the fields and v/c, verify (2–44) for a plane polarized wave in which $\mathbf{E} = \mathbf{E}_0(\mathbf{r}) \cos \omega t$.

2-12 The components of the relativistic equation of motion for a charged particle in the magnetic field $\mathbf{B} = B_0\hat{\mathbf{k}}$ are[16]

$$\frac{d}{dt}\left(\frac{\dot{x}}{(1 - \beta^2)^{1/2}}\right) = \frac{eB_0}{mc}\dot{y}, \quad \frac{d}{dt}\left(\frac{\dot{y}}{(1 - \beta^2)^{1/2}}\right) = -\frac{eB_0}{mc}\dot{x},$$

$$\frac{d}{dt}\left(\frac{\dot{z}}{(1 - \beta^2)^{1/2}}\right) = \frac{d}{dt}\left(\frac{c}{(1 - \beta^2)^{1/2}}\right) = 0$$

Prove that the velocity of the particle is again given by (2–9) but with

$$\Omega = \Omega_0(1 - \beta^2)^{1/2}, \quad \Omega_0 = eB_0/mc.$$

3

Macroscopic equations

3–1 Introduction
Discussion of the behaviour of an ionized gas in terms of particle orbit theory proves useful in certain situations and provides a convenient starting point for a development of plasma dynamics. However, it is adequate only when the density of charged particles is low enough for their interaction to be ignored, and under other conditions we need a different approach. In this chapter we discuss the hydrodynamic model of plasma behaviour. The value of the hydrodynamic approach lies in the fact that the dynamics of non-conducting fluids has been studied exhaustively and many aspects of their behaviour are now well understood. When the fluid is electrically conducting its motion is, in general, more complex and it is often useful to be able to draw analogies with the behaviour of a non-conducting fluid.

The study of the motion of electrically conducting fluids in electromagnetic fields is known variously as hydromagnetics or magnetohydrodynamics (MHD). The distinguishing feature of hydromagnetics lies in the interaction of electric currents with the magnetic field. Currents due to the motion of the fluid

(i) interact with the magnetic field producing forces which modify the motion of the fluid; and
(ii) induce magnetic fields which alter the existing field.

Hydromagnetics has developed extensively, the impetus coming first from astrophysics, and later from the quest for a controlled thermonuclear reaction. The modern development of the subject began with the work of Alfvén who, in 1942, stated the basic theorem that the magnetic lines of force in a perfectly conducting fluid are 'frozen' into the fluid. In other words, the currents induced in a perfectly conducting fluid moving in a magnetic field act to prevent the motion of the fluid relative to the field. These ideas are discussed in Chapter 4; for the present, we shall establish the equations relating to a fluid description of a plasma (Sections 3–3 to 3–5) and assess their applicability to plasma phenomena (Section 3–6).

3–2 Fluid model of a plasma
Before deriving the hydromagnetic equations from the more fundamental kinetic theory, we shall extrapolate the hydrodynamic equations to describe the behaviour of conducting fluids. In hydrodynamics, the equations of

motion of a fluid are often obtained by plausible appeals to the laws of conservation of mass, momentum, and energy.[20] However, to *derive* these equations one must have recourse to a more fundamental theory—the kinetic theory of fluids. The first equation of hydrodynamics, the continuity equation, expresses the conservation of mass:

$$\frac{\partial \rho}{\partial t} + \mathbf{\nabla}.(\rho \mathbf{U}) = 0 \qquad (3\text{--}1)$$

in which ρ is the mass density and $\mathbf{U}$ the macroscopic flow velocity. No assumptions about the nature of the fluid are used to arrive at (3–1); it is valid for viscous and inviscid fluids alike and its validity may be extrapolated to include electrically conducting fluids. The second equation, the hydrodynamic equation of motion, follows from momentum conservation. For simplicity, confining attention to the case of an inviscid fluid, one gets Euler's equation

$$\rho \left[\frac{\partial \mathbf{U}}{\partial t} + (\mathbf{U}.\mathbf{\nabla})\mathbf{U} \right] = -\mathbf{\nabla} P + \mathbf{F} \qquad (3\text{--}2)$$

in which P is the fluid pressure and $\mathbf{F}$ represents external forces. Thus if the fluid is in a gravitational field then a force $\rho \mathbf{g}$ acts on a unit volume of the fluid, $\mathbf{g}$ being the acceleration due to gravity. From (3–1) and (3–2), we see that an ideal fluid in motion is characterized by the quantities ρ, $\mathbf{U}$, and P; so one further relation is needed for the system to be fully determined. In discussing the flow of an ideal fluid, not only are viscous effects ignored but the effects of thermal conductivity as well. This means that there is no exchange of heat between fluid elements—that is, the motion of the fluid is *adiabatic*. For a compressible fluid the adiabatic condition is simply

$$P\rho^{-\gamma} = \text{const.} \qquad (3\text{--}3)$$

where γ is the ratio of the specific heat at constant pressure to that at constant volume.

When the fluid is electrically conducting there will be an electromagnetic contribution $\mathbf{F}_{em}$ to the force term, given by

$$\mathbf{F}_{em} = q\mathbf{E} + \frac{\mathbf{j} \times \mathbf{B}}{c}$$

where q is the charge density, $\mathbf{j}$ the current density, and $\mathbf{E}$, $\mathbf{B}$ the electric and magnetic fields, respectively. In practice, $\mathbf{F}_{em}$ is usually the dominant contribution to $\mathbf{F}$ (though exceptions to this arise in some astrophysical applications of hydromagnetics) so that (3–2) becomes

$$\rho \left[\frac{\partial \mathbf{U}}{\partial t} + (\mathbf{U}.\mathbf{\nabla})\mathbf{U} \right] = -\mathbf{\nabla} P + q\mathbf{E} + \frac{1}{c}\mathbf{j} \times \mathbf{B} \qquad (3\text{--}4)$$

There will again be an adiabatic condition (3–3). However, one sees at once from (3–1), (3–3), and (3–4) that with the appearance of q, $\mathbf{j}$, $\mathbf{E}$, and $\mathbf{B}$ the number of unknowns in the hydromagnetic equations has risen to 15. The two vector Maxwell equations (2–2), (2–3) determine six unknowns; thus, four further equations are needed to specify the system completely. The equation of conservation of charge provides one of these:

$$\frac{\partial q}{\partial t} + \mathbf{\nabla}.\mathbf{j} = 0 \qquad (3\text{–}5)$$

The remaining three come from the equation relating $\mathbf{j}$ and the electric field, namely Ohm's law; for a moving conductor the electric field[16] is given by $\mathbf{E} + \mathbf{U} \times \mathbf{B}/c$, so that Ohm's law reads

$$\mathbf{j} = \sigma\left(\mathbf{E} + \frac{\mathbf{U} \times \mathbf{B}}{c}\right) \qquad (3\text{–}6)$$

in which σ is the electrical conductivity of the fluid.

In the following sections, (3–1) and (3–3) to (3–6) will be derived from the kinetic equations.

3–3 The moment equations

The kinetic equation

$$\frac{\partial f}{\partial t} + \mathbf{v}.\frac{\partial f}{\partial \mathbf{r}} + \frac{\mathbf{F}}{m}.\frac{\partial f}{\partial \mathbf{v}} = \left(\frac{\partial f}{\partial t}\right)_c \qquad (3\text{–}7)$$

is derived and discussed in Chapters 1 and 10. The moment equations are obtained from (3–7) through multiplying by various functions of velocity, $\psi(\mathbf{v})$, and integrating over velocity space. Essentially, $\psi(\mathbf{v})$ takes the values 1, $\mathbf{v}$ (or v_i), $\mathbf{vv}$ (or v_iv_j), . . ., thus giving rise to the zero-order, first-order, second-order, . . ., *moment equations* respectively. The procedure is straightforward, though tedious for higher orders. It replaces an equation for the distribution function, $f(\mathbf{r}, \mathbf{v}, t)$, by equations for quantities which are functions of $\mathbf{r}$ and t only. The particular choices of $\psi(\mathbf{v})$ ensure that these quantities are related to local mass density ($\psi = m$), momentum ($\psi = m\mathbf{v}$), energy ($\psi = \tfrac{1}{2}mv^2$), and so on.

The average over velocity space of an arbitrary function $\phi(\mathbf{r}, \mathbf{v}, t)$ is defined by

$$\langle\phi\rangle = \frac{1}{n(\mathbf{r}, t)}\int \phi(\mathbf{r}, \mathbf{v}, t)\, f(\mathbf{r}, \mathbf{v}, t)\, d\mathbf{v} \qquad (3\text{–}8)$$

where

$$n(\mathbf{r}, t) = \int f(\mathbf{r}, \mathbf{v}, t)\, d\mathbf{v} \qquad (3\text{–}9)$$

is the particle number density in the spatial volume element $d\mathbf{r}$ at position $\mathbf{r}$ and time t. Hence, multiplying (3–7) by $\psi(\mathbf{v})$ and integrating over velocity space, the first term in the general moment equation is (since ψ is independent of time)

$$\int \psi(\mathbf{v}) \frac{\partial f}{\partial t} d\mathbf{v} = \frac{\partial}{\partial t} \int \psi f \, d\mathbf{v} = \frac{\partial}{\partial t}(n\langle\psi\rangle) \tag{3–10}$$

Similarly, since ψ is also independent of $\mathbf{r}$, the second term

$$\int \psi(\mathbf{v}) \, \mathbf{v} \cdot \frac{\partial f}{\partial \mathbf{r}} \, d\mathbf{v} = \frac{\partial}{\partial \mathbf{r}} \cdot \int \psi \, \mathbf{v} \, f \, d\mathbf{v} = \frac{\partial}{\partial \mathbf{r}} \cdot (n\langle\psi\mathbf{v}\rangle) \tag{3–11}$$

For those applied forces, $\mathbf{F}$, which are independent of velocity

$$\frac{1}{m} \int \psi(\mathbf{v}) \, \mathbf{F} \cdot \frac{\partial f}{\partial \mathbf{v}} d\mathbf{v} = \frac{\mathbf{F}}{m} \cdot \int \psi \frac{\partial f}{\partial \mathbf{v}} d\mathbf{v} = \frac{\mathbf{F}}{m} \cdot \int \left[\frac{\partial(\psi f)}{\partial \mathbf{v}} - f \frac{\partial \psi}{\partial \mathbf{v}} \right] d\mathbf{v} \tag{3–12a}$$

$$= -\frac{n}{m} \mathbf{F} \cdot \left\langle \frac{\partial \psi}{\partial \mathbf{v}} \right\rangle \tag{3–12}$$

The first term in (3–12a) vanishes since we assume that $f \to 0$ sufficiently rapidly as $|\mathbf{v}| \to \infty$ so that for all $\psi(\mathbf{v})$

$$\lim_{|\mathbf{v}|\to\infty} (\psi f) = 0$$

In the case of the Lorentz force, which is velocity-dependent, one has (using the summation convention),

$$\frac{e}{mc} \int \psi(\mathbf{v})(\mathbf{v} \times \mathbf{B})_i \frac{\partial f}{\partial v_i} d\mathbf{v}$$

in which e is the charge on the particle. Since $(\mathbf{v} \times \mathbf{B})_i$ does not contain v_i this may be rewritten

$$\frac{e}{mc} \int \left[\frac{\partial}{\partial v_i}(\psi f(\mathbf{v} \times \mathbf{B})_i) - f(\mathbf{v} \times \mathbf{B})_i \frac{\partial \psi}{\partial v_i} \right] d\mathbf{v}$$

As in (3–12a), the first of these terms is zero, giving

$$\frac{e}{mc} \int \psi(\mathbf{v})(\mathbf{v} \times \mathbf{B}) \cdot \frac{\partial f}{\partial \mathbf{v}} d\mathbf{v} = -\frac{ne}{mc} \left\langle (\mathbf{v} \times \mathbf{B}) \cdot \frac{\partial \psi}{\partial \mathbf{v}} \right\rangle \tag{3–13}$$

From (3–10) to (3–13), therefore, it follows that the *general moment equation* may be written

$$\frac{\partial(n\langle\psi\rangle)}{\partial t} + \frac{\partial}{\partial \mathbf{r}}\cdot(n\langle\mathbf{v}\psi\rangle) - \frac{n}{m}\mathbf{F}\cdot\left\langle\frac{\partial\psi}{\partial\mathbf{v}}\right\rangle - \frac{ne}{mc}\left\langle(\mathbf{v}\times\mathbf{B})\cdot\frac{\partial\psi}{\partial\mathbf{v}}\right\rangle$$

$$= \int \psi(\mathbf{v})\left(\frac{\partial f}{\partial t}\right)_c d\mathbf{v} \qquad (3\text{--}14)$$

where **F** now refers to velocity-independent forces only. To evaluate the moment of the collision term (the right-hand side of (3–14)), one needs to know the form of $(\partial f/\partial t)_c$. It is necessary, therefore, to adopt some model for describing the particle interactions. Unfortunately, for any realistic model $(\partial f/\partial t)_c$ is, in general, a complicated, non-linear, integral function of f. Rather than discuss particular models for the collision term, assumptions will be made concerning its moments. All the assumptions to be used are valid for the Boltzmann collision integral discussed in Chapter 1. Remembering that $(\partial f/\partial t)_c$ represents the rate of change of f due to collisions, a comparison of (3–7) and (3–14) suggests that the moment of the collision term might represent the rate of change of $n\langle\psi\rangle$ due to collisions, i.e.,

$$\int \psi(\mathbf{v})\left(\frac{\partial f}{\partial t}\right)_c d\mathbf{v} = \left(\frac{\partial}{\partial t}(n\langle\psi\rangle)\right)_c \qquad (3\text{--}15)$$

This will be taken as an assumption, the consequences of which can always be checked once the form of the collision term is given. Also, since our principal interest is in electromagnetic forces, **F** will be replaced by $e\mathbf{E}$, so that (3–14) becomes

$$\frac{\partial}{\partial t}(n\langle\psi\rangle) + \frac{\partial}{\partial \mathbf{r}}\cdot(n\langle\mathbf{v}\psi\rangle) - \frac{ne}{m}\mathbf{E}\cdot\left\langle\frac{\partial\psi}{\partial\mathbf{v}}\right\rangle$$

$$- \frac{ne}{mc}\left\langle(\mathbf{v}\ \mathbf{B})\cdot\frac{\partial\psi}{\partial\mathbf{v}}\right\rangle = \left(\frac{\partial}{\partial t}(n\langle\psi\rangle)\right)_c \qquad (3\text{--}16)$$

This general moment equation is also known as the *transfer equation* since, depending on the choice of ψ, it describes the transfer of mass, momentum, energy, and so on.

Thus the kinetic equation for the distribution function f is replaced by a set of equations containing averaged quantities. It is necessarily a *set* of equations. For suppose one took only the zero-order moment equation, $\psi = 1$; this contains n and $\langle\mathbf{v}\rangle$. Since it is a scalar equation it may be regarded as an equation for n; thus, a vector equation for $\langle\mathbf{v}\rangle$ is required. This comes from the first-order moment equation with $\psi = \mathbf{v}$. However, this in turn contains $\langle\mathbf{vv}\rangle$ as well as $\langle\mathbf{v}\rangle$ and a further six scalar equations are necessary ($\langle v_i v_j\rangle$ contains only six independent elements since $\langle v_i v_j\rangle = \langle v_j v_i\rangle$). Clearly, each moment equation introduces the next higher-order velocity moment, and some method of breaking the chain of equations and restricting the variables to a manageable number must be devised. This question is discussed further in Section 3–5.

Before writing down the first few moment equations specifically, we will define some of the physical quantities occurring in them. The first of these, the local particle number density, is already defined by (3–9). Remember that there exists a distribution function, kinetic equation, and a set of moment equations for each type of particle present in the plasma. For simplicity, we consider a plasma consisting of protons and electrons only and distinguish proton and electron densities, velocities, and so on by $+$ and $-$ superscripts respectively. Wherever possible, we shall write an equation applying equally to both types of particle without superscripts to save writing two separate equations. This is consistent with what has been done so far in this section; for example, (3–9) applies to both types of particle and, therefore, represents the two equations

$$n^+ = \int f^+ \, d\mathbf{v} \qquad n^- = \int f^- \, d\mathbf{v} \qquad (3\text{–}17)$$

Similarly,
$$\mathbf{u} \equiv \langle \mathbf{v} \rangle = \frac{1}{n} \int \mathbf{v} f \, d\mathbf{v} \qquad (3\text{–}18)$$

represents two equations defining the local mean velocities $\mathbf{u}^+$ and $\mathbf{u}^-$. With the definitions (3–17) and (3–18) a local centre of mass velocity may be defined:

$$\mathbf{U} = (m^+ n^+ \mathbf{u}^+ + m^- n^- \mathbf{u}^-)/(m^+ n^+ + m^- n^-) \qquad (3\text{–}19)$$

This is also called the *stream* or *flow velocity*. The velocity of a particle relative to the centre of mass

$$\mathbf{w} = \mathbf{v} - \mathbf{U} \qquad (3\text{–}20)$$

is called the *random* or *peculiar velocity*. Pressure and heat tensors are usually defined in terms of this velocity.† The pressure tensors are given by

$$p_{ij} = mn\langle w_i w_j \rangle = m \int w_i w_j f \, d\mathbf{v} = p_{ji} \qquad (3\text{–}21)$$

and the heat tensors by

$$q_{ijk} = mn\langle w_i w_j w_k \rangle = m \int w_i w_j w_k f \, d\mathbf{v} \qquad (3\text{–}22)$$

The zero-order moment equation is obtained from (3–16) by putting

† Alternative definitions are given in Section 3–4.

$\psi = 1$; hence, using (3–18),

$$\frac{\partial n}{\partial t} + \mathbf{\nabla}.(n\mathbf{u}) = \left(\frac{\partial n}{\partial t}\right)_c \qquad (3\text{–}23)$$

Similarly for the first-order equation, putting $\psi = mv_i$ we get for the ith component of the equation

$$\frac{\partial(mnu_i)}{\partial t} + \mathbf{\nabla}.(mn\langle v v_i\rangle) - neE_i - \frac{ne}{c}(\mathbf{u} \times \mathbf{B})_i = \left(\frac{\partial}{\partial t}(mnu_i)\right)_c \qquad (3\text{–}24)$$

For the purposes of this discussion, in the general second-order moment equation $[\psi = (m/2)v_i v_j]$ it is adequate to take the cases $i = j$ and sum over $i = 1, 2, 3$. Then, $\psi = \frac{1}{2}mv^2$, and this gives the *energy equation*,

$$\frac{\partial}{\partial t}(n\langle \tfrac{1}{2}mv^2\rangle) + \mathbf{\nabla}.(n\langle \tfrac{1}{2}mv^2\mathbf{v}\rangle) - ne\mathbf{E}.\mathbf{u} = \left(\frac{\partial}{\partial t}(n\langle \tfrac{1}{2}mv^2\rangle)\right) \qquad (3\text{–}25)$$

In dealing with the collision terms in (3–23) to (3–25), we use certain properties of individual collisions. Since the plasma consists of protons and electrons only, collision processes which change the number of particles (such as recombination and ionization) are taken to be negligible; in other words, collisions conserve numbers of particles. Further, in a collision involving only electrons the total electron momentum and energy must be conserved, and similarly for proton–proton collisions. These considerations lead one to suspect that the right-hand side of (3–23) is zero, and those of (3–24) and (3–25) represent the exchange (between electrons and protons) of momentum and energy respectively. That this is so may be shown for certain forms of the collision integral (for example, the Boltzmann form). However, the right-hand sides of these equations represent the rates of change, due to collisions, of the number, momentum, and energy densities *within the volume element* dr. If an electron within dr interacts with another electron outside dr it is clear that, although total electron density, momentum, and energy are conserved, the electron density, momentum, and energy within dr need not be conserved by this collision. One must assume, therefore, that collisions take place only with neighbouring particles (i.e., within dr). In situations where particle interactions are mainly due to short-range, large-angle collisions (as in hydrodynamics), this assumption can be justified by making the scale length of the volume element dr much bigger than the mean free path. For a plasma, where interactions are predominantly long-range and weak, a significant deflection of a particle is caused by many small deflections. Particle trajectories are thus relatively smooth and it is not obvious that the same justification will suffice. However, assuming that the above condition on the size of dr is adequate for plasmas, too, it is clear that a second averaging has been introduced. The quantities, $n(\mathbf{r}, t)$, $\mathbf{u}(\mathbf{r}, t)$, and so on appearing in the moment equations are averages over a small, but finite, volume around the position $\mathbf{r}$.

With these assumptions concerning the moments of the collision terms, (3–23) to (3–25) become

$$\frac{\partial n}{\partial t} + \mathbf{V}.(n\mathbf{u}) = 0 \tag{3–26}$$

$$\frac{\partial(mnu_i)}{\partial t} + \mathbf{V}.(mn\langle vv_i\rangle) - neE_i - \frac{ne}{c}(\mathbf{u} \times \mathbf{B})_i = \pm K_i \tag{3–27}$$

$$\frac{\partial}{\partial t}(n\langle\tfrac{1}{2}mv^2\rangle) + \mathbf{V}.(n\langle\tfrac{1}{2}mv^2\mathbf{v}\rangle) - ne\mathbf{E}.\mathbf{u} = \pm H \tag{3–28}$$

where

$$\mathbf{K} = \int m^+\mathbf{v}\left(\frac{\partial f^+}{\partial t}\right)_c dv = -\int m^-\mathbf{v}\left(\frac{\partial f^-}{\partial t}\right)_c dv \tag{3–29}$$

and

$$H = \int \tfrac{1}{2}m^+v^2\left(\frac{\partial f^+}{\partial t}\right)_c dv = -\int \tfrac{1}{2}m^-v^2\left(\frac{\partial f^-}{\partial t}\right)_c dv \tag{3–30}$$

represent the rate of transfer of momentum and energy from electrons to protons. The $+$ and $-$ signs preceding K and H in (3–27) and (3–28) refer to the proton and electron equations respectively, since if the protons in $d\mathbf{r}$ are gaining momentum (energy) due to collisions at a rate given by $\mathbf{K}(H)$, the electrons must be losing momentum (energy) at the same rate. This is also stated in (3–29) and (3–30). Note that in those plasmas where ionization and recombination processes are non-negligible (3–26) also has source and sink terms on the right-hand side.

Now, using the definitions (3–18) to (3–22), $\langle v_i v_j\rangle$ and $\langle v^2 v_i\rangle$ may be rewritten

$$\langle v_i v_j\rangle = \langle(w_i + U_i)(w_j + U_j)\rangle = \frac{p_{ij}}{mn} + u_i U_j + u_j U_i - U_i U_j \tag{3–31}$$

and

$$\langle v^2 v_i\rangle = \langle v_j v_j v_i\rangle = \langle(w_j + U_j)(w_j + U_j)(w_i + U_i)\rangle$$

$$= \frac{q_{jji}}{mn} + \frac{p_{ji}}{mn}U_i + \frac{2p_{ij}}{mn}U_j + 2u_j U_j U_i + u_i U_j U_j - 2U_i U_j U_j \tag{3–32}$$

Substituting (3–31) and (3–32) in (3–27) and (3–28), we get

$$\frac{\partial}{\partial t}(mnu_i) + \frac{\partial}{\partial r_j}[p_{ij} + mn(u_i U_j + u_j U_i - U_i U_j)] - neE_i - \frac{ne}{c}(\mathbf{u} \times \mathbf{B})_i$$

$$= \pm K_i \tag{3–27a}$$

and

$$\frac{\partial}{\partial t}[\tfrac{1}{2}p_{ii} + mn(\mathbf{u}.\mathbf{U} - \tfrac{1}{2}U^2)] + \frac{\partial}{\partial r_j}[\tfrac{1}{2}q_{iij} + \tfrac{1}{2}p_{ii}U_j + p_{ij}U_i$$

$$+ mn(\mathbf{u}.\mathbf{U}U_j + \tfrac{1}{2}u_jU^2 - U_jU^2)] - ne\mathbf{E}.\mathbf{u} = \pm H \quad (3\text{--}28\text{a})$$

The description of the plasma in terms of (3–26), (3–27a), and (3–28a) is a two-fluid model since these equations refer to both protons and electrons. It is often more desirable to have a one-fluid description in which proton and electron quantities are combined by the following definitions:

Total number density

$$N = n^+ + n^- \quad (3\text{--}33)$$

Mass density

$$\rho = m^+n^+ + m^-n^- \quad (3\text{--}34)$$

Charge density

$$q = e^+n^+ + e^-n^- \quad (3\text{--}35)$$

Current density

$$\mathbf{j} = e^+n^+\mathbf{u}^+ + e^-n^-\mathbf{u}^- \quad (3\text{--}36)$$

Total pressure tensor

$$P_{ij} = p_{ij}^+ + p_{ij}^- = P_{ji} \quad (3\text{--}37)$$

It is also convenient to introduce the concept of kinetic temperature, which is defined in terms of the mean random energy of the plasma. From elementary gas kinetic theory, the energy associated with each degree of freedom of a gas is $\tfrac{1}{2}\kappa T$, where κ is Boltzmann's constant and T is the kinetic temperature of the gas. That is,

$$3N\frac{\kappa T}{2} = \tfrac{1}{2}m^+ \int w^2 f^+ \, d\mathbf{v} + \tfrac{1}{2}m^- \int w^2 f^- \, d\mathbf{v}$$

Using (3–21)

$$3N\kappa T = p_{ii}^+ + p_{ii}^- = P_{ii}$$

i.e.,

$$T = \frac{P_{ii}}{3N\kappa} \quad (3\text{--}38)$$

Finally, the components of the heat flux vector are defined as

$$Q_i = \tfrac{1}{2}m^+ \int w^2 w_i f^+ \, d\mathbf{v} + \tfrac{1}{2}m^- \int w^2 w_i f^- \, d\mathbf{v}$$

$$= \tfrac{1}{2}(q_{ijj}^+ + q_{ijj}^-) \tag{3-39}$$

by (3–22).

Applying (3–26) to electrons and protons, respectively:

$$\frac{\partial n^-}{\partial t} + \nabla \cdot (n^- \mathbf{u}^-) = 0 \tag{3-26)$^-$}$$

$$\frac{\partial n^+}{\partial t} + \nabla \cdot (n^+ \mathbf{u}^+) = 0 \tag{3-26)$^+$}$$

Multiplying (3–26)$^-$ by m^- and (3–26)$^+$ by m^+ and adding gives, using (3–19) and (3–34),

$$\frac{\partial \rho}{\partial t} + \nabla \cdot (\rho \mathbf{U}) = 0 \tag{3-40}$$

which is the *equation of continuity* or *mass conservation*. If (3–26)$^-$ and (3–26)$^+$ are multiplied by e$^-$ and e$^+$, respectively, then adding we get, using (3–35) and (3–36),

$$\frac{\partial q}{\partial t} + \nabla \cdot \mathbf{j} = 0 \tag{3-41}$$

the *equation of charge conservation*.

Likewise, adding the electron and proton versions of (3–27a) gives, using (3–19) and (3–34) to (3–37),

$$\frac{\partial}{\partial t}(\rho U_i) + \frac{\partial}{\partial r_j}(P_{ij} + \rho U_i U_j) - qE_i - \frac{1}{c}(\mathbf{j} \times \mathbf{B})_i = 0$$

Using (3–40) this reduces to

$$\rho\frac{\partial U_i}{\partial t} + \frac{\partial P_{ij}}{\partial r_j} + \rho U_j \frac{\partial U_i}{\partial r_j} - qE_i - \frac{1}{c}(\mathbf{j} \times \mathbf{B})_i = 0 \tag{3-42a}$$

or, in vector notation,

$$\rho\frac{D\mathbf{U}}{Dt} + \nabla \cdot \mathbf{P} - q\mathbf{E} - \frac{1}{c}(\mathbf{j} \times \mathbf{B}) = 0 \tag{3-42b}$$

where $\mathbf{P}$ is the dyadic form of the pressure tensor (cf. Appendix 1) and

$$\frac{D}{Dt} \equiv \frac{\partial}{\partial t} + \mathbf{U}.\nabla \tag{3-43}$$

Treating (3–28a) in the same way, and using (3–19) and (3–34) to (3–39), gives the *energy equation*

$$\frac{\partial}{\partial t}(\tfrac{1}{2}P_{ii} + \tfrac{1}{2}\rho U^2) + \nabla.[(\tfrac{1}{2}P_{ii} + \tfrac{1}{2}\rho U^2)\mathbf{U} + \mathbf{Q} + \mathbf{P}.\mathbf{U}] - \mathbf{j}.\mathbf{E} = 0 \tag{3-44a}$$

or, in view of (3–38),

$$\frac{\partial}{\partial t}(\tfrac{3}{2}N\kappa T + \tfrac{1}{2}\rho U^2) + \nabla.[(\tfrac{3}{2}N\kappa T + \tfrac{1}{2}\rho U^2)\mathbf{U} + \mathbf{Q} + \mathbf{P}.\mathbf{U}] - \mathbf{j}.\mathbf{E} = 0 \tag{3-44b}$$

In these equations,

$$\nabla.(\mathbf{P}.\mathbf{U}) \equiv \frac{\partial}{\partial r_j}(P_{ji}U_i)$$

To see the physical significance of the terms in (3–44b), integrate over a small volume. The first term expresses the rate of change of random or internal energy $(\tfrac{3}{2}N\kappa T)$ and ordered or flow energy $(\tfrac{1}{2}\rho U^2)$ with time. Applying Gauss' theorem to the second term, we see that $(\tfrac{3}{2}N\kappa T + \tfrac{1}{2}\rho U^2)\mathbf{U}$ is the flux of random and ordered energy across the surface bounding the volume, $\mathbf{Q}$ is the heat flux across this surface, while $\mathbf{P}.\mathbf{U}$ represents the rate at which the pressure does work on the surface. The final term in (3–44b) is the rate of Joule heating.

An alternative form of (3–44) is obtained by using (3–40), (3–42b), and (3–43):

$$\frac{D}{Dt}(\tfrac{1}{2}P_{ii}) + \tfrac{1}{2}P_{ii}\nabla.\mathbf{U} + \mathbf{P}:\nabla\mathbf{U} + \nabla.\mathbf{Q} + (q\mathbf{U} - \mathbf{j}).\left(\mathbf{E} + \frac{\mathbf{U} \times \mathbf{B}}{c}\right) = 0$$

$$\tag{3-44c}$$

where $\qquad\qquad\qquad \mathbf{P}:\nabla\mathbf{U} \equiv P_{ij}\dfrac{\partial U_i}{\partial r_j}$

3–4 Alternative form of the moment equations

In the previous section, we defined the pressure and the heat tensors, (3–21) and (3–22), in terms of the random velocity, $\mathbf{w}$, that is, the velocity relative to the centre of mass velocity $\mathbf{U}$. These definitions are not unique and sometimes the tensors are defined in terms of the random velocity relative to the local mean velocity $\mathbf{u}$ (for electrons relative to $\mathbf{u}^-$ and for protons relative to $\mathbf{u}^+$). Thus

$$p_{ij} = mn\langle(v_i - u_i)(v_j - u_j)\rangle = m\int(v_i - u_i)(v_j - u_j)f \, d\mathbf{v} = p_{ji} \tag{3-45}$$

and
$$q_{ijk} = mn\langle(v_i - u_i)(v_j - u_j)(v_k - u_k)\rangle$$

$$= m\int (v_i - u_i)(v_j - u_j)(v_k - u_k)\, f\, d\mathbf{v} \qquad (3\text{--}46)$$

Since p_{ij}^+, p_{ij}^- (and q_{ijk}^+, q_{ijk}^-) are defined relative to different mean velocities $\mathbf{u}^+$, $\mathbf{u}^-$, there are no physically meaningful definitions equivalent to (3–37) to (3–39). (Partial pressures can be added to get a total pressure only if they are defined relative to the same flow velocity.) However, a temperature and a heat flux vector may be defined for protons and electrons separately; by analogy with (3–38), (3–39)

$$T = \frac{p_{ii}}{3n\kappa} \qquad (3\text{--}47)$$

and
$$q_i = \tfrac{1}{2}q_{ijj} \qquad (3\text{--}48)$$

The moment equations are derived in the same way. The zero-order equations are unaltered:

$$\frac{\partial n^+}{\partial t} + \mathbf{V}.(n^+\mathbf{u}^+) = 0 \qquad (3\text{--}26)^+$$

$$\frac{\partial n^-}{\partial t} + \mathbf{V}.(n^-\mathbf{u}^-) = 0 \qquad (3\text{--}26)^-$$

The higher-order equations are a little easier to evaluate with the alternate definitions (3–45) and (3–46). For example,

$$\langle v_i v_j\rangle = \langle\overline{(v_i - u_i + u_i)(v_j - u_j + u_j)}\rangle = \frac{p_{ij}}{mn} + u_i u_j$$

where the cross-terms have vanished since

$$\langle \mathbf{v} - \mathbf{u}\rangle = 0$$

The momentum equations now read

$$n^+m^+\left(\frac{\partial}{\partial t} + \mathbf{u}^+.\mathbf{V}\right)\mathbf{u}^+ + \mathbf{V}.\mathbf{p}^+ - n^+e^+\left(\mathbf{E} + \frac{1}{c}(\mathbf{u}^+ \times \mathbf{B})\right) = \mathbf{K} \quad (3\text{--}49)^+$$

$$n^-m^-\left(\frac{\partial}{\partial t} + \mathbf{u}^-.\mathbf{V}\right)\mathbf{u}^- + \mathbf{V}.\mathbf{p}^- - n^-e^-\left(\mathbf{E} + \frac{1}{c}(\mathbf{u}^- \times \mathbf{B})\right) = -\mathbf{K} \quad (3\text{--}49)^-$$

for protons and electrons, respectively. Similarly, the energy equations are

$$\left(\frac{\partial}{\partial t} + \mathbf{u}^+.\nabla\right)(\tfrac{3}{2}n^+\kappa T^+) + \tfrac{3}{2}n^+\kappa T^+\nabla.\mathbf{u}^+ + \mathbf{p}^+ : \nabla\mathbf{u}^+ + \nabla.\mathbf{q}^+$$

$$= -\mathbf{u}^+.\mathbf{K} + H \quad (3\text{--}50)^+$$

and

$$\left(\frac{\partial}{\partial t} + \mathbf{u}^-.\nabla\right)(\tfrac{3}{2}n^-\kappa T^-) + \tfrac{3}{2}n^-\kappa T^-\nabla.\mathbf{u}^- + \mathbf{p}^- : \nabla\mathbf{u}^- + \nabla.\mathbf{q}^-$$

$$= \mathbf{u}^-.\mathbf{K} - H \quad (3\text{--}50)^-$$

The counterparts of (3–49) and (3–50) in Section 3–3 are (3–27a) and (3–28a). This alternative form of the moment equations is appropriate whenever proton and electron temperatures differ appreciably. Since collisions are responsible for the exchange of energy between protons and electrons, such a temperature difference *may* occur when collisions are sufficiently infrequent.

Since we cannot combine p_{ij}^+ and p_{ij}^- to form a physically meaningful total pressure tensor as in Section 3–3, we cannot usefully add the species equations in (3–49) and (3–50) to give equations analogous to (3–42) and (3–44). Accordingly, this alternative form of the moment equations is properly regarded as a *two-fluid* description of the plasma. In the development of hydromagnetics, the one-fluid treatment of Section 3–3 is usually the appropriate description and this will be discussed in the remainder of this chapter.

3–5 Hydromagnetic equations

We now summarize the development of the moment equations given in Section 3–3:

(i) Zero-, first-, and second-order moment equations were derived for protons and electrons separately.

(ii) The zero-order equations were combined to give equations of mass and charge conservation, (3–40) and (3–41).

(iii) The first-order equations were added to give the momentum equation, (3–42).

(iv) The second-order equations were added to give the energy equation, (3–44).

In the discussion following the derivation of the general moment equation, (3–16), we saw that the set of moment equations is not closed. Since only one of a possible six, independent, second-order equations has been derived, it follows that, for the set of equations derived in Section 3–3, five of the six independent elements of p_{ij} must remain indeterminate; so also must q_{ijj}. The equations referred to in (i) are to be regarded as equations for n^+, n^-, $\mathbf{u}^+$, $\mathbf{u}^-$, p_{ii}^+, and p_{ii}^-.†

† Instead of treating (3–28a) as an equation for p_{11}, p_{22}, or p_{33} it may be treated as an equation for $p_{11} + p_{22} + p_{33}$. There are then only five independent elements of p_{ij}, all of which are indeterminate as far as the set of equations is concerned.

Combining proton and electron equations, as in (ii) to (iv), introduced new equations in which the variables, ρ, q, $\mathbf{U}$, $\mathbf{j}$, and P_{ii} have replaced n^+, n^-, $\mathbf{u}^+$, $\mathbf{u}^-$, p_{ii}^+, and p_{ii}^-. However, this transformation of equations is incomplete since in both (iii) and (iv) two equations have been combined to yield only one. In hydromagnetics (iv) is satisfactory: this is a property of the hydromagnetic approximation and has already been anticipated in that only one variable P_{ii} has been defined in terms of the two variables p_{ii}^+ and p_{ii}^- (cf. definitions of ρ and q in terms of n^+, n^-, or $\mathbf{U}$ and $\mathbf{j}$ in terms of $\mathbf{u}^+$, $\mathbf{u}^-$). The further vector equation, to be obtained from (3–27a), cannot be written in terms of ρ, q, $\mathbf{U}$, $\mathbf{j}$, and P_{ii} without approximation. Before getting this equation, therefore, we shall discuss the further approximations involved in the hydromagnetic equations. The need for these is twofold: (a) closure of the equations; (b) simplification of the equations, in particular, so that one has a one-fluid description (involving ρ, q, $\mathbf{U}$, $\mathbf{j}$, and P_{ii}) rather than a two-fluid description (involving n^+, n^-, $\mathbf{u}^+$, $\mathbf{u}^-$, p_{ii}^+, and p_{ii}^-).

There are two general situations in which the moment equations may be closed. One of these is the so-called 'cold plasma' limit; here the thermal, or random, energies are small enough for the pressure and heat tensors to be neglected:

$$\left.\begin{array}{ll} p_{ij} = 0 & \text{all } i, j \\ q_{ijk} = 0 & \text{all } i, j, k \end{array}\right\} \tag{3–51}$$

The cold plasma is a highly idealized model. However, in view of the simplification of the equations, it has been given much consideration. The cold plasma equations and the precise nature of the limit on the thermal velocities is discussed in Section 3–6.

The other situation in which the moment equations may be closed arises when collisions are dominant. Collisions, being a random process, tend to smooth out any anisotropies so that within a few collision times the local distribution functions approach Maxwellian distributions

$$f_M(n, \mathbf{U}, T) = n\left(\frac{m}{2\pi\kappa T}\right)^{3/2} \exp\left(-\frac{m(\mathbf{v} - \mathbf{U})^2}{2\kappa T}\right) \tag{3–52}$$

In this case, both q_{ijk} and the non-diagonal elements of p_{ij} are small and $p_{11} \approx p_{22} \approx p_{33}$. Further, since strong electric fields oppose charge separation (cf. Chapter 1)

$$n^+ \approx n^- \approx N/2 \quad \text{and} \quad \mathbf{u}^+ \approx \mathbf{u}^- \approx \mathbf{U} \quad (\text{since } m^- \ll m^+) \tag{3–52a}$$

This is the basis of the Chapman–Enskog method [21] for truncating the moment equations in which the distribution function is expanded about $f_M(N/2, \mathbf{U}, T)$. To derive the hydromagnetic equations, however, we can neglect q_{ijk} completely and assume that the pressure tensor is exactly isotropic. That is,

$$q_{ijk} = 0 \qquad \text{all } i, j, k$$

$$\left. \begin{array}{l} p_{11} = p_{22} = p_{33} = \tfrac{1}{3}p_{ii} = p \\ p_{ij} = 0 \qquad i \neq j \end{array} \right\} \tag{3-53}$$

Assumption (3-53) is equivalent to neglect of viscosity (cf. Section 5-2). Substituting (3-53) into (3-42b) reduces the momentum equation to (3-4)

$$\rho \frac{DU}{Dt} = -\nabla P + q\mathbf{E} + \frac{1}{c}(\mathbf{j} \times \mathbf{B}) \tag{3-54}$$

where, by analogy with (3-53), $P = P_{ii}/3$.

Similarly, the energy equation, (3-44c), becomes

$$\frac{D}{Dt}(\tfrac{3}{2}P) + \tfrac{5}{2}P\nabla.\mathbf{U} = (\mathbf{j} - q\mathbf{U}).\left(\mathbf{E} + \frac{\mathbf{U} \times \mathbf{B}}{c}\right) \tag{3-55}$$

or, substituting for $\nabla.\mathbf{U}$ from (3-40),

$$\frac{D}{Dt}(\tfrac{3}{2}P) - \frac{5}{2}\frac{P}{\rho}\frac{D\rho}{Dt} = (\mathbf{j} - q\mathbf{U}).\left(\mathbf{E} + \frac{\mathbf{U} \times \mathbf{B}}{c}\right)$$

i.e.,

$$\frac{D}{Dt}(P\rho^{-5/3}) = \tfrac{2}{3}\rho^{-5/3}(\mathbf{j} - q\mathbf{U}).\left(\mathbf{E} + \frac{\mathbf{U} \times \mathbf{B}}{c}\right) \tag{3-56}$$

Consider now the derivation of a second equation from the two equations contained in (3-27a). We do this by first multiplying the proton and electron equations by e^+/m^+ and e^-/m^-, respectively, and then adding. Provided that terms quadratic in $\mathbf{U}$ and $\mathbf{j}$ may be neglected, we get (cf. Problem 3-3)

$$\frac{\partial j_k}{\partial t} + \frac{\partial}{\partial r_i}\left[e\left(\frac{p_{ik}^+}{m^+} - \frac{p_{ik}^-}{m^-}\right)\right] - e^2 E_k\left(\frac{n^+}{m^+} + \frac{n^-}{m^-}\right) - \frac{e^2}{c}\left[\left(\frac{n^-}{m^-}\mathbf{u}^- + \frac{n^+}{m^+}\mathbf{u}^+\right) \times \mathbf{B}\right]_k$$

$$= eK_k\left(\frac{1}{m^+} + \frac{1}{m^-}\right) \tag{3-57}$$

where e is the charge on the proton. From (3-52), (3-52a), and the definition of pressure (3-21) it follows that

$$p^+ \simeq p^- \simeq P/2 \tag{3-58}$$

Furthermore, since $m^+ \gg m^-$, (3-57) becomes

$$\frac{\partial \mathbf{j}}{\partial t} - \frac{e}{2m^-}\nabla P - \frac{\rho e^2}{m^- m^+}\left(\mathbf{E} + \frac{\mathbf{U} \times \mathbf{B}}{c}\right) + \frac{e}{m^- c}(\mathbf{j} \times \mathbf{B}) = \frac{e}{m^-}\mathbf{K} \tag{3-59}$$

The final step needed to close the set of moment equations is to write the collision moment $\mathbf{K}$ in terms of one-fluid variables. The assumption made takes the momentum exchange between protons and electrons as proportional to their relative mean velocities,† i.e.,

$$\mathbf{K} \propto (\mathbf{u}^- - \mathbf{u}^+) = -\frac{m^+\mathbf{j}}{\rho e} \qquad (3\text{-}60)$$

It is convenient to choose the constant of proportionality such that

$$\mathbf{K} = -\frac{\rho e}{\sigma m^+}\mathbf{j} \qquad (3\text{-}60a)$$

since, if $\partial\mathbf{j}/\partial t$, ∇P, and $\mathbf{B}$ are all zero, (3–59) becomes Ohm's law with σ the electrical conductivity. Thus (3–59) becomes

$$\frac{m^+m^-}{\rho e^2}\cdot\frac{\partial\mathbf{j}}{\partial t} = \mathbf{E} + \frac{\mathbf{U}\times\mathbf{B}}{c} - \frac{m^+}{\rho ec}(\mathbf{j}\times\mathbf{B}) + \frac{m^+}{2\rho e}\nabla P - \frac{\mathbf{j}}{\sigma} \qquad (3\text{-}61)$$

Since $\mathbf{E}$ and $\mathbf{B}$ are also variables, the complete, closed set of equations for ρ, q, $\mathbf{U}$, $\mathbf{j}$, P, $\mathbf{E}$, and $\mathbf{B}$ are the two vector Maxwell equations

$$\nabla\times\mathbf{E} = -\frac{1}{c}\frac{\partial\mathbf{B}}{\partial t} \qquad (3\text{-}62)$$

$$\nabla\times\mathbf{B} = \frac{1}{c}\frac{\partial\mathbf{E}}{\partial t} + \frac{4\pi}{c}\mathbf{j} \qquad (3\text{-}63)$$

the mass and charge conservation equations, (3–40) and (3–41), the momentum and energy equations, (3–54) and (3–56), together with (3–61), which is known as the *generalized Ohm's law*.

To reduce (3–61) further, we consider the relative magnitude of the terms in it. As a preliminary to this, a dimensional analysis‡ of (3–62) shows that

$$\frac{E}{B} \sim \frac{\omega L}{c} \qquad (3\text{-}64)$$

where L and ω^{-1} are, respectively, a length and a time over which the fields change appreciably. Hydromagnetics involves the interaction between flow and fields and for this to be significant one would anticipate that

$$U \sim \omega L \qquad (3\text{-}65)$$

---

† This assumption may be justified using the Boltzmann collision integral.

‡ In this and the following analyses, the vector character of the equations is ignored; this *may* be a dangerous simplification and should be checked in particular instances.

which implies that $\omega L/c \ll 1$. From (3–63), it follows that

$$\left|\frac{1}{c}\frac{\partial \mathbf{E}}{\partial t}\right| \bigg/ |\nabla \times \mathbf{B}| \sim (\omega L/c)^2 \ll 1 \qquad (3\text{–}66)$$

Hence, in hydromagnetics, (3–63) may be replaced by

$$\nabla \times \mathbf{B} = \frac{4\pi}{c}\mathbf{j} \qquad (3\text{–}67)$$

Now, if we compare the magnitudes of all the terms in the generalized Ohm's law, taking them in the same order as they appear in (3–61), the result is

$$\left(\frac{\omega}{\omega_p}\right)^2\left(\frac{c}{U}\right)^2 : 1 : 1 : \left(\frac{\omega}{\omega_p}\right)\left(\frac{\Omega_e}{\omega_p}\right)\left(\frac{c}{U}\right)^2 : \left(\frac{\omega}{\Omega_i}\right)\left(\frac{c_s}{U}\right)^2 : \left(\frac{\omega}{\omega_p}\right)\left(\frac{\nu_c}{\omega_p}\right)\left(\frac{c}{U}\right)^2 \qquad (3\text{–}68)$$

where $\qquad \Omega_e = \dfrac{e^-B}{m^-c} \qquad \Omega_i = \dfrac{e^+B}{m^+c} \qquad c_s \sim (P/\rho)^{1/2} \qquad (3\text{–}69)$

are the cyclotron frequencies and sound speed, ω_p is the electron plasma frequency, and ν_c is the electron-proton collision frequency. In comparing magnitudes (3–64), (3–65), and (3–67) have been used. For the last term in (3–68), (3–60a) was used to replace j/σ by $m^+K/\rho e$; also

$$K \sim nm^- |\mathbf{u}^- - \mathbf{u}^+| \nu_c = \frac{m^-j\nu_c}{e} \qquad (3\text{–}60b)$$

Thus the term in $\partial \mathbf{j}/\partial t$ may be neglected if ω is such that

$$\frac{\omega}{\omega_p} \ll \frac{U}{c} \qquad (3\text{–}70)$$

The term in $\mathbf{j} \times \mathbf{B}$, the *Hall term*, may be neglected provided

$$\frac{\omega\Omega_e}{\omega_p^2} \ll \left(\frac{U}{c}\right)^2 \qquad (3\text{–}71)$$

as may the pressure term if

$$\frac{\omega}{\Omega_i} \ll \left(\frac{U}{c_s}\right)^2 \qquad (3\text{–}72)$$

If conditions (3–70) to (3–72) are all satisfied, (3–61) becomes

$$\mathbf{j} = \sigma\left(\mathbf{E} + \frac{\mathbf{U} \times \mathbf{B}}{c}\right) \qquad (3\text{–}6)$$

in which case σ may be interpreted as the electrical conductivity as in Section 3–2. Sometimes the assumption of *infinite* or *perfect conductivity* is made, so that (3–6) becomes

$$\mathbf{E} + \frac{\mathbf{U} \times \mathbf{B}}{c} = 0 \tag{3–73}$$

The condition for this is, from (3–68),

$$\frac{\omega \nu_c}{\omega_p^2} \ll \left(\frac{U}{c}\right)^2 \tag{3–74}$$

For a perfectly conducting plasma, the right-hand side of (3–56) vanishes and the adiabatic condition is retrieved, that is

$$P\rho^{-5/3} = \text{const.} \tag{3–3}$$

For consistency, the order of magnitude treatment should be applied to *all* the hydromagnetic equations. This has the effect of removing q altogether so that (3–41) may be omitted from the set of equations. A comparison of magnitudes in (3–54) and (3–56) gives, using $\nabla \cdot \mathbf{E} = 4\pi q$,

$$\frac{qE}{(jB/c)} \sim \frac{E^2}{B^2} \sim \left(\frac{U}{c}\right)^2 \tag{3–75}$$

and

$$\frac{qU}{j} \sim \frac{UE}{cB} \sim \left(\frac{U}{c}\right)^2 \tag{3–76}$$

In conclusion, therefore, the hydromagnetic equations (i.e., under conditions (3–66), (3–70) to (3–72)) are:

Mass conservation equation, (3–40)

$$\frac{\partial \rho}{\partial t} + \nabla \cdot (\rho \mathbf{U}) = 0 \tag{3–77}$$

Momentum equation, from (3–54) and (3–75)

$$\rho \frac{D\mathbf{U}}{Dt} = -\nabla P + \frac{1}{c}(\mathbf{j} \times \mathbf{B}) \tag{3–78}$$

Energy equation, from (3–56), (3–6), and (3–76)

$$\frac{D}{Dt}(P\rho^{-5/3}) = \frac{2}{3}\frac{\rho^{-5/3}}{\sigma}j^2 \tag{3–79}†$$

† For generalizations of (3–79) and (3–85) see Summary.

Ohm's law, (3-6)

$$\mathbf{j} = \sigma\left(\mathbf{E} + \frac{\mathbf{U} \times \mathbf{B}}{c}\right) \tag{3-80}$$

and *Maxwell equations* (3-62), (3-67)

$$\nabla \times \mathbf{E} = -\frac{1}{c}\cdot\frac{\partial \mathbf{B}}{\partial t} \tag{3-81}$$

$$\nabla \times \mathbf{B} = \frac{4\pi}{c}\mathbf{j} \tag{3-82}$$

This is a closed set of fourteen scalar equations for ρ, $\mathbf{U}$, $\mathbf{j}$, P, $\mathbf{E}$, and $\mathbf{B}$.

For a perfectly conducting plasma (condition (3-74)), (3-79) and (3-80) become (3-3) and (3-73), respectively, and (3-73) may be used to eliminate $\mathbf{E}$ giving the *ideal hydromagnetic equations*:

$$\frac{\partial \rho}{\partial t} + \nabla\cdot(\rho\mathbf{U}) = 0 \tag{3-83}$$

$$\rho\frac{D\mathbf{U}}{Dt} = -\nabla P + \frac{1}{c}(\mathbf{j} \times \mathbf{B}) \tag{3-84}$$

$$P\rho^{-5/3} = \text{const.} \tag{3-85}†$$

$$\nabla \times (\mathbf{U} \times \mathbf{B}) = \frac{\partial \mathbf{B}}{\partial t} \tag{3-86}$$

$$\nabla \times \mathbf{B} = \frac{4\pi}{c}\mathbf{j} \tag{3-87}$$

3-6 Criteria for applicability of a fluid description

We will now discuss the important question of the *validity* of a fluid description of a plasma.

To start, we introduce a length λ and a time τ characteristic of distances and times over which plasma quantities change appreciably. For a fluid description we must consider a fluid element $d\mathbf{r}$ such that $d\mathbf{r} \ll \lambda$, but large enough for most of the particles within it at a time t also to be present at time $t + \tau$; that is, the fluid element *persists*. In hydrodynamics, this persistence is assured by collisions; particles are prevented from leaving the fluid element by collisions with neighbouring particles. This implies that $d\mathbf{r} \gg \lambda_c$, the mean free path. For a collision-dominated plasma this argument is less clear-cut, since the nature of collisions is different from that in hydrodynamics; small-angle collisions are now dominant, so that a significant

† See footnote on p. 50.

change in the direction of the particle is more often the result of many small deflections rather than a single impact.† Nevertheless, an essential requirement for the validity of the hydromagnetic equations is still

$$\lambda \gg \lambda_c \tag{3-88}$$

Moreover, the relaxation of the particle distribution function to a Maxwellian takes place in a time of the order of a collision time τ_c. Hence an additional requirement is

$$\tau \gg \tau_c \tag{3-89}$$

In practice, it turns out that in very many situations of interest these conditions are violated. The hydrodynamic description would be circumscribed indeed if one had to insist that (3–88) and (3–89) were satisfied. In fact, the hydrodynamic approach may sometimes be used when the plasma, far from being collision-dominated, is indeed collisionless! Consider the velocity $\mathbf{v}$ of a particle in $\mathbf{dr}$; from (3–20) it is apparent that it has a random component $\mathbf{w}$, and a flow component $\mathbf{U}$. Since $\mathbf{U}$ is the same for *all* particles within $\mathbf{dr}$, the concept of a fluid element is still meaningful provided

$$|\mathbf{U}| \gg |\mathbf{w}| \tag{3-90}$$

Since collisions are responsible for the random velocities, this condition means that the effect of collisions is negligible compared with the coherence produced by the self-consistent fields. In other words, particles constituting the fluid element are now held together by these fields alone. From the definitions (3–21), (3–37), and (3–38), neglect of $\mathbf{w}$ clearly means neglect of pressure and temperature; this is the *cold plasma limit*. A plasma in this limit is very different from a collisionless *neutral* gas, where there is no equivalent of self-consistent fields to bring about coherence so that the particles move freely and a hydrodynamic description is meaningless. The set of *cold plasma equations*, from (3–40), (3–41), (3–42b), (3–59), (3–62), and (3–63), is:

Mass conservation equation

$$\frac{\partial \rho}{\partial t} + \nabla.(\rho\mathbf{U}) = 0 \tag{3-91}$$

Charge conservation equation

$$\frac{\partial q}{\partial t} + \nabla.\mathbf{j} = 0 \tag{3-92}$$

† Cf. discussion following (3–25).

Momentum equation

$$\rho \frac{D\mathbf{U}}{Dt} = q\mathbf{E} + \frac{1}{c}\mathbf{j} \times \mathbf{B} \tag{3-93}$$

Generalized Ohm's law

$$\frac{m^+ m^-}{\rho e^2} \frac{\partial \mathbf{j}}{\partial t} = \mathbf{E} + \frac{\mathbf{U} \times \mathbf{B}}{c} - \frac{m^+}{\rho e c}\mathbf{j} \times \mathbf{B} \tag{3-94}$$

Maxwell equations

$$\nabla \times \mathbf{E} = -\frac{1}{c}\frac{\partial \mathbf{B}}{\partial t} \tag{3-95}$$

$$\nabla \times \mathbf{B} = \frac{1}{c}\frac{\partial \mathbf{E}}{\partial t} + \frac{4\pi}{c}\mathbf{j} \tag{3-96}$$

This gives us a set of 14 equations for ρ, q, $\mathbf{U}$, $\mathbf{j}$, $\mathbf{B}$, and $\mathbf{E}$. The displacement current term is now retained in (3–96) as opposed to (3–82) in the hydro-magnetic equations, since the inequality (3–66) may no longer be valid in the cold plasma limit (cf. the discussion of electromagnetic wave propagation using these equations in Section 7–3). Consequently the term $q\mathbf{E}$ in (3–54) is now retained in the momentum equation, which in turn requires the charge conservation equation to determine q.

Consider, finally, the behaviour of a plasma in a strong magnetic field; suppose now that the zero temperature, collisionless condition is relaxed to admit a finite pressure and a finite, though very large, mean free path, λ_c. By 'strong magnetic field' is meant one for which the particle Larmor radius, r_L, is such that

$$\lambda_c \gg r_L \tag{3-97}$$

We introduce a scale length $\lambda_\perp$ characteristic of distances over which the macroscopic plasma quantities change appreciably in directions perpendicular to the magnetic field. Then the condition analogous to (3–88) (for collision-dominated plasmas) is

$$\lambda_\perp \gg r_L \tag{3-98}$$

This is a necessary prerequisite for a fluid description in so far as motion perpendicular to the field is concerned. This means, physically, that the magnetic field now assumes the role previously played by collisions and acts to bind the particles tightly to the lines of force so that, in the motion of a fluid element, transverse movements of the particles are constrained within the element. This inhibition of particle motion by the field does not of itself produce a set of fluid equations: in a sense the plasma is only two thirds of the way towards behaving like a fluid.

In terms of macroscopic quantities, a strong field means that the pressure is not—as in the collision-dominated case—isotropic, but differs for directions perpendicular $(P_\perp)$ and parallel $(P_{||})$ to the magnetic field direction $(\hat{z})$. In deriving the moment equations, one must now distinguish between the moments $\langle v_\perp^2 \rangle = \langle v_1^2 + v_2^2 \rangle$ and $\langle v_{||}^2 \rangle$; that is, instead of the energy equation, (3–44c), one now gets an equation for $P_\perp$ and one for $P_{||}$. The problem of truncating the moment equations must now be faced. For perpendicular directions, the heat flow tensor may be neglected because of the effect of the magnetic field in inhibiting motion across the field. For the parallel direction—since collisions are infrequent—the only possibility for closing the set of equations is to make the low-temperature approximation; that is, the temperature is small, but finite, and one retains a pressure, $P_{||}$, but ignores the heat flow along the lines of force. Thus

$$P_{ij} = \begin{pmatrix} P_\perp & 0 & 0 \\ 0 & P_\perp & 0 \\ 0 & 0 & P_{||} \end{pmatrix} \tag{3–99}$$

and $$Q_{ijk} = 0 \qquad \text{all } i, j, k \tag{3–100}$$

From Section 3–5 we recall that the energy equation gave rise to an adiabatic condition under certain assumptions. In this case, two adiabatic conditions arise from the equations for $P_\perp$ and $P_{||}$, namely

$$\frac{D}{Dt}\left(\frac{P_\perp}{\rho B}\right) = 0 \tag{3–101}$$

$$\frac{D}{Dt}\left(\frac{P_{||}B^2}{\rho^3}\right) = 0 \tag{3–102}$$

(cf. Problem 3–6). For this reason, the theory is sometimes known as the *double-adiabatic approximation*, but we shall not pursue it further here.

Summary

1. *Conditions for truncation of the moment equations*
   (a) *Collisions dominant*

$$\omega \ll \nu_c \rightarrow \begin{cases} f \approx f_M \\ n^+ \approx n^- \approx \tfrac{1}{2}N \\ \mathbf{u}^+ \approx \mathbf{u}^- \approx \mathbf{U} \end{cases} \rightarrow \begin{cases} q_{ijk} \approx 0 \\ P_{ij} \approx \begin{cases} 0 & i \neq j \\ p & i = j \end{cases} \\ p^+ \approx p^- \approx \tfrac{1}{2}P \end{cases}$$

(b) *Cold plasma model*

$$\text{thermal velocity} \ll U \to \begin{cases} q_{ijk} \approx 0 \\ p_{ij} \approx 0 \end{cases} \quad \text{all } i, j, k$$

2. *Conditions for validity of hydromagnetic equations*

(i) $\dfrac{U}{c} \sim \dfrac{\omega L}{c} \ll 1.$

(ii) $\dfrac{\omega}{\omega_p} \ll \dfrac{U}{c}.$

(iii) $\dfrac{\omega \Omega_e}{\omega_p^2} \ll \left(\dfrac{U}{c}\right)^2.$

(iv) $\dfrac{\omega}{\Omega_i} \ll \left(\dfrac{U}{c_s}\right)^2.$

(v) $\dfrac{\omega \nu_c}{\omega_p^2} \ll \left(\dfrac{U}{c}\right)^2$ (*ideal* hydromagnetic equations *only*).

3. *Generalization of energy equation*
More generally, (3–79) and (3–85) may be written

$$\frac{D}{Dt}(P\rho^{-\gamma}) = \frac{(\gamma - 1)j^2}{\sigma\rho^\gamma} \tag{3–79}$$

$$P\rho^{-\gamma} = \text{const.} \tag{3–85}$$

where $\gamma = (n + 2)/n$ and n is the number of degrees of freedom (cf. Problem 3–7).

Problems

3–1 Explain why there is no term containing **B** in the energy equation (3–44).

3–2 Derive (3–49) and (3–50).

3–3 In the derivation of (3–57) show that the conditions for the neglect of terms quadratic in **U** and **j** are

$$\left(\frac{\omega}{\omega_p}\right)^2 \ll \left(\frac{U}{c}\right)^2, 1$$

[Use the same dimensional analysis as for (3–68).] Compare (3–70) and comment.

3-4 Show that the moment equations may be written in the form of conservation equations:

$$\frac{\partial \rho}{\partial t} = -\nabla.(\rho U)$$

$$\frac{\partial G}{\partial t} = -\nabla.\Pi$$

$$\frac{\partial W}{\partial t} = -\nabla.S$$

where

$$G = \rho U + \frac{1}{4\pi c}(E \times B) \equiv \text{total momentum density}$$

$$W = \tfrac{3}{2}N\kappa T + \tfrac{1}{2}\rho U^2 + \frac{1}{8\pi}[E^2 + B^2] \equiv \text{total energy density}$$

$$\Pi_{ij} = \rho U_i U_j + P_{ij} - T_{ij} \equiv \text{total momentum flux}$$

$$T_{ij} = \frac{1}{4\pi}(B_i B_j + E_i E_j) - \frac{\delta_{ij}}{8\pi}(B^2 + E^2)$$

$$\equiv \text{electromagnetic stress tensor}$$

$$S = \frac{c}{4\pi}(E \times B) + Q + P.U + U[\tfrac{3}{2}N\kappa T + \tfrac{1}{2}\rho U^2] \equiv \text{total energy flux}$$

3-5 Consider the generalized Ohm's law in the form

$$E + \frac{U \times B}{c} - \frac{m^+}{\rho ec}(j \times B) - \frac{j}{\sigma} = 0$$

and express it in terms of components parallel and perpendicular to **B**. In particular, show that

$$j_\perp \left(1 + \frac{\Omega_e^2}{v_c^2} \right) = \sigma \left[E_\perp^* - \frac{|\Omega_e|}{v_c} \frac{(E^* \times B)}{|B|} \right]$$

where

$$E^* = E + \frac{U \times B}{c}$$

Discuss this result.

3-6 Derive (3–101) and (3–102).

Hint: use $\psi = mv_i^2$ (no summation over i) in (3–16), then take $i = 3$ to derive (3–102) and the summation of $i = 1, 2$ for (3–101). Use the assumption of infinite conductivity to obtain

$$\frac{\partial U_3}{\partial r_3} = \frac{1}{B}\frac{DB}{Dt} - \frac{1}{\rho}\frac{D\rho}{Dt}$$

3-7 Verify the generalizations of (3–79) and (3–85) contained in section 3 of the Summary.

Hint: use (3–44c) and observe that for n degrees of freedom the generalization of (3–38) is $nN\kappa T = nP = P_{ii}$.

3-8 Consider the motion of electrons in an electric field **E**. Assume that the electric force is balanced by a frictional force $m^-\nu_c(\mathbf{u}^- - \mathbf{u}^+)$, due to the relative motion between ions and electrons and deduce the relationship

$$\sigma = ne^2/m^-\nu_c$$

which is implied by (3–60a, b).

4

Hydromagnetics

4–1 Introduction

In this chapter we apply hydromagnetic equations to a variety of problems. Certain aspects of hydromagnetics, such as the pinch effect (Sections 4–3, 4–5) and Alfvén waves (Section 4–7), are considered in some detail whereas others, such as hydromagnetic stability, are discussed only briefly. Still others, such as hydromagnetic turbulence, are not treated at all; these latter are no less important in hydromagnetics but, because of their essentially non-linear nature, there is still a wide gap between theory and observation. Two special classes of problem (flow problems and hydromagnetic shocks) will be treated separately in Chapters 5 and 6.

4–2 Kinematics

First, consider the effect of a moving, electrically conducting fluid on a magnetic field, neglecting the effect of the magnetic forces on the motion. This is sometimes called the kinematic—as opposed to the dynamic—aspect of hydromagnetics; for the time being the equation of motion is ignored. Then, using Maxwell's equations and Ohm's law, (3–80) to (3–82),

$$\mathbf{\nabla} \times \mathbf{B} = \frac{4\pi}{c}\mathbf{j} = \frac{4\pi\sigma}{c}\left[\mathbf{E} + \frac{\mathbf{U} \times \mathbf{B}}{c}\right]$$

i.e.,

$$\mathbf{\nabla} \times \mathbf{\nabla} \times \mathbf{B} = \frac{4\pi\sigma}{c}\left[-\frac{1}{c}\frac{\partial \mathbf{B}}{\partial t} + \frac{1}{c}\mathbf{\nabla} \times (\mathbf{U} \times \mathbf{B})\right]$$

supposing that σ is spatially uniform. Thus,

$$\frac{4\pi\sigma}{c^2}\frac{\partial \mathbf{B}}{\partial t} = \nabla^2\mathbf{B} + \frac{4\pi\sigma}{c^2}\mathbf{\nabla} \times (\mathbf{U} \times \mathbf{B})$$

Denoting $c^2/4\pi\sigma$ by η, this becomes

$$\frac{\partial \mathbf{B}}{\partial t} = \eta\,\nabla^2\mathbf{B} + \mathbf{\nabla} \times (\mathbf{U} \times \mathbf{B}) \qquad (4\text{--}1)$$

This equation is analogous to the hydrodynamic equation describing the evolution of the vorticity $\boldsymbol{\omega}$ of an incompressible fluid with kinematic viscosity ν,

$$\frac{\partial \boldsymbol{\omega}}{\partial t} = \nu \, \nabla^2 \boldsymbol{\omega} + \nabla \times (\mathbf{U} \times \boldsymbol{\omega}) \qquad (4\text{--}2)$$

in which the first term on the right-hand side represents the effect of *diffusion*, while the second describes the *convection* of the vorticity. By analogy, therefore, one may interpret η as the *magnetic viscosity*. One should be careful not to infer that **B** is the complete hydromagnetic analogue of $\boldsymbol{\omega}$; the vorticity is defined as $\boldsymbol{\omega} \equiv \nabla \times \mathbf{U}$ while **B** is not intrinsically related to the flow velocity. Note, however, that by definition $\boldsymbol{\omega}$ is solenoidal ($\nabla . \boldsymbol{\omega} = 0$), as is **B**.

It is simpler to consider the diffusion and convection terms separately. Suppose the fluid is at rest; then (4–1) becomes

$$\frac{\partial \mathbf{B}}{\partial t} = \eta \, \nabla^2 \mathbf{B} \qquad (4\text{--}3)$$

This is simply a diffusion equation; in the theory of fluids such an equation states that the fluid density changes by transfer of molecules from one region in the fluid to another. In the context of hydromagnetics, (4–3) states that the magnetic field changes by diffusing through the electrically conducting fluid, and the rate of change is given in terms of a parameter characteristic of the medium; thus η is known sometimes as the *magnetic diffusivity*. The field decays in a characteristic time, τ, given by

$$\tau \sim \frac{L^2}{\eta} = \frac{4\pi\sigma L^2}{c^2}$$

where L is a measure of the distance over which B changes appreciably. Some typical decay times for magnetic fields in a variety of conductors are shown in Table 4–1. Because of the L^2 dependence, magnetic fields clearly persist for long times when L is large. However, one must not conclude that τ is necessarily a measure of the *true* lifetime of the magnetic fields involved;

Table 4–1   Characteristic diffusion times and magnetic Reynolds' numbers

	L (cm)	τ (s)	R_M
Mercury	10	10^{-2}	10^{-1}
Arc discharge	10	10^{-3}	1
Earth's core	10^8	10^{12}	10^7
Sunspot	10^9	10^{14}	10^9
Solar corona	10^{11}	10^{18}	10^{12}

thus sunspot fields persist for far shorter times than 10^{14} seconds, while the reverse is true for the field in the earth's core. This simply means that with the former the field does not diffuse through the photosphere but decays as a result of some much more explosive process, while in the latter case the implication is that some mechanism regenerating the field is at work.

Now consider a fluid which is in motion but which is a perfect conductor; then from (4–1)

$$\frac{\partial \mathbf{B}}{\partial t} = \mathbf{\nabla} \times (\mathbf{U} \times \mathbf{B}) \tag{4–4}$$

The analogy between (4–4) and the hydrodynamic equation

$$\frac{\partial \boldsymbol{\omega}}{\partial t} = \mathbf{\nabla} \times (\mathbf{U} \times \boldsymbol{\omega}) \tag{4–5}$$

leads one to expect that there should be hydromagnetic counterparts to the theorems due to Kelvin and Helmholtz in inviscid hydrodynamics. These state that:[20]

(i) The velocity circulation around a closed contour moving with the fluid is constant in time; $\oint \mathbf{U}.\mathbf{dr} = \text{const.} = \int_S \boldsymbol{\omega}.\mathbf{dS}$ by Stokes' theorem. In other words, the flux of vorticity through any closed contour moving with the fluid is constant.

(ii) Fluid elements which lie initially on a vortex line continue to lie on a vortex line.

Thus, by analogy, one might expect the following theorems to be true in hydromagnetics:

(a) The magnetic flux through any closed contour moving with a perfectly conducting fluid is constant.

(b) Fluid elements which lie initially on a magnetic line of force continue to lie on a line of force in a perfectly conducting fluid.

To establish the first theorem, introduce the magnetic flux Φ through a closed surface S. Thus

$$\Phi = \int_S \mathbf{B}.\mathbf{dS} = \oint \mathbf{A}.\mathbf{dr} \tag{4–6}$$

using Stokes' theorem; $\mathbf{A}$ is the vector potential.

Then

$$\frac{D\Phi}{Dt} = \oint \left[\frac{D\mathbf{A}}{Dt}.\mathbf{dr} + \mathbf{A}.\frac{D}{Dt}(\mathbf{dr}) \right] \tag{4–7}$$

Writing (cf. Problem 4–2)

$$\frac{D}{Dt}(d\mathbf{r}) = (d\mathbf{r}.\nabla)\mathbf{v} \tag{4-8}$$

(4–7) becomes

$$\frac{D\Phi}{Dt} = \oint \left[\frac{D\mathbf{A}}{Dt}.d\mathbf{r} + \mathbf{A}.(d\mathbf{r}.\nabla)\mathbf{v} \right]$$

Using the vector identity (cf. Problem 4–2)

$$\mathbf{A}.(d\mathbf{r}.\nabla)\mathbf{v} \equiv d\mathbf{r}.[(\mathbf{A}.\nabla)\mathbf{v} + \mathbf{A} \times (\nabla \times \mathbf{v})]$$

one finds

$$\frac{D\Phi}{Dt} = \oint \left[\frac{\partial \mathbf{A}}{\partial t} + (\mathbf{v}.\nabla)\mathbf{A} + (\mathbf{A}.\nabla)\mathbf{v} + \mathbf{A} \times (\nabla \times \mathbf{v}) \right].d\mathbf{r}$$

$$= \oint \left[\frac{\partial \mathbf{A}}{\partial t} + \nabla(\mathbf{v}.\mathbf{A}) - \mathbf{v} \times (\nabla \times \mathbf{A}) \right].d\mathbf{r}$$

$$= -c \oint \left[\mathbf{E} + \frac{\mathbf{v} \times \mathbf{B}}{c} + \nabla\left(\phi - \frac{\mathbf{v}.\mathbf{A}}{c}\right) \right].d\mathbf{r}$$

where ϕ is the scalar potential. It follows that

$$\frac{D\Phi}{Dt} = -c \oint \left[\mathbf{E} + \frac{\mathbf{v} \times \mathbf{B}}{c} \right].d\mathbf{r} \tag{4-9}$$

Thus for a closed circuit moving with the fluid the integrand becomes $\mathbf{E} + (\mathbf{U} \times \mathbf{B})/c$ and for a perfectly conducting fluid this vanishes; so Φ is constant for any closed contour moving with the fluid. An alternative proof is the subject of Problem 4–3.

To establish the second theorem, we expand the right-hand side of (4–4):

$$\frac{\partial \mathbf{B}}{\partial t} = (\mathbf{B}.\nabla)\mathbf{U} - (\mathbf{U}.\nabla)\mathbf{B} - \mathbf{B}(\nabla.\mathbf{U})$$

since $\mathbf{B}$ is solenoidal, which gives on rearranging

$$\frac{D\mathbf{B}}{Dt} = (\mathbf{B}.\nabla)\mathbf{U} - \mathbf{B}(\nabla.\mathbf{U}) \tag{4-10}$$

Using the continuity equation, (3–40), (4–10) becomes

$$\frac{D\mathbf{B}}{Dt} = (\mathbf{B}.\nabla)\mathbf{U} + \frac{\mathbf{B}}{\rho}\frac{D\rho}{Dt}$$

i.e.,
$$\frac{D}{Dt}\left(\frac{\mathbf{B}}{\rho}\right) = \left(\frac{\mathbf{B}}{\rho}\cdot\nabla\right)\mathbf{U} \qquad (4\text{--}11)$$

a result first derived by Walén. From (4–8), using the fact that we are now considering a fluid element of length dr in motion,

$$\frac{D}{Dt}(\mathbf{dr}) = (\mathbf{dr}\cdot\nabla)\mathbf{U} \qquad (4\text{--}12)$$

and so the evolution of $\mathbf{B}/\rho$ and dr are described by the same equation. If at $t = 0$ these quantities are parallel, so that

$$\varepsilon\mathbf{B_0} = \rho_0\,\mathbf{dr_0}$$

where ε is infinitesimal, then from (4–11) and (4–12)

$$\frac{D}{Dt}\left(\mathbf{dr} - \varepsilon\frac{\mathbf{B}}{\rho}\right) = 0 \qquad \text{at } t = 0$$

This equation is sometimes taken to imply that $[\mathbf{dr} - \varepsilon(\mathbf{B}/\rho)]$ remains zero and hence that a line element of the fluid lying originally along a field line always lies along a field line. Although this conclusion is correct, the argument thus far is insufficient, as was pointed out by Larmor in connection with the proof of Helmholtz's theorem in hydrodynamics. Goldstein[22] illustrates the fallacy by observing that, reduced to its simplest form, this argument requires that a function $f(t)$, such that $f'(t) = 0$ when $f(t) = 0$, be identically zero, and quotes $f(t) \equiv 1 - \cos t$ as a simple case in which this is certainly not true. A rigorous discussion is given by Goldstein, who integrates (4–11) and finds

$$\frac{\mathbf{B}}{\rho} = \left(\frac{\mathbf{B_0}}{\rho_0}\cdot\nabla_0\right)\mathbf{r} \qquad (4\text{--}13)$$

in which the subscript denotes initial values.

Consider the situation in Fig. 4–1 in which a point in the fluid, originally at P_0, moves to some position P at a later time t. P_0' is a point on the same line of force as P_0, and so $\mathbf{dr_0}$ is in the direction of the magnetic field. Similarly, P_0' moves to the position P'. Then,

$$\mathbf{dr} = \mathbf{dr_0} + \mathbf{s}'(\mathbf{r_0} + \mathbf{dr_0}, t) - \mathbf{s}(\mathbf{r_0}, t) \qquad (4\text{--}14)$$

Using Taylor's theorem,

$$\mathbf{s}'(\mathbf{r_0} + \mathbf{dr_0}) = \mathbf{s}(\mathbf{r_0}) + (\mathbf{dr_0}\cdot\nabla_0)\mathbf{s}$$

(4–14) becomes

$$\mathbf{dr} = \mathbf{dr_0} + (\mathbf{dr_0}\cdot\nabla_0)\mathbf{s} = (\mathbf{dr_0}\cdot\nabla_0)\mathbf{r} \qquad (4\text{--}15)$$

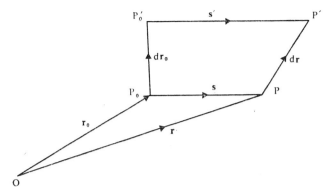

Fig. 4-1 Fluid element originally at P_0P_0' moves to PP' at a later time t

Now dr_0 has been chosen so that it is oriented along a field line, i.e., $dr_0 = \varepsilon(B_0/\rho_0)$. Then, by (4–15),

$$dr = \left(\varepsilon\frac{B_0}{\rho_0}\cdot\nabla_0\right)r = \varepsilon\frac{B}{\rho} \tag{4-16}$$

by (4–13); that is, dr is also oriented along a field line. Thus a fluid element which lies initially on a magnetic line of force continues to lie on a magnetic line of force in a perfectly conducting fluid.

It follows directly from (4–16) that, if $dr = dr_0$, B/ρ is constant along the path of the fluid element. Also, if the fluid element in moving from P_0P_0' to PP' is stretched, so that $|dr| > |dr_0|$, then, if the fluid is incompressible, (4–16) implies that the magnetic field intensity is increased.

Alfvén summed up these results for the behaviour of the magnetic field in a perfectly conducting fluid by stating that the field behaved as if it were 'frozen' into the fluid; in other words, any fluid motion transverse to the lines of force transports them along with the fluid. This statement condenses the results of the two theorems which have been proved. For a fluid element which lies along a line of force is by definition also a tube of force, and the first theorem then asserts that the flux threading such a tube is constant if it moves with the fluid. Hence Alfvén's remark that the field behaves as though it were frozen into the fluid. The fluid may, of course, flow freely *along* the lines of force.

In the general case, in which the motion is not negligible and the fluid not a perfect conductor, one must revert to (4–1). Physically one expects that, while the lines of force are still transported by the moving fluid, they are no longer frozen into the fluid but may now 'leak' through it; that is, the lines slip relative to the fluid. From (4–1), if L is a characteristic length, we see that convection of the lines of force dominates diffusion, provided

$$LU \gg \eta \tag{4-17}$$

By analogy with the definition of the Reynolds number in hydrodynamics, $R = LU/\nu$ (ν is the kinematic viscosity), one may introduce here the *magnetic Reynolds number*

$$R_M = \frac{LU}{\eta} \qquad (4\text{--}18)$$

Thus convection is dominant provided $R_M \gg 1$. Table 4–1 shows values for R_M characteristic of a variety of conducting fluids. From this it is apparent that the requirement $R_M \gg 1$, while not met in laboratory situations, is easily satisfied in naturally occurring plasmas.

4–3 Static problems

Thus far, we have confined discussion to kinematics and have ignored the effect of forces on the motion. Ultimately, we wish to examine the effect of the magnetic field on the motion of the fluid through the $\mathbf{j} \times \mathbf{B}$ term, but we begin with a specially simple case in which the fluid is at rest. Sometimes a close parallel is drawn between magnetohydrostatics and the steady flow of an inviscid, incompressible fluid. (By *steady flow* is meant one in which the flow velocity $\mathbf{U}$ is not time-dependent.) As this comparison can be rather misleading it is worth clarifying the analogy. The hydrodynamic equation describing steady inviscid flow (assuming that flow does not take place in a gravitational field) is

$$\mathbf{\nabla}(P + \tfrac{1}{2}\rho U^2) + \rho(\mathbf{\omega} \times \mathbf{U}) = 0 \qquad (4\text{--}19)$$

Using the identity

$$\mathbf{\nabla}\left(\frac{U^2}{2}\right) = \mathbf{U} \times (\mathbf{\nabla} \times \mathbf{U}) + (\mathbf{U}.\mathbf{\nabla})\mathbf{U}$$

(4–19) becomes for constant ρ

$$\mathbf{\nabla}P + \rho(\mathbf{U}.\mathbf{\nabla})\mathbf{U} = 0 \qquad (4\text{--}20)$$

Steady flow in hydromagnetics is described by (3–78) with $\partial\mathbf{U}/\partial t = 0$,

$$\rho(\mathbf{U}.\mathbf{\nabla})\mathbf{U} = -\mathbf{\nabla}P + \frac{1}{c}\mathbf{j} \times \mathbf{B} = -\mathbf{\nabla}P + \frac{(\mathbf{\nabla} \times \mathbf{B}) \times \mathbf{B}}{4\pi}$$

i.e.,

$$\mathbf{\nabla}\left(P + \frac{B^2}{8\pi}\right) + \rho(\mathbf{U}.\mathbf{\nabla})\mathbf{U} - \frac{(\mathbf{B}.\mathbf{\nabla})\mathbf{B}}{4\pi} = 0$$

This equation may be rewritten for constant ρ

$$\mathbf{\nabla}\left(P + \frac{B^2}{8\pi}\right) + \rho[(\mathbf{V}.\mathbf{\nabla})\mathbf{W} + (\mathbf{W}.\mathbf{\nabla})\mathbf{V}] = 0 \qquad (4\text{--}21)$$

where

$$\left.\begin{array}{c} \mathbf{V} = \dfrac{1}{\sqrt{2}}(\mathbf{U} + \mathbf{v}_A) \\[2ex] \mathbf{W} = \dfrac{1}{\sqrt{2}}(\mathbf{U} - \mathbf{v}_A) \\[2ex] \mathbf{v}_A = \dfrac{\mathbf{B}}{(4\pi\rho)^{1/2}} \end{array}\right\} \qquad (4\text{--}22)$$

This form provides a direct comparison with (4–20). For magnetohydrostatic problems, $\mathbf{U} = 0$ and (4–21) becomes

$$\nabla\!\left(P + \frac{B^2}{8\pi}\right) - \rho(\mathbf{v}_A . \nabla)\mathbf{v}_A = 0 \qquad (4\text{--}23)$$

The hydrostatic equation corresponding to (4–23) is properly

$$\nabla P = 0$$

The basis for the comparison between hydrodynamic steady flow and mag-netohydrostatics lies in the similarity between (4–20) and (4–23), with $B^2/8\pi$ interpreted as a 'magnetic pressure' and $\mathbf{v}_A$ as a velocity characteristic of the magnetic field. The obvious dissimilarity between (4–20) and (4–23) lies in the difference in sign in the second terms and this can only be understood by reference to the proper analogy, namely by comparing (4–20) and (4–21).

In magnetohydrostatic situations, from (4–1)

$$\eta\, \nabla^2 \mathbf{B} = 0 \qquad (4\text{--}24)$$

and, of course, $\nabla . \mathbf{B} = 0$. In the case of a perfectly conducting fluid, (4–24) is identically true and does not imply a condition on $\mathbf{B}$. A special situation arises whenever $\nabla \times \mathbf{B}$ and $\mathbf{B}$ are everywhere parallel; such fields are called *force-free* and are discussed briefly in Section 4–4. For the present, we shall only consider static equilibria in which the magnetic force $[(\nabla \times \mathbf{B}) \times \mathbf{B}]/4\pi$ is balanced by the fluid pressure gradient. Rewriting,

$$\frac{(\nabla \times \mathbf{B}) \times \mathbf{B}}{4\pi} = \frac{1}{4\pi}(\mathbf{B}.\nabla)\mathbf{B} - \frac{1}{8\pi}\nabla B^2 = \nabla.\left(\frac{\mathbf{B}\mathbf{B}}{4\pi}\right) - \nabla\!\left(\frac{B^2}{8\pi}\right) \qquad (4\text{--}25)$$

since $\mathbf{B}$ is solenoidal; $\mathbf{B}\mathbf{B}$ is a dyadic form (cf. Appendix 1). In tensor notation, (4–25) reads

$$\left(\frac{(\nabla \times \mathbf{B}) \times \mathbf{B}}{4\pi}\right)_i = \frac{\partial}{\partial r_k}\!\left(\frac{B_i B_k}{4\pi}\right) - \frac{\partial}{\partial r_k}\!\left(\frac{B^2}{8\pi}\delta_{ik}\right) \qquad (4\text{--}25a)$$

From electrodynamics,[23] the definition of the Maxwell stress tensor is

$$T_{ik} = \frac{1}{4\pi}\left[E_i E_k + B_i B_k - \frac{\delta_{ik}}{2}(E^2 + B^2)\right]$$

and, dropping the terms involving the electric stresses (for consistency, since the electric force term has been dropped from the equation of motion, (3–54)), one sees at once that (4–25a) is just $\partial T_{ik}/\partial r_k$. The static equation reads

$$\frac{\partial}{\partial r_k}\left[P\delta_{ik} - T_{ik}\right] = 0$$

i.e.,

$$\frac{\partial}{\partial r_k}\mathfrak{I}_{ik} = 0 \qquad (4\text{–}26)$$

The total stress tensor

$$\mathfrak{I}_{ik} = \left[\left(P + \frac{B^2}{8\pi}\right)\delta_{ik} - \frac{B_i B_k}{4\pi}\right] \qquad (4\text{–}27)$$

may be reduced to diagonal form by transformation to the principal axes. The eigenvalues may be obtained from the secular equation

$$|\,\mathfrak{I}_{ik} - \delta_{ik}\lambda\,| = 0 \qquad (4\text{–}28)$$

the solution being

$$\lambda_1 = P + \frac{B^2}{8\pi} = \lambda_2 \qquad \lambda_3 = P - \frac{B^2}{8\pi}$$

Thus, referred to the principal axes, the total stress tensor takes the form

$$\begin{bmatrix} P + \dfrac{B^2}{8\pi} & 0 & 0 \\[2ex] 0 & P + \dfrac{B^2}{8\pi} & 0 \\[2ex] 0 & 0 & P - \dfrac{B^2}{8\pi} \end{bmatrix}$$

The principal axes are oriented so that the axis corresponding to λ_3 is parallel to **B** and the axes corresponding to λ_1, λ_2 are perpendicular to **B**. From this, we see that the stress caused by the magnetic field amounts to a pressure $B^2/8\pi$ in directions transverse to the field and a tension $B^2/8\pi$ along the lines of force. In other words, the total stress amounts to an isotropic pressure which is the sum of the fluid pressure and the *magnetic pressure* $B^2/8\pi$, and a *tension* $B^2/4\pi$ along the lines of force. This is illustrated in Fig. 4–2.

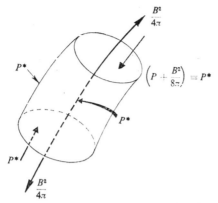

Fig. 4–2 Total stress composed of an isotropic pressure P^* and a tension $B^2/4\pi$ along the lines of force

The ratio of fluid pressure to magnetic pressure, $8\pi P/B^2$, is an important parameter commonly denoted by β. In hydromagnetics, it is often convenient to picture a *tube* of force (a tube with walls made up of lines of force) behaving like an elastic string under tension. Thus stretching the tube of force increases the tension, which means that the field is increased (cf. Section 4–2).

Returning to the static condition

$$\nabla P = \frac{\mathbf{j} \times \mathbf{B}}{c} \tag{4–29}$$

it follows that

$$\mathbf{B}.\nabla P = 0 \tag{4–30a}$$

and

$$\mathbf{j}.\nabla P = 0 \tag{4–30b}$$

That is, both $\mathbf{B}$ and $\mathbf{j}$ lie on surfaces of constant pressure. Supposing the constant pressure (isobaric) surfaces are closed, then, since (4–30a) states that no magnetic field line passes through the surface, one may picture the surface as made up from a winding of field lines. Likewise, from (4–30b), the same isobaric surface is made up from lines of current density; these lines will, in general, intersect the field lines at any angle. The cross-section in Fig. 4–3 shows a set of nested surfaces on which the pressure increases in passing from the outside towards the axis; the currents are such that $\mathbf{j} \times \mathbf{B}$ points towards the axis. The important implication here is that a plasma may be contained entirely by the magnetic force, an arrangement referred to as *magnetic confinement*.

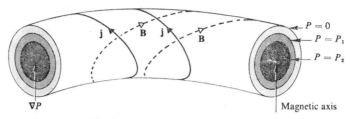

Fig. 4–3 Set of nested isobaric surfaces with pressure increasing towards the axis

A specially simple instance arises whenever the magnetic lines of force are straight and parallel; then, since there is no variation in the direction of the field, $(\mathbf{B} \cdot \nabla)\mathbf{B} = 0$ and so, from (4–23),

$$P^* \equiv P + \frac{B^2}{8\pi} = \text{const.} \qquad (4\text{–}31)$$

Such a field may be produced by currents flowing azimuthally; devices designed to contain plasma in this configuration are known as *theta pinches* or *thetatrons* (since θ is used to denote the azimuthal coordinate). Azimuthal currents are induced by discharging a current suddenly in a metal conductor enclosing the discharge tube, as in Fig. 4–4. The induced currents flow in the

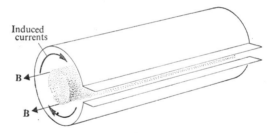

Fig. 4–4 Theta pinch

opposite direction, and an axial magnetic field is generated in the region between; for perfect conductivity this field is restricted to the region between the plasma and the metal conductor, since it cannot penetrate the plasma. The $\mathbf{j} \times \mathbf{B}$ force acts to push the plasma towards the axis until the magnetic pressure is balanced by the plasma pressure. Since there is no field within the plasma, nor is there any plasma in the region occupied by the field, condition (4–31) for pressure balance is simply

$$P = \frac{B^2}{8\pi} \qquad (4\text{–}32)$$

Another magnetostatic configuration important for plasma containment is the *linear pinch*. Here the $\mathbf{j}$, $\mathbf{B}$ lines of the theta pinch are interchanged so that **j** is now axial and **B** is azimuthal; the **j** × **B** force is again directed towards the axis. Consider a fully ionized gas contained in a cylindrical discharge tube with a current flowing parallel to the axis of the tube (Fig. 4–5);

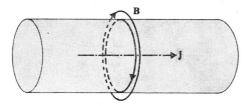

Fig. 4–5   Linear pinch

under the action of the **j** × **B** force the plasma is squeezed or 'pinched' into a filament along the axis of the tube. The static condition (4–29) gives

$$\frac{\mathrm{d}P}{\mathrm{d}r} = -\frac{B}{4\pi r}\frac{\mathrm{d}}{\mathrm{d}r}(rB) \tag{4–33}$$

If the radius of the pinch is r_0, then multiplying (4–33) by r^2 and integrating gives

$$\int_0^{r_0} r^2 \frac{\mathrm{d}P}{\mathrm{d}r}\,\mathrm{d}r = -\frac{1}{4\pi}\int_0^{r_0}(rB)\,\mathrm{d}(rB)$$

i.e.,

$$\left[r^2 P\right]_0^{r_0} - 2\int_0^{r_0} rP\,\mathrm{d}r = -\frac{1}{8\pi}(rB)_{r=r_0}^2$$

If we suppose that the plasma pressure vanishes at $r = r_0$ and assume that both ions and electrons have the same temperature T which is constant across the pinch, then

$$2\kappa T\int_0^{r_0} n\,.\,2r\,\mathrm{d}r = \frac{1}{8\pi}(rB)_{r=r_0}^2$$

Introducing the number of electrons per unit length of the plasma column $N = \int_0^{r_0} n\,.\,2\pi r\,\mathrm{d}r$ one finds

$$2N\kappa T = \frac{1}{8}(rB)_{r=r_0}^2 \tag{4–34}$$

Now from the Maxwell equation

$$\frac{1}{r}\frac{\mathrm{d}}{\mathrm{d}r}(rB) = \frac{4\pi}{c}j$$

one has
$$(rB)_{r=r_0} = \frac{2}{c} \int_0^{r_0} j \cdot 2\pi r \, dr$$

Defining the total current flowing in the plasma column $I = \int_0^{r_0} j \cdot 2\pi r \, dr$ gives, on substituting for $(rB)_{r=r_0}$ in (4–34),

$$2N\kappa T = \frac{I^2}{2c^2}$$

i.e.,
$$I^2 = 4N\kappa T c^2 \qquad (4\text{–}35)$$

which is known as *Bennett's relation* for the linear pinch (after W. H. Bennett who examined the pinch effect in 1934). It shows that for a stable pinch configuration the plasma temperature is proportional to the square of the discharge current and inversely proportional to the number of particles per unit length (the particle *line density*). Note that for this analysis we have assumed that the pinched plasma exists in a stable configuration (in fact the simple linear pinch is unstable, cf. Section 4–5) and have regarded the plasma temperature as constant.

One further point about magnetically contained plasmas arises from (4–31). Denoting the magnetic field inside and outside the plasma by B_i, B_e, respectively,

$$\frac{B_e^2}{8\pi} = P + \frac{B_i^2}{8\pi}$$

that is, $B_e > B_i$ so that the plasma is diamagnetic (cf. Section 2–2).

4–4 Force-free fields

The special case in which the current and magnetic field are parallel, resulting in the vanishing of the electromagnetic body force, is of interest in certain astrophysical situations. Such magnetic field configurations are said to be *force-free*. The problem arose originally from a situation in which a very tenuous ionized gas carried a strong magnetic field, so that the gradient of the fluid pressure was at least an order of magnitude smaller than the magnetic pressure gradient. From (4–23)

$$\nabla\left(\frac{B^2}{8\pi}\right) \simeq \frac{(\mathbf{B}.\nabla)\mathbf{B}}{4\pi}$$

or
$$\mathbf{B} \times (\nabla \times \mathbf{B}) \simeq 0 \qquad (4\text{–}36)$$

There is once again a parallel with hydrodynamics where velocity fields exist for which

$$\mathbf{U} \times (\nabla \times \mathbf{U}) = 0 \qquad (4\text{–}37)$$

Such vector fields are known as *Beltrami fields* and flows satisfying (4–37) are called *Beltrami flows*. These flows are of interest in certain practical hydrodynamic and aerodynamic applications.

Apart from the trivial case in which $\nabla \times \mathbf{B} = 0$, magnetic fields for which

$$\nabla \times \mathbf{B} = \alpha\mathbf{B}$$

will be force-free; in the most general case, α will be spatially dependent. Some theorems relating to force-free fields have been established by L. Woltjer.[24] In particular, he showed that:

(i) Force-free fields with α constant represent the state of minimal magnetic energy in a closed system.

(ii) Hydromagnetic equilibrium in the absence of flow requires a force-free field with α constant. (Cf. Problem 4–11.)

4–5 Hydromagnetic stability

The most striking experimental feature of the linear pinch is its marked predilection for twisting or wriggling prior to breaking up (cf. Fig. 4–10); linear pinches appear to be inherently unstable dynamical systems. Stability is, therefore, of vital importance for plasma containment. The terms *stable*, *unstable* describe the behaviour of dynamical systems towards small perturbations about equilibrium situations. If forces act on the system tending to restore it to its equilibrium configuration as a result of this perturbation, the system is said to be in *stable equilibrium* (with respect to the class of perturbations considered). If, on the other hand, the system tends to depart further and further from the equilibrium configuration as a result of the perturbation, it is in *unstable equilibrium*.

In general, plasma instabilities may be broadly categorized as macroscopic (hydromagnetic) or microscopic. The first class may be discussed within the framework of hydromagnetics and involves the physical (spatial) displacement of plasma or conducting fluid, as the case may be. To describe microscopic instabilities we need kinetic theory, since there are changes in the velocity distribution functions and this information is lost in hydromagnetics.

In this and the following section, the hydromagnetic stability of some static configurations is discussed. Since we are concerned with small departures from equilibrium we can apply perturbation theory to the hydromagnetic equations. For simplicity, we assume perfect conductivity, so that the appropriate set of equations is (3–83) to (3–87). In a perturbation approach one writes

$$\left.\begin{aligned}
\rho &= \rho_0 + \rho' \\
\mathbf{U} &= \mathbf{U}_0 + \mathbf{U}' = \mathbf{U}' \\
P &= P_0 + P' \\
\mathbf{j} &= \mathbf{j}_0 + \mathbf{j}' \\
\mathbf{B} &= \mathbf{B}_0 + \mathbf{B}'
\end{aligned}\right\} \qquad (4\text{–}40)$$

where the subscript 0 denotes equilibrium values and primed variables are the perturbations. ($U_0 = 0$ since the equilibrium is static.) Substituting (4–40) in this set of equations and ignoring products of the perturbations (since these are second order of smallness), we arrive at the linearized equations. The zero-order (equilibrium) equations read

$$\nabla P_0 = \frac{1}{c}(\mathbf{j}_0 \times \mathbf{B}_0) \qquad (4\text{–}41a)$$

$$P_0 \rho_0^{-\gamma} = \text{const.} \qquad (4\text{–}41b)$$

$$\nabla \times \mathbf{B}_0 = \frac{4\pi}{c}\mathbf{j}_0 \qquad (4\text{–}41c)$$

and the first-order equations are

$$\frac{\partial \rho'}{\partial t} + \nabla \cdot (\rho_0 \mathbf{U}') = 0 \qquad (4\text{–}42)$$

$$\rho_0 \frac{\partial \mathbf{U}'}{\partial t} = -\nabla P' + \frac{1}{c}[\mathbf{j}' \times \mathbf{B}_0 + \mathbf{j}_0 \times \mathbf{B}'] \qquad (4\text{–}43)$$

$$P'\rho_0 = \gamma \rho' P_0 \qquad (4\text{–}44)$$

$$\frac{\partial \mathbf{B}'}{\partial t} = \nabla \times (\mathbf{U}' \times \mathbf{B}_0) \qquad (4\text{–}45)$$

$$\nabla \times \mathbf{B}' = \frac{4\pi}{c}\mathbf{j}' \qquad (4\text{–}46)$$

Using (4–44) and (4–46), ρ' and $\mathbf{j}'$ may be eliminated from the set of equations; (4–42) becomes

$$\frac{\rho_0}{P_0}\frac{\partial P'}{\partial t} + \gamma \nabla \cdot (\rho_0 \mathbf{U}') = 0 \qquad (4\text{–}42a)$$

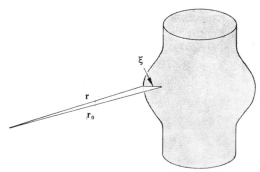

Fig. 4–6 Perturbed linear pinch plasma

Conveniently, we work in terms of the displacement of a fluid element from its equilibrium position $\mathbf{r}_0$; that is, $\xi(\mathbf{r}_0, t) = \mathbf{r} - \mathbf{r}_0$ (Fig. 4–6). Then

$$\mathbf{U}' = \frac{D\mathbf{r}}{Dt} = \frac{\partial \xi}{\partial t}$$

since $\mathbf{r}_0$ is not a function of t; thus, (4–42a) may be integrated to give†
(since initially the perturbations are zero except for $\mathbf{U}'$)

$$P' = -\gamma \frac{P_0}{\rho_0} \nabla \cdot (\rho_0 \xi) = -\gamma P_0 \nabla \cdot \xi - \gamma \frac{P_0}{\rho_0} \xi \cdot \nabla \rho_0$$
$$= -\gamma P_0 \nabla \cdot \xi - \xi \cdot \nabla P_0 \tag{4–47}$$

where the gradient of (4–41b) has been used. Integrating (4–45),

$$\mathbf{B}' = \nabla \times (\xi \times \mathbf{B}_0) \tag{4–48}$$

Substituting (4–41c), (4–46) to (4–48) in (4–43) then gives

$$\rho_0 \frac{\partial^2 \xi}{\partial t^2} = \nabla(\xi \cdot \nabla P_0 + \gamma P_0 \nabla \cdot \xi)$$
$$+ \frac{1}{4\pi} \Big[(\nabla \times \nabla \times [\xi \times \mathbf{B}_0]) \times \mathbf{B}_0 + (\nabla \times \mathbf{B}_0) \times (\nabla \times [\xi \times \mathbf{B}_0]) \Big]$$
$$\equiv \mathbf{Q}(\xi(\mathbf{r}, t)) \tag{4–49}$$

Since ρ_0, P_0, and $\mathbf{B}_0$ are given by the equilibrium configuration, (4–49) together with appropriate boundary conditions determines the displacement vector ξ. The usual technique for solving such a linear equation is Fourier analysis; one writes

$$\xi(\mathbf{r}, t) = \sum_n \xi(\mathbf{r}, \omega_n) \exp i\omega_n t \tag{4–50}$$

in which case (4–49) reduces to the set of *normal mode equations*

$$-\rho_0 \omega_n^2 \xi(\mathbf{r}, \omega_n) = \mathbf{Q}(\xi(\mathbf{r}, \omega_n)) \tag{4–51}$$

It is clear from (4–50) that if for all the *normal frequencies* ω_n, $\omega_n^2 > 0$ then all the modes are periodic. This means that the system oscillates about the equilibrium position and the initial configuration is stable. On the other hand, if at least one of the normal frequencies is such that $\omega_n^2 < 0$ the perturbation will grow exponentially and the equilibrium configuration is unstable.

---

† Throughout the perturbed equations, (4–42) to (4–46), $\mathbf{r}$ is written as $\mathbf{r}_0 + \xi(\mathbf{r}_0, t)$. However, these equations are already of first-order in small quantities so this merely has the effect of replacing $\mathbf{r}$ by $\mathbf{r}_0$ and ∇ by ∇_0. Since $\mathbf{r}$ and ∇ no longer appear in the equations, the suffix on $\mathbf{r}_0$ and ∇_0 may be dropped.

The boundary conditions to be applied in solving (4–51) depend, of course, on the nature of the problem. The situation met in the linear pinch is typical of magnetically contained plasmas: the plasma is confined by a magnetic field so that it is separated from the walls by a vacuum. Denoting the vacuum fields by the subscript e, and the plasma fields by i, one may write for the vacuum fields

$$\mathbf{E}_e = \mathbf{E}_{0e} + \mathbf{E}'_e = \mathbf{E}'_e = -\frac{1}{c}\frac{\partial \mathbf{A}'}{\partial t} \tag{4-52}$$

$$\mathbf{B}_e = \mathbf{B}_{0e} + \mathbf{B}'_e = \mathbf{B}_{0e} + \nabla \times \mathbf{A}'$$

where $\mathbf{A}'$ denotes the perturbation of the vacuum vector potential. In the vacuum, (4–46) gives

$$\nabla \times \nabla \times \mathbf{A}' = 0 \tag{4-53}$$

It may be shown (cf. Problem 4–12) that the boundary condition on the electric field is

$$\hat{\mathbf{n}} \times \left[\mathbf{E}'_i + \frac{1}{c}(\mathbf{U}' \times \mathbf{B}_{0i})\right] = \hat{\mathbf{n}} \times \left[\mathbf{E}'_e + \frac{1}{c}(\mathbf{U}' \times \mathbf{B}_{0e})\right] \tag{4-54}$$

where $\hat{\mathbf{n}}$ is a unit vector normal to the plasma surface. However, since perfect conductivity has been assumed

$$\mathbf{E}'_i + \frac{1}{c}(\mathbf{U}' \times \mathbf{B}_{0i}) = 0$$

and (using (4–52)), (4–54) becomes

$$\hat{\mathbf{n}} \times \frac{\partial \mathbf{A}'}{\partial t} = \hat{\mathbf{n}} \times \left(\frac{\partial \xi}{\partial t} \times \mathbf{B}_{0e}\right) \tag{4-55}$$

On integration,†

$$\hat{\mathbf{n}} \times \mathbf{A}' = \hat{\mathbf{n}} \times (\xi \times \mathbf{B}_{0e}) = -(\hat{\mathbf{n}}.\xi)\mathbf{B}_{0e} \tag{4-56}$$

since the vacuum field is initially parallel to the surface.

A further boundary condition is provided by the continuity of the total pressure across the surface. This is required since a discontinuity would give rise to an infinite acceleration. Thus, remembering that the condition must be applied at the perturbed boundary,

$$(P_0 + \xi.\nabla P_0) + P' + \frac{1}{8\pi}[(\mathbf{B}_{0i} + (\xi.\nabla)\mathbf{B}_{0i}) + \mathbf{B}'_i]^2$$

$$= \frac{1}{8\pi}[(\mathbf{B}_{0e} + (\xi.\nabla)\mathbf{B}_{0e}) + \mathbf{B}'_e]^2 \tag{4-57}$$

† Since (4–55) is of first-order, one should read $\hat{\mathbf{n}}_0$ (the unit normal to the equilibrium surface) for $\hat{\mathbf{n}}$.

Hence, using the equilibrium condition

$$P_0 + \frac{B_{0i}^2}{8\pi} = \frac{B_{0e}^2}{8\pi}$$

(4–57) becomes on linearization, and using (4–47) for P',

$$-\gamma P_0 \nabla \cdot \boldsymbol{\xi} + \frac{\mathbf{B}_{0i}}{4\pi} \cdot [\mathbf{B}_i' + (\boldsymbol{\xi} \cdot \nabla)\mathbf{B}_{0i}] = \frac{\mathbf{B}_{0e}}{4\pi} \cdot [\mathbf{B}_e' + (\boldsymbol{\xi} \cdot \nabla)\mathbf{B}_{0e}] \qquad (4\text{–}58)$$

4-5-1 Stability of the linear pinch

As an illustration, we apply the theory to the stability of the linear pinch.[25] Note, however, that the situation now differs from that in Section 4–3. The assumption of perfect conductivity means that the current is not distributed throughout the plasma cross-section but flows in a thin 'skin' on the surface of the column. We shall assume that an imposed, constant axial field exists inside the plasma; perfect conductivity then means that this field is frozen inside the plasma (cf. Section 4–2). For simplicity, assume that no axial field is imposed outside the plasma. Initially, therefore, the external field is the azimuthal field due to the current. Then, in cylindrical polar coordinates,

$$\mathbf{B}_{0i} = (0, 0, B_{0z} = \text{const.}) \qquad (4\text{–}59)$$

$$\mathbf{B}_{0e} = (0, B_\theta(r), 0) \qquad (4\text{–}60)$$

If P_0 is also constant, (4–51) (dropping the subscript on ω) reduces to

$$-\rho_0 \omega^2 \boldsymbol{\xi} = \gamma P_0 \nabla(\nabla \cdot \boldsymbol{\xi}) + \frac{1}{4\pi}[\nabla \times \nabla \times (\boldsymbol{\xi} \times \mathbf{B}_{0i})] \times \mathbf{B}_{0i} \qquad (4\text{–}61)$$

The discussion is restricted to perturbations which retain cylindrical symmetry. Then

$$\boldsymbol{\xi}(\mathbf{r}) = \boldsymbol{\xi}(r) \exp{(im\theta + ikz)} = (\xi_r(r), \xi_\theta(r), \xi_z(r)) \exp{(im\theta + ikz)}$$

and to lighten the algebra we examine only $m = 0$ modes, a further restriction to axially symmetric perturbations. In this case, (4–61) becomes, after some straightforward reduction,

$$-\rho_0 \omega^2 \xi_r = \gamma P_0 e^{-ikz} \frac{\partial}{\partial r}(\nabla \cdot \boldsymbol{\xi}) + \frac{B_{0z}}{4\pi}\left[\frac{\partial}{\partial r}\left(\frac{B_{0z}}{r}\frac{\partial}{\partial r}(r\xi_r)\right) - k^2 B_{0z}\xi_r\right]$$

$$-\rho_0 \omega^2 \xi_\theta = -\frac{k^2 B_{0z}^2}{4\pi}\xi_\theta$$

$$-\rho_0 \omega^2 \xi_z = ik\gamma P_0 e^{-ikz}(\nabla \cdot \boldsymbol{\xi})$$

Writing $(\gamma P_0/\rho_0)^{1/2} = c_s$, the speed of sound, and $(B_{0z}^2/4\pi\rho_0)^{1/2} = v_A$, the Alfvén speed (cf. Section 4–7), these equations become

$$(k^2 v_A^2 - \omega^2)\xi_r = (c_s^2 + v_A^2)\frac{d}{dr}\left[\frac{1}{r}\frac{d}{dr}(r\xi_r)\right] + ikc_s^2\frac{d\xi_z}{dr} \qquad (4\text{–}61a)$$

$$(k^2 v_A^2 - \omega^2)\xi_\theta = 0 \qquad (4\text{–}61b)$$

$$(k^2 c_s^2 - \omega^2)\xi_z = c_s^2\left[\frac{ik}{r}\frac{d}{dr}(r\xi_r)\right] \qquad (4\text{–}61c)$$

from which, eliminating ξ_r,

$$\frac{d^2\xi_z}{dr^2} + \frac{1}{r}\frac{d\xi_z}{dr} - \frac{(k^2 c_s^2 - \omega^2)(k^2 v_A^2 - \omega^2)}{k^2 c_s^2 v_A^2 - \omega^2(c_s^2 + v_A^2)}\xi_z = 0 \qquad (4\text{–}62)$$

This is Bessel's equation having the solution (cf. Appendix 2)

$$\xi_z = I_0(Kr) \qquad (4\text{–}63)$$

where I_0 is the modified Bessel function of the first kind (of order zero),[†] and

$$K^2 = \frac{(k^2 c_s^2 - \omega^2)(k^2 v_A^2 - \omega^2)}{k^2 c_s^2 v_A^2 - \omega^2(c_s^2 + v_A^2)} = k^2\left[1 + \frac{(\omega/k)^4}{c_s^2 v_A^2 - (\omega/k)^2(c_s^2 + v_A^2)}\right]$$

Using (4–63) in (4–61a, c),

$$\xi_r = \frac{K}{ik}\left[\frac{c_s^2(k^2 v_A^2 - \omega^2) - \omega^2 v_A^2}{c_s^2(k^2 v_A^2 - \omega^2)}\right]I_0'(Kr) \qquad (4\text{–}64)$$

where the prime on I_0 denotes differentiation with respect to its argument.

We now apply the boundary conditions appropriate to the problem. Using (4–48) for $\mathbf{B}_i'$,

$$B_{iz}' = -\frac{B_{0z}}{r}\frac{d}{dr}(r\xi_r)$$

we obtain from the boundary condition (4–58)

$$-\gamma P_0 e^{-ikz}\nabla.\boldsymbol{\xi} - \frac{B_{0z}^2}{4\pi r}\frac{d(r\xi_r)}{dr} = \frac{B_{0e}}{4\pi}.\left[\nabla \times \mathbf{A}' + \xi_r\frac{dB_{0e}}{dr}\right] \qquad (4\text{–}65)$$

From the boundary condition (4–56) we find (since $\hat{\mathbf{n}}$ is in the direction of the radius vector, $\mathbf{r}$)

$$\mathbf{A}' = \xi_r B_\theta\hat{\mathbf{k}} + \alpha\hat{\mathbf{r}}$$

[†] The modified Bessel function of the second kind which is part of the general solution of (4–62) is inadmissible because of its behaviour for small values of r.

where α is an arbitrary function. Hence,

$$\mathbf{\nabla} \times \mathbf{A}' = \hat{\mathbf{\theta}}\left[\frac{\partial \alpha}{\partial z} - \frac{\partial}{\partial r}(\xi_r B_\theta)\right] \tag{4-66}$$

so that from (4–53)

$$\mathbf{\nabla} \times \mathbf{\nabla} \times \mathbf{A}' = \hat{\mathbf{r}}\left[-\frac{\partial^2 \alpha}{\partial z^2}\right] + \hat{\mathbf{k}}\left[\frac{\partial^2 \alpha}{\partial r\,\partial z} - \frac{\partial^2}{\partial r^2}(\xi_r B_\theta)\right] = 0$$

This gives two differential equations from which, on integrating once,

$$\frac{\partial \alpha}{\partial z} = f(r) \tag{4-67a}$$

$$\frac{\partial \alpha}{\partial z} - \frac{\partial}{\partial r}(\xi_r B_\theta) = g(z) \tag{4-67b}$$

where $f(r)$ and $g(z)$ are arbitrary functions. Using (4–67a) in (4–67b), and the fact that $\xi_r B_\theta$ is a function of r only, implies that $g(z) = \text{const}$. Thus, from (4–66)

$$\mathbf{\nabla} \times \mathbf{A}' = g(z)\hat{\mathbf{\theta}} = 0 \qquad \text{(since } \mathbf{B}' = 0 \text{ initially)}$$

Then (4–65) becomes, on using (4–61c) and the fact that $B_\theta \propto r^{-1}$,

$$\frac{\gamma\omega^2 P_0\xi_z}{ikc_s^2} - \frac{B_{0z}^2(k^2c_s^2 - \omega^2)\xi_z}{4\pi ikc_s^2} = -\frac{B_\theta^2\xi_r}{4\pi r}$$

Substituting for ξ_z and ξ_r from (4–63) and (4–64), this boundary condition becomes

$$[c_s^2\omega^2 - v_A^2(k^2c_s^2 - \omega^2)]I_0(Kr) = -\frac{B_\theta^2}{B_{0z}^2}v_A^2\frac{K}{r}I_0'(Kr)\left[\frac{c_s^2(k^2v_A^2 - \omega^2) - \omega^2v_A^2}{k^2v_A^2 - \omega^2}\right]$$

This expression should be regarded as a relation between ω and k (known as a *dispersion relation*, cf. Chapter 7) and may be written

$$\omega^2 = k^2v_A^2 - \frac{B_\theta^2}{B_{0z}^2}v_A^2\frac{K}{r}\frac{I_0'(Kr)}{I_0(Kr)}$$

$$= \frac{k^2B_{0z}^2}{4\pi\rho_0} - \frac{B_\theta^2}{4\pi\rho_0 r^2}\left[\frac{KrI_0'(Kr)}{I_0(Kr)}\right] \tag{4-68}$$

Since K involves ω^2, the general solution of the dispersion relation must be obtained numerically. However, it may be shown that ω^2 takes only real values and so the condition for the pinch to be unstable to $m = 0$ perturbations is

$$B_{0z}^2 \leqslant \frac{B_\theta^2}{(kr)^2}\left[\frac{KrI_0'(Kr)}{I_0(Kr)}\right] \qquad (4\text{--}69)$$

Moreover, for real values of K the function $[KrI_0'(Kr)/I_0(Kr)]$ is positive and increases monotonically with $|K|$. Clearly, from the definition of K^2, in the absence of the field B_{0z} (i.e., $v_A = 0$) there always exists a solution of (4–68) (such that $\omega^2 < 0$) which satisfies (4–69) and hence under these circumstances the pinch is always unstable. The introduction of the axial field B_{0z} in the plasma stabilizes the plasma for a certain range of values of B_{0z} (such that (4–69) is *not* satisfied). As one might guess from (4–69), there is a finite minimum value of B_{0z} for stability. Moreover, a finite maximum value of B_{0z} exists, since the equilibrium pressure condition

$$P_0 + \frac{B_{0z}^2}{8\pi} = \frac{B_\theta^2}{8\pi}$$

implies that $B_{0z}^2 < B_\theta^2$.

These results may be explained physically in the following way. Consider Fig. 4–7, which shows an axially symmetric perturbation of the equilibrium configuration of the linear pinch. Since $B_\theta \propto r^{-1}$, the external magnetic

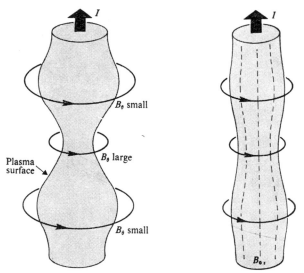

Fig. 4–7 Sausage ($m = 0$) instability of the plasma in a linear pinch; the second figure indicates the stabilizing effect of an axial magnetic field

pressure on the plasma surface is increased where the perturbation squeezes the plasma into a neck and is decreased where the perturbation fattens the plasma into a bulge. This gradient in the external magnetic field causes the perturbation to grow and, without an internal field, the necks contract to the axis, giving a sausage-like appearance to the plasma; hence the name *sausage instability*. Applying a similar argument to the internal field, B_{0z}, one sees that constricting the plasma increases B_{0z} and hence the internal magnetic pressure opposes the perturbation.

Similar arguments may be applied to other modes ($m \neq 0$). For example, the *kink instability* ($m = 1$) arises from the perturbation shown in Fig. 4–8. The distortion grows because the magnetic pressure on the concave side of

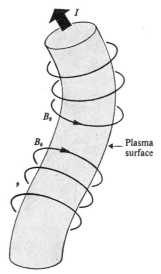

Fig. 4-8 Kink ($m = 1$) instability of the plasma in a linear pinch

the kink is increased (the B_θ lines are closer together), while that on the convex side is decreased (the B_θ lines are further apart). Again, the introduction of an internal field is favourable to stability since the tension in the lines of force caused by their stretching (cf. Section 4–3) tries to restore the pinch to its equilibrium position.

These unfavourable properties of the azimuthal field, B_θ, have a more general interpretation. Thus (cf. Problem 4–13) whenever the external field (the field containing the plasma) is concave towards the plasma any ripple on the plasma will tend to grow for the same reasons that the sausage and kink perturbations grow. Conversely, containing fields which are convex towards the plasma tend to smooth out ripple perturbations. For plasma containment, therefore, it is highly desirable to construct such fields. An example, shown in Fig. 4–9, is the *cusp field*.

Returning to the linear pinch, there are two points in particular which, for purposes of simplification, we have so far neglected. Commonly, a discharge tube has conducting walls since this enhances pinch stabilization. (One can demonstrate this by the same arguments used above concerning the internal field B_{0z}.) The presence of conducting walls means that the external field B_θ is also trapped between two conductors—namely, the walls and the plasma. Thus, plasma distortions towards the walls of the tube are repelled by the increased pressure of the external field.

The second point concerns relaxation of the perfect conductivity assumption. Although plasmas are highly conducting (the more so the hotter they

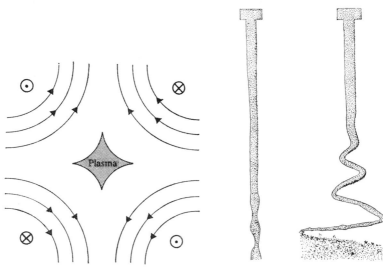

Fig. 4–9   Containment of plasma in a cusp field

Fig. 4–10   Experimental observation of sausage and kink instabilities in a mercury column carrying an electric current (after Alfvén and Fälthammer[29])

are) it is a well observed fact that currents do not flow in infinitesimally thin surface layers and the assumption of perfect conductivity is often very much a matter of mathematical convenience. The most obvious effect of finite conductivity considerations is a decrease in stability, since the lower the conductivity the more easily can magnetic fields penetrate (or leave) the plasma. Put the other way round, plasma motion is less dominated by the field configurations and the flow of plasma distortions across field lines is facilitated.

The sausage and kink instabilities predicted by linear perturbation theory have been observed in laboratory experiments with both a plasma and a conducting fluid. Figure 4–10 shows both the $m = 0$, $m = 1$ modes for a mercury column carrying an electric current.

4–6   Interchange instabilities

The sausage and kink (or wriggling) instabilities are illustrations of two general types of hydromagnetic instability endemic in plasma configurations.

The sausage mode belongs to the family of *interchange instabilities* and this wider genre will now be discussed briefly using a hydrodynamic analogy.

From hydrodynamics it is well known that an unstable fluid configuration exists when a fluid of density ρ_1 is supported by one less dense (ρ_2) in a gravitational field. Assuming that initially the boundary between the fluids is a horizontal plane, then any perturbation which produces a ripple on this boundary will grow in amplitude; the instability resulting is known as the *Rayleigh–Taylor instability*. Physically, what happens is that in displacing some of the uppermost (denser) fluid downwards the potential energy is, of course, decreased; this decrease is not balanced by the simultaneous upward displacement of an equal volume of the less dense fluid (both fluids are assumed to be

Fig. 4-11 Perturbation of the plasma–vacuum interface for a plasma supported against gravity by a magnetic field; the magnetic field is normal to the page

incompressible). An instability develops and a mode of wave number k grows in time according to $\exp\left[(kg)^{1/2}t\right]$ (cf. Problem 4–14). As one might anticipate, surface tension tends to inhibit the onset and development of instability; perturbations with wavelengths below a certain critical wavelength are stabilized by the surface tension.

What happens when the fluids are electrically conducting and a magnetic field is present? This general case is rather complicated, so we examine a rather special case (which has no counterpart in hydrodynamics) in which a perfectly conducting plasma is supported by a magnetic field against gravity, i.e., $\rho_2 = 0$. Suppose the plasma density and the magnetic field components are functions of z (cf. Fig. 4–11), that is, $\rho(z)$,† $B_x(z)$, $B_y(z)$, and initially the plasma boundary is the plane $z = 0$. Then, from the static condition (4–23), modified to take account of the gravitational force,

$$\frac{d}{dz}\left(P + \frac{B^2}{8\pi}\right) = -\rho g$$

Now suppose (as in Problem 4–14 for the Rayleigh–Taylor instability) that the perturbation of the plasma–vacuum boundary may be written

$$\xi(\mathbf{r}, t) = \xi(z) \exp\left[i\omega t - i(k_x x + k_y y)\right]$$

Then, for an incompressible plasma, it may be shown (cf. Problem 4–15)

† The density $\rho(z)$ is a step function given by ρ_1 within the plasma and equal to zero elsewhere.

that the linearized hydromagnetic equations reduce to

$$\frac{d}{dz}\left[\omega^2\rho - \frac{(\mathbf{k}\cdot\mathbf{B})^2}{4\pi}\right]\frac{d\xi_z}{dz} + k^2\left[-\rho\omega^2 + \frac{(\mathbf{k}\cdot\mathbf{B})^2}{4\pi} - g\frac{d\rho}{dz}\right]\xi_z = 0 \quad (4\text{-}70)$$

where $k^2 = k_x^2 + k_y^2$.

In the special case, $B_y = 0$, $k_x = 0$ (see Fig. 4–11), (4–70) reduces to the equation governing the Rayleigh–Taylor instability (cf. Problem 4–14). Consequently, the hydrodynamic growth rate applies in this special case to the hydromagnetic instability—that is, the mode grows in time according to $\exp[(k_y g)^{1/2}t]$. One would have anticipated no departure from the hydrodynamic result in this case on physical grounds since the perturbation does not distort the field lines and so leads to no change in the magnetic energy of the system.

In the general case, $B_y \neq 0$ and the perturbation *will* distort the field lines and alter the magnetic energy. The presence of a B_y (shear) component will

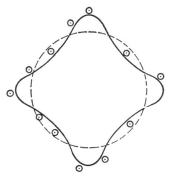

oppose the deformation of the interface, so tending to stabilize the system. In this sense, the shear component of the magnetic field fills a role analogous to surface tension in the hydrodynamic case; it may be shown[26] that the shear field is equivalent to a surface tension $T = (\lambda/\pi)(B_y^2/4\pi)$ where λ is the wavelength of the ripples.

This analogue of the Rayleigh–Taylor instability is known in hydromagnetics as the *Kruskal–Schwarzschild instability*.[27] The configuration of the Kruskal–Schwarzschild instability is not particularly relevant in practice for laboratory plasmas, but an interchange

Fig. 4–12 Flute instability in a magnetically contained plasma

instability does play an important role in the containment of a plasma between mirrors (cf. Section 2–8). There, the interchange instability is often known as the *flute instability*, due to the resemblance between the shape of the cross-section of the perturbed plasma (Fig. 4–12) and a fluted column. The interchange of plasma and magnetic field is illustrated in Fig. 4–12.

To determine whether or not the plasma in a mirror trap is stable to such a perturbation one may calculate the energy change δQ resulting from the interchange. Then, provided $\delta Q > 0$, such an interchange will not occur spontaneously nor, if imposed on the plasma–field configuration as a perturbation, will it develop into an instability. The stability condition $\delta Q > 0$† may be shown to reduce in this geometry to

$$\delta\int\frac{dl}{B} < 0 \quad (4\text{-}71)$$

† The stability condition that the energy change due to a perturbation of a dynamical

in which the integral is taken over the entire length of a flux tube. Further, using the fact that $dl \simeq -R\,d\theta$, where R is the radius of curvature and θ the angle the flux line makes with the z-axis, the requirement for stability becomes

$$\int \frac{d\theta}{B^{3/2}} > 0 \qquad (4\text{-}72)$$

The field in an adiabatic mirror trap does not satisfy (4–72) and so in this configuration a contained plasma is unstable towards a perturbation interchanging plasma and magnetic field on this simple analysis.†

Finally, all of this discussion in Sections 4–5 and 4–6 has been concerned with a linear analysis. Clearly, if a system is unstable this description is no longer valid once the amplitude of the growing perturbation becomes large and we must include non-linear effects in the analysis.

4–7 Alfvén waves

One of the most interesting aspects of an electrically conducting fluid in a magnetic field is the great variety of waves which it can support. A more complete treatment of waves in plasmas is deferred to Chapters 7 and 8. However, a discussion of hydromagnetics at this point would be incomplete without some mention of the hydromagnetic waves known as *Alfvén waves*. In Section 4–3, we saw that the stresses caused in a fluid by a magnetic field were equivalent to an isotropic pressure $B^2/8\pi$ together with a tension $B^2/4\pi$ along the lines of force, and we found that it was often convenient to picture tubes of force behaving like elastic strings under tension. Alfvén, using the analogy with elastic strings, recognized that transverse waves propagating along the lines of force might be generated by 'plucking' the tubes of force. If the analogy is valid then the velocity of these transverse waves, v_A, would be just $(T/\rho)^{1/2}$, T being the tension in the 'magnetic string' and ρ the density of the fluid; that is,

$$v_A = \frac{B}{\sqrt{(4\pi\rho)}}$$

Alfvén showed in 1942 that these waves should exist in a conducting fluid in a magnetic field. The discussion presented here follows an argument due to Walén (1944).[29] To make things as simple as possible, assume that the fluid is incompressible and perfectly conducting; then, from (3–83) and (3–86)

$$\nabla . \mathbf{U} = 0$$

system should be positive is a general one widely used to determine whether or not particular configurations are stable.

† The fact that plasmas contained in such traps are not always unstable depends on other features omitted here, such as the effect of a finite Larmor radius and of conducting end plates.[28]

and
$$\frac{\partial \mathbf{B}}{\partial t} = \mathbf{\nabla} \times (\mathbf{U} \times \mathbf{B}) = (\mathbf{B}.\mathbf{\nabla})\mathbf{U} - (\mathbf{U}.\mathbf{\nabla})\mathbf{B} \qquad (4\text{–}73)$$

The momentum equation, (3–84), is

$$\rho \frac{D\mathbf{U}}{Dt} = -\mathbf{\nabla}P + \frac{1}{c}(\mathbf{j} \times \mathbf{B}) = -\mathbf{\nabla}P - \frac{\mathbf{B} \times (\mathbf{\nabla} \times \mathbf{B})}{4\pi} \qquad (4\text{–}74)$$

Assume that the flow velocity is zero in the unperturbed fluid and that a static, spatially uniform magnetic field $\mathbf{B}_0$ exists throughout the fluid. Then, consider a small transverse disturbance of the equilibrium situation (such as that caused by 'plucking' a tube of magnetic force),

$$\mathbf{B} = \mathbf{B}_0 + \mathbf{B}' \qquad (4\text{–}75)$$

$$\mathbf{U} = \mathbf{U}' \qquad (4\text{–}76)$$

$$P = P_0 + P' \qquad (4\text{–}77)$$

Neglecting second-order terms, (4–74) becomes

$$\rho \frac{\partial \mathbf{U}'}{\partial t} = -\mathbf{\nabla}P' - \frac{\mathbf{B}_0}{4\pi} \times (\mathbf{\nabla} \times \mathbf{B}')$$
$$= -\mathbf{\nabla}\left(P' + \frac{\mathbf{B}_0.\mathbf{B}'}{4\pi}\right) + \left(\frac{\mathbf{B}_0}{4\pi}.\mathbf{\nabla}\right)\mathbf{B}' \qquad (4\text{–}78)$$

Since $\mathbf{\nabla}.\mathbf{U}' = 0 = \mathbf{\nabla}.\mathbf{B}'$, it follows that

$$\nabla^2\left(P' + \frac{\mathbf{B}_0.\mathbf{B}'}{4\pi}\right) = 0$$

Now, away from the perturbed zone of the fluid, P' and $\mathbf{B}'$ are zero. From potential theory we know that, since $P' + (\mathbf{B}_0.\mathbf{B}')/4\pi$ is zero outside the region of the perturbation and is harmonic, it must be everywhere zero.[30] So (4–78) becomes simply

$$\rho \frac{\partial \mathbf{U}'}{\partial t} = \left(\frac{\mathbf{B}_0}{4\pi}.\mathbf{\nabla}\right)\mathbf{B}' \qquad (4\text{–}79)$$

Also, using (4–75) and (4–76), (4–73) becomes

$$\frac{\partial \mathbf{B}'}{\partial t} = (\mathbf{B}_0.\mathbf{\nabla})\mathbf{U}' \qquad (4\text{–}80)$$

Combining (4–79) and (4–80) gives

$$\rho \frac{\partial^2 \mathbf{U}'}{\partial t^2} = \frac{1}{4\pi}(\mathbf{B}_0.\mathbf{\nabla})^2\mathbf{U}' \qquad (4\text{–}81)$$

For $\mathbf{B}_0$ oriented along the z-axis, (4–81) becomes

$$\frac{\partial^2 \mathbf{U}'}{\partial t^2} = \frac{B_0^2}{4\pi\rho} \frac{\partial^2 \mathbf{U}'}{\partial z^2} = v_A^2 \frac{\partial^2 \mathbf{U}'}{\partial z^2} \qquad (4\text{–}82)$$

Similarly,

$$\frac{\partial^2 \mathbf{B}'}{\partial t^2} = v_A^2 \frac{\partial^2 \mathbf{B}'}{\partial z^2} \qquad (4\text{–}83)$$

From (4–79) and (4–80) it follows that $\mathbf{U}$ and $\mathbf{B}'$ are parallel, and so dynamically the system resembles transverse waves on a stretched elastic string, thus bearing out the result anticipated at the beginning of this section. The propagation velocity of these transverse waves, v_A, is known as the *Alfvén velocity*, and the waves as *Alfvén waves*.[31] Here again, as in Section 4–3, where it became clear that magnetohydrostatics was much richer in physical phenomena than hydrostatics, one sees that the $\mathbf{j} \times \mathbf{B}$ force gives rise to a phenomenon without parallel in hydrodynamics. In a non-conducting fluid —if it is incompressible—the energy associated with a perturbation in the fluid remains localized; it may only move with the bulk flow velocity and there is no true wave motion. If, on the other hand, the fluid is compressible then this perturbation energy may radiate from its source in the form of sound waves (which are, of course, compression waves). For a conducting fluid, however, even if it is incompressible, the existence of the transverse Alfvén waves means that energy may be transported with a velocity $\mathbf{v}_A$ from its source region even though there is no bulk flow velocity $\mathbf{U}_0$. Note that *equipartition* of the perturbation energy exists between the fluid motion and the perturbed field since $\frac{1}{2}\rho U'^2 \equiv B'^2/8\pi$ (cf. Problem 4–16).

In deriving (4–78) the fluid was taken to be incompressible; since we shall be interested mainly in ionized gases rather than conducting liquids like mercury this assumption may not be tenable. Rather than give a full analysis here one may compare this with the situation in aerodynamics where the modifications introduced into hydrodynamic results are not serious provided the appropriate characteristic velocities are very much less than the speed of sound. So, by analogy, one would expect the incompressible fluid result derived above should be valid in the case of an ionized gas provided

$$v_A \ll \sqrt{\left(\frac{\gamma P}{\rho}\right)}$$

i.e.,

$$\frac{B_0^2}{4\pi} \ll \gamma P$$

(cf. Problem 4–17).

The other restriction on (4–78) is that no account has been taken of dissipative effects, such as viscosity or electrical resistivity. Only the neglect of resistivity may be serious in practice. To see what effect it has, in place of (4–73) we must use (4–1) which, for an incompressible fluid, becomes

$$\frac{\partial \mathbf{B'}}{\partial t} = \eta \nabla^2 \mathbf{B'} + (\mathbf{B_0} \cdot \nabla)\mathbf{U'} \tag{4-84}$$

while (4-79) remains unaffected. Combining (4-79) and (4-84) now gives

$$\frac{\partial}{\partial t}\left(\frac{\partial}{\partial t} - \eta\nabla^2\right)\mathbf{B'} = \frac{(\mathbf{B_0} \cdot \nabla)^2\mathbf{B'}}{4\pi\rho} \tag{4-85}$$

In this case, too, $\mathbf{U'}$ satisfies the same differential equation as $\mathbf{B'}$. Taking $\mathbf{B_0}$ to be the direction of the z-axis as before, (4-85) becomes

$$\frac{\partial}{\partial t}\left(\frac{\partial}{\partial t} - \eta\nabla^2\right)\mathbf{B'} = v_A^2 \frac{\partial^2\mathbf{B'}}{\partial z^2} \tag{4-86}$$

Suppose now that $\mathbf{B'} = \mathbf{b}\exp i(kz - \omega t)$; then, substituting in (4-86) gives $\omega^2 + ik^2\eta\,\omega - k^2 v_A^2 = 0$, from which

$$\omega = -\tfrac{1}{2}ik^2\eta \pm \tfrac{1}{2}(4k^2 v_A^2 - k^4\eta^2)^{1/2} \tag{4-87}$$

As one would have expected, taking account of the electrical conductivity of the fluid leads to a *damping* of the Alfvén wave. Assuming $2v_A > k^2\eta$, the waves are damped by the factor $\exp(-\tfrac{1}{2}k^2\eta t)$. Clearly, if $2v_A < k^2\eta$, there is no wave form and the perturbation is damped monotonically (rather than as a damped oscillation).

Some time after Alfvén's prediction of these transverse waves, attempts were made to detect them experimentally. First, Lundquist[32] tried to generate torsional Alfvén waves in mercury using magnetic fields of the order of 10,000 G; he succeeded in exciting waves but his results did not show quantitative agreement with theoretical predictions. This was not surprising since mercury is hardly suited to this purpose; its conductivity is not very high, while it has a high density. Some time later Lehnert[33] tried the experiments again; in place of mercury he used liquid sodium, which has a lower density and higher electrical conductivity—combined with the less desirable property of being highly inflammable. These experiments, in their turn, produced qualitative rather than quantitative agreement with theory, although they did improve on the earlier work. Later experiments on Alfvén waves have been carried out in plasmas—both gaseous and solid-state—and satisfactory agreement with theory has been reached (cf. Chapter 7).

Summary

1. *Kinematics*

(i) $\dfrac{\partial \mathbf{B}}{\partial t} = \eta\nabla^2\mathbf{B} + \nabla \times (\mathbf{U} \times \mathbf{B})$

            ↓   ↓

   diffusion   convection

(dominant if $R_M = \dfrac{LU}{\eta} \gg 1$)

(ii) Magnetic flux through any closed contour moving with a perfectly conducting fluid is constant;

(iii) Fluid elements initially on a line of force continue to lie on a line of force in a perfectly conducting fluid;

From (ii) and (iii), the field behaves as though it were frozen into the fluid.

2. Magnetohydrostatics

(i) $\dfrac{\partial \Im_{ik}}{\partial r_k} = 0$ $\Im_{ik} = \left(P + \dfrac{B^2}{8\pi}\right)\delta_{ik} - \dfrac{B_i B_k}{4\pi}$

(ii) The total stress in a conducting fluid amounts to an isotropic pressure which is the sum of the fluid and magnetic pressures together with a tension $B^2/4\pi$ along the lines of force.

(iii) Closed isobaric surfaces may be pictured as windings of $\mathbf{B}$ and $\mathbf{j}$ lines; a set of nested surfaces on which pressure increases inwards ($\mathbf{j} \times \mathbf{B}$ points inwards) makes possible magnetic containment (such as linear pinch, theta pinch).

3. Hydromagnetic stability

(i) A magnetohydrostatic system in equilibrium has to be examined for stability; if forces act on the system tending to restore it to its equilibrium configuration after it has suffered a small perturbation, the system is stable towards such perturbations.

(ii) The linear pinch configuration in which the conducting fluid is free from magnetic fields and currents (the containing field being generated by a skin current) is, for $m = 0$, unstable for all k. A suitable axial magnetic field resists the tendency towards sausage instability. Conducting walls also have a stabilizing effect on the pinch.

(iii) A plasma in a gravitational field supported by a magnetic field is unstable (Kruskal–Schwarzschild instability); a shear component in the $\mathbf{B}$ field has a stabilizing effect.

(iv) Containing fields concave (convex) towards the plasma permit (inhibit) the growth of a ripple on the plasma surface.

4. Alfvén waves

(i) The analogy between magnetic tubes of force under tension and elastic strings suggests the existence of transverse waves with velocity

$$v_A = \sqrt{\left(\frac{T}{\rho}\right)} = \frac{B}{\sqrt{(4\pi\rho)}}$$

(ii) The effect of finite conductivity is to damp the wave; the damping factor is $\exp\left(-\tfrac{1}{2}k^2\eta t\right)$ (if $2v_A > k^2\eta$).

P D—G

Problems

4-1 Consider the equation describing diffusion of the magnetic field. For a one-dimensional field $B(z, t)$, show that a solution of this equation which is continuous and has continuous derivatives for all z when $t > 0$ and is such that $B(z, 0) = f(z)$, is given by

$$B(z, t) = (4\pi\eta t)^{-1/2} \int_{-\infty}^{\infty} f(s) \exp\left[-\frac{(s - z)^2}{4\eta t}\right] ds$$

where η is the magnetic diffusivity.

Next consider a half-space problem in which a conductor fills the space $x > 0$ and is bounded by the plane $x = 0$. Examine how a uniform magnetic field in the region $x < 0$ and parallel to the boundary diffuses through the conductor. Assume that $B(x > 0, t = 0) = 0$ and that $B(x < 0, t) = $ constant.

Give a physical interpretation of these results.

4-2 Establish

(i) $\dfrac{D}{Dt}(dr) = (dr . \nabla)v$

(ii) $A.(dr.\nabla)v \equiv dr.[(A.\nabla)v + A \times (\nabla \times v)]$

Hint: start with the term $dr.A \times (\nabla \times v)$ and use the permutation tensor (cf. Appendix 1).

4-3 Devise an alternative proof of the theorem that the magnetic flux through any closed contour moving with a perfectly conducting fluid is constant.

Hint: consider how Faraday's law

$$\frac{d\Phi}{dt} = -\oint .Edr$$

may be represented in situations in which the closed circuit over which the integration is carried out is no longer stationary. Allow for the additional contribution to $d\Phi/dt$ and show that it amounts to

$$-\oint \frac{(v \times B)}{c} . dr$$

4-4 Using the result established in (4-3), construct a plausibility argument to support the statement that fluid elements which lie initially on a magnetic line of force continue to lie on a line of force in a perfectly conducting fluid.

4-5 Consider a rotating inviscid fluid of uniform conductivity and a magnetic field with no azimuthal component, i.e., $B = (B_r, 0, B_z)$. Show that along a line of force the fluid rotates with uniform angular velocity.

Hint:
$$\nabla \times j = \nabla \times \left\{\sigma\left(E + \frac{U \times B}{c}\right)\right\}$$

$$= \frac{\sigma}{c} \nabla \times (U \times B)$$

Observe that since $B_\theta = 0$, the current is wholly azimuthal, i.e., $[\nabla \times (U \times B)]_\theta = 0$. Using the null divergence of B the result $rB.\nabla (U_\theta/r) = 0$ follows.

4-6 By transforming to principal axes show that the total stress tensor

$$\mathfrak{I}_{ik} = \left[\left(P + \frac{B^2}{8\pi}\right)\delta_{ik} - \frac{B_iB_k}{4\pi}\right]$$

may be reduced to the diagonal form

$$\begin{bmatrix} P + \dfrac{B^2}{8\pi} & 0 & 0 \\[2mm] 0 & P + \dfrac{B^2}{8\pi} & 0 \\[2mm] 0 & 0 & P - \dfrac{B^2}{8\pi} \end{bmatrix}$$

4-7 (i) Choosing $n \sim 7.10^{15}$ particles/cm³ obtain estimates of the magnetic field strength needed to contain plasmas with temperatures of 1 keV and 100 keV respectively.

(ii) Bennett's relation, (4–35), predicts that the plasma temperature in a pinch discharge should vary as the square of the current. As the temperature increases, the loss of energy from the plasma due to radiation increases (cf. Chapter 9); this means that there is a limit to the heating of the discharge and to the current that may be passed. If the limiting current is assumed to be 10^6 A, calculate the highest temperature attainable for line densities of 10^{18} particles/cm and 10^{20} particles/cm.

4-8 Consider a linear pinch in which the proton and electron temperatures are T^+, T^- and assume that the electron axial velocities are very much greater than those of the protons, in addition to being independent of position. Show that if $n = n_0$ on the axis of the discharge, the particle density distribution is given by

$$n(r) = n_0\left[1 + \frac{\pi n_0 e^2 u^2 r^2}{2c^2\kappa(T^+ + T^-)}\right]^{-2}$$

where u_z is the electron axial velocity.

4-9 From (4–1), together with the hydromagnetic Navier–Stokes equation (cf. Section 5–2)

$$\rho\frac{DU}{Dt} = -\nabla P + \frac{1}{c}(\mathbf{j} \times \mathbf{B}) + \rho\nu\,\nabla^2\mathbf{U}$$

show by introducing new variables

$$\mathbf{Q}_1 = \mathbf{U} + (4\pi\rho)^{-1/2}\mathbf{B} \qquad \mathbf{Q}_2 = \mathbf{U} - (4\pi\rho)^{-1/2}\mathbf{B}$$

that these equations may be rewritten in the form

$$\frac{\partial\mathbf{Q}_1}{\partial t} + (\mathbf{Q}_2\cdot\nabla)\mathbf{Q}_1 = -\frac{1}{\rho}\nabla\left(P + \frac{B^2}{8\pi}\right) + \nu_1\,\nabla^2\mathbf{Q}_1 + \nu_2\,\nabla^2\mathbf{Q}_2$$

$$\frac{\partial \mathbf{Q}_2}{\partial t} + (\mathbf{Q}_1 \cdot \nabla)\mathbf{Q}_2 = -\frac{1}{\rho} \nabla\left(P + \frac{B^2}{8\pi}\right) + \nu_2 \nabla^2 \mathbf{Q}_1 + \nu_1 \nabla^2 \mathbf{Q}_2$$

where $\nu_{1,2} = \frac{1}{2}(\nu \pm \eta)$.

Comment on this form of the equations, which is due to Elsasser.

4-10 Show that within a finite region a magnetic field cannot be force-free everywhere.

4-11 Establish Woltjer's result that force-free fields with α constant represent the state of minimal magnetic energy in a closed system.

Hint: Assuming perfect conductivity and using $\mathbf{B} = \nabla \times \mathbf{A}$, where $\mathbf{A}$ is the vector potential, one has

$$\frac{\partial \mathbf{A}}{\partial t} = \mathbf{v} \times \nabla \times \mathbf{A}$$

Using this show that $\int_V \mathbf{A} \cdot \nabla \times \mathbf{A}\, d\tau = $ const. for all $\mathbf{A}$ which are constant on the boundary of V. Then examine the stationary values of the magnetic energy

$$\int \frac{B^2}{8\pi} = d\tau \int \frac{(\nabla \times \mathbf{A})^2}{8\pi} d\tau$$

Introduce a Lagrangian multiplier $\alpha/8\pi$ and obtain the following condition for stationary values of the magnetic energy

$$\delta \int [(\nabla \times \mathbf{A})^2 - \alpha \mathbf{A} \cdot \nabla \times \mathbf{A}]\, d\tau = 0$$

Performing the variation, the condition $\nabla \times \mathbf{B} = \alpha\mathbf{B}$ follows (where α is constant).

4-12 Establish the boundary condition on the tangential electric field given in (4-54).

Also consider more generally the boundary conditions on the normal component of an electric field and on the normal and tangential magnetic fields. Establish that the normal electric field E_n is discontinuous on account of surface charge, and that B_n is continuous, while B_t is discontinuous due to surface currents.

4-13 Construct a physical argument to show that when the magnetic field containing a plasma is concave towards the plasma, any disturbance in the form of a ripple on the surface of the plasma will tend to grow in amplitude.

4-14 Consider an ideal fluid of density ρ_1 supported in a gravitational field by another ideal fluid of density ρ_2; neither fluid is an electrical conductor. Initially the boundary between the fluids is a horizontal plane; this is then perturbed so that gravity waves propagate on the surface of separation. Show that the relation between frequency and wave number for these waves is $\omega^2 = kg(\rho_2 - \rho_1)/(\rho_2 + \rho_1)$, where g is the acceleration due to gravity. For $\rho_1 \gg \rho_2$ the waves are unstable, growing according to exp $[(kg)^{1/2}t]$; this is the Rayleigh–Taylor instability.

4-15 Consider a perfectly conducting plasma supported against gravity by a magnetic field. For a perturbation of the plasma–vacuum boundary of the form

$$\xi(\mathbf{r}, t) = \xi(z) \exp\left[i\omega t - i(k_x x + k_y y)\right]$$

show that the linearized hydromagnetic equations reduce to (4–70).

4–16 Confirm the observation made in Section 4–7 that equipartition of energy between fluid motion and wave magnetic field occurs for Alfvén waves.

4–17 Show that the neglect of compressibility in considering Alfvén wave propagation is unimportant provided the fluid pressure is very much greater than the magnetic pressure.

4–18 Consider a plane surface $x = 0$ across which a discontinuity in the density of a conducting fluid exists. A uniform magnetic field is applied parallel to the x-axis. An Alfvén wave is incident normally on the surface $x = 0$ from the left. Calculate the amplitudes of the transmitted and reflected Alfvén waves.

4–19 Let $\mathbf{B} = \mathbf{B}_0 + \mathbf{b}$ where $\mathbf{B}_0$ is a uniform magnetic field and *no* restriction is placed on the magnitude of $\mathbf{b}$. Show that the equation corresponding to (4–74) is now

$$\rho \frac{D\mathbf{U}}{Dt} = -\nabla\left(P + \frac{B^2}{8\pi}\right) + \left[\frac{(\mathbf{B}_0 + \mathbf{b})}{4\pi} \cdot \nabla\right]\mathbf{b}$$

and that if $P + B^2/8\pi = \text{const.}$ and $\mathbf{U} = \pm(4\pi\rho)^{-1/2}\mathbf{b}$, this equation together with that for the magnetic field gives

$$\frac{\partial^2 \mathbf{U}}{\partial t^2} = v_A^2 \frac{\partial^2 \mathbf{U}}{\partial z^2}$$

Observe that since $\mathbf{b}$ is unrestricted this result implies that non-linear Alfvén waves may propagate in a uniform, perfectly conducting, incompressible fluid; it was first derived by Walén.[34]

5

Hydromagnetic flows

5–1 Introduction

In general, a mathematical analysis of hydromagnetic flows is complicated by the non-linearity of the equations; usually one must resort to approximate or numerical methods of solution. To avoid this complication, we concentrate mainly on flows which may be solved analytically, such as flows along ducts of constant cross-section in which two of the dimensions (including the length) are very much greater than the third. This has the effect of making the problem one-dimensional, since the flow variables no longer depend on the two coordinates corresponding to the larger dimensions,† thus permitting the calculation of exact solutions. Although such solutions can only be found for simplified problems, they not only serve as approximate solutions for more realistic problems but demonstrate clearly the basic flow dynamics. This in turn enables one to discuss, qualitatively at least, various applications of hydromagnetic flows.

A brief, qualitative discussion of the stability of flows[35] is given in Section 5–5. A valuable exploitation of the similarity between the hydromagnetic and hydrodynamic equations is demonstrated in Sections 5–6 and 5–7.

5–2 Hydromagnetic Navier–Stokes equation

The derivation of the hydromagnetic equations in Chapter 3 was for an inviscid fluid. In discussing flows this is a restriction which it is desirable to remove. However, since the algebraic manipulation involved in deriving the viscous, hydromagnetic equations is out of all proportion to the use which will be made of them, the procedure is merely outlined.‡

In Section 3–5 we saw that the Chapman–Enskog method for closing the set of moment equations involved the expansion of the proton and electron distribution functions, $f^{\pm}$, about local Maxwellians, $f_M^{\pm}$. The lowest-order approximation, $f = f_M$, then led to the inviscid, hydromagnetic equations. In this approximation, the *unknown*§ quantities $q_{ijk}^{\pm}$ and the off-diagonal elements of $p_{ij}^{\pm}$ were all zero and the diagonal elements of $p_{ij}^{\pm}$ were all equal (to one third of the trace of $p_{ij}^{\pm}$); see (3–53). The viscous terms appear in first-order theory for which one writes $f^{\pm} = f_M^{\pm}(1 + \phi^{\pm})$, where $\phi^{\pm}$ are the first-

---

† Strictly, this may not be true for the pressure, cf. Section 5–3.

‡ For a further discussion see reference 4.

§ The term *unknown* is used here to denote those quantities for which no moment equation has been derived.

order corrections to the Maxwellians. The next step is to calculate these first-order corrections, as functionals of $n^{\pm}$, **U**, and T, from the proton and electron kinetic equations. This done, the unknowns may be evaluated to first-order, giving the closed set of *Navier–Stokes equations*. The complete set of Navier–Stokes equations is very complicated but reduces considerably for collision-dominated situations (such that $\Omega_e \ll \nu_c$) and one finds for the pressure tensor

$$P_{ij} = P\delta_{ij} - \eta\left(\frac{\partial U_i}{\partial r_j} + \frac{\partial U_j}{\partial r_i}\right) + \tfrac{2}{3}\eta\frac{\partial U_k}{\partial r_k}\delta_{ij}$$

where, by analogy with hydrodynamics, η may be identified with the coefficient of viscosity and is related to the kinematic viscosity ν by $\eta = \rho\nu$. Although η is a fairly complex expression involving the temperature one may usually treat it as constant. This leads to the momentum equation

$$\rho\frac{D\mathbf{U}}{Dt} = -\nabla P + \frac{1}{c}(\mathbf{j} \times \mathbf{B}) + \frac{\eta}{3}\nabla(\nabla.\mathbf{U}) + \eta\nabla^2\mathbf{U} \tag{5-1}$$

or, in tensor notation,

$$\rho\frac{DU_i}{Dt} = -\frac{\partial P}{\partial r_i} + \frac{1}{c}(\mathbf{j} \times \mathbf{B})_i + \frac{\eta}{3}\frac{\partial^2 U_j}{\partial r_i\partial r_j} + \eta\frac{\partial^2 U_i}{\partial r_j^2} \tag{5-1a}$$

which is the hydromagnetic *Navier–Stokes equation*. Since we are not here concerned with the energy equation, its first-order form will not be reproduced. The mass conservation equation is unchanged, as is Ohm's law under conditions (3–70) to (3–72). The set of equations to be used in this chapter, therefore, is (5–1) together with (3–77) and (3–80) to (3–82), which are repeated for convenience:

$$\frac{\partial\rho}{\partial t} + \nabla.(\rho\mathbf{U}) = 0 \tag{5-2}$$

$$\mathbf{j} = \sigma\left(\mathbf{E} + \frac{\mathbf{U} \times \mathbf{B}}{c}\right) \tag{5-3}$$

$$\nabla \times \mathbf{E} = -\frac{1}{c}\frac{\partial\mathbf{B}}{\partial t} \tag{5-4}$$

$$\nabla \times \mathbf{B} = \frac{4\pi}{c}\mathbf{j} \tag{5-5}$$

5–3 Hartmann flow

Consider the steady flow of a viscous, conducting fluid along a duct of constant cross-section, illustrated in Fig. 5–1. Cartesian axes are chosen so

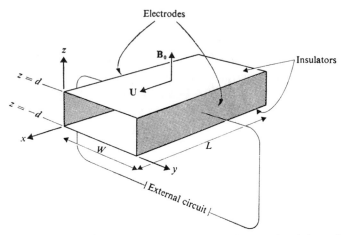

Fig. 5-1 Rectangular duct for Hartmann flow. If electrodes are placed down the sides of
the duct, current may be extracted from, or supplied to, the conducting fluid

that the flow velocity $\mathbf{U}$ and the constant, uniform applied magnetic field $\mathbf{B_0}$
are given by

$$\mathbf{U} = (U, 0, 0) \tag{5-6a}$$

and $$\mathbf{B_0} = (0, 0, B_0) \tag{5-6b}$$

If the width, W, and length, L, of the duct are such that

$$W \gg d \tag{5-7a}$$

$$L \gg d \tag{5-7b}$$

then the problem may be approximated to flow between infinite, parallel
planes at $z = \pm d$. In this approximation, the flow does not depend on x or y
so that, with the exception of the pressure P, all variables may be taken as
functions of z only. Since the flow is caused by a pressure gradient in the
x direction, P must be a function of x as well as z although, as the solution
confirms (cf. (5–13)), ∇P is independent of x. (It is ∇P which occurs in the
equation of motion so this is consistent with the one-dimensional nature of
the flow.) This then corresponds to classical *Hartmann flow*. Relaxation of
condition (5–7a) introduces y dependence and may lead to *secondary flows*.
Similarly, relaxation of (5–7b) introduces x dependence—that is, end-effects.
These will not be considered here.[36]
    Substituting (5–6a) in (5–1) to (5–5) we see that (5–2) is satisfied identically
and (5–5) gives

$$\mathbf{j} = \frac{c}{4\pi}\boldsymbol{\nabla} \times \mathbf{B} = \frac{c}{4\pi}\left(-\frac{dB_y}{dz}, \frac{dB_x}{dz}, 0\right) \tag{5-8}$$

Also, since $\nabla.\mathbf{B} = 0$,

$$B_z = \text{const.} = B_0 \qquad (5\text{-}9)$$

and (5–1) gives

$$\frac{\partial P}{\partial x} = \frac{B_0}{4\pi}\frac{dB_x}{dz} + \eta\frac{d^2U}{dz^2} \qquad (5\text{-}10a)$$

$$\frac{dB_y}{dz} = 0 \qquad (5\text{-}10b)$$

$$\frac{\partial P}{\partial z} = -\frac{B_x}{4\pi}\frac{dB_x}{dz} \qquad (5\text{-}10c)$$

Substituting (5–10b) in (5–8)

$$\mathbf{j} = \left(0, \frac{c}{4\pi}\frac{dB_x}{dz}, 0\right) \qquad (5\text{-}8a)$$

so that (5–3) gives (cf. Problem 5–1)

$$E_x = 0 = E_z$$

$$\frac{j_y}{\sigma} = \frac{c}{4\pi\sigma}\frac{dB_x}{dz} = E_y - \frac{UB_0}{c} \qquad (5\text{-}11)$$

Lastly, from (5–4)

$$E_y = \text{const.} = E_0 \qquad (5\text{-}12)$$

the induced electric field.

Differentiating (5–10a, c) with respect to x shows that

$$\frac{\partial P}{\partial x} = \text{const.} = -P_0$$

i.e., $$P(x, z) = -P_0 x + P_1(z) \qquad (5\text{-}13)$$

(where P_0 is proportional to the constant pressure head maintaining the flow) and

$$-P_0 = \frac{B_0}{4\pi}\frac{dB_x}{dz} + \eta\frac{d^2U}{dz^2}$$

Substituting from (5–11) and (5–12), this last equation becomes

$$\frac{d^2U}{dz^2} - \frac{\sigma B_0^2}{\eta c^2}U = -\frac{(cP_0 + \sigma B_0 E_0)}{\eta c} \qquad (5\text{-}14)$$

Taking the velocity boundary conditions as $U(\pm d) = 0$, the solution of (5–14) is

$$U(z) = \frac{c(cP_0 + \sigma B_0 E_0)}{\sigma B_0^2}\left[1 - \frac{\cosh(Mz/d)}{\cosh M}\right] \tag{5–15}$$

where the *Hartmann number* M is given by

$$M = \frac{B_0 d}{c}\left(\frac{\sigma}{\eta}\right)^{1/2} \tag{5–16}$$

In this solution, all the parameters are given constants with the exception of the electric field E_0 induced in the y direction. If, as in Fig. 5–1, electrodes are placed on either side of the duct some external circuit may be connected across them so that the fluid is part of a continuous electrical circuit. E_0 will then be determined by the current which flows in this circuit. Integrating (5–11) with respect to z and using (5–12) and (5–15) gives for the mean current density

$$j_0 \equiv \frac{1}{2d}\int_{-d}^{d} j_y\, dz = \sigma\left(E_0 - \frac{B_0 U_0}{c}\right) \tag{5–17}$$

where the mean flow velocity

$$U_0 = \frac{1}{2d}\int_{-d}^{d} U\, dz = \frac{c(cP_0 + \sigma B_0 E_0)}{M\sigma B_0^2}[M - \tanh M] \tag{5–18}$$

Substituting (5–18) in (5–17) and rearranging gives E_0 in terms of j_0:

$$E_0 = \frac{(cP_0 + j_0 B_0)}{\sigma B_0}M\coth M - \frac{cP_0}{\sigma B_0} \tag{5–19}$$

Hence,

$$U_0 = \frac{c(cP_0 + B_0 j_0)}{\sigma B_0^2}[M\coth M - 1] \tag{5–20}$$

and

$$U(z) = \frac{cM(cP_0 + B_0 j_0)}{\sigma B_0^2 \sinh M}[\cosh M - \cosh(Mz/d)] \tag{5–21}$$

$$= U_0 M\frac{[\cosh M - \cosh(Mz/d)]}{(M\cosh M - \sinh M)} \tag{5–22}$$

It is of physical significance whether the mean current is positive, negative, or zero and the various possibilities lead to different technological applications. Suppose first that the electrodes are short-circuited. Then $E_0 = 0$ and it follows from (5–17) that $j_0 < 0$ for all positive B_0. Now introduce some resistance in the external circuit and gradually increase its value. The mean current will continue to flow in the negative y direction, decreasing in magnitude as the circuit resistance increases, until at infinite resistance, or open circuit, $j_0 = 0$. It is easy to verify for all $j_0 < 0$ that the mean Lorentz force, $(\mathbf{j_0} \times \mathbf{B_0})/c$, acts *against* the flow. The action of the applied magnetic field on the flow therefore is that of a brake. This braking action is maximum when the current is maximum, that is when the external resistance is zero. The short-circuit condition, therefore, corresponds to the *electromagnetic brake*. For a *finite* external resistance, electrical power is extracted from the fluid and this is the basic action of the *magnetohydrodynamic generator* (cf. Section 5–9). Note that in the electromagnetic brake, flow energy is dissipated wholly within the conducting fluid.

The open-circuit condition, $j_0 = 0$, corresponds to a *flowmeter* since, from (5–17), $E_0 = B_0 U_0/c$, so that by measuring E_0 the fluid flux may be determined.

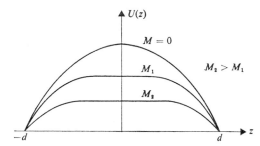

Fig. 5–2 Velocity profile for various values of Hartmann number keeping P_0 and j_0 fixed

Now suppose that instead of a resistance the external circuit contains a power supply. A current may then be driven through the fluid in the positive y direction and the Lorentz force acts in the direction of flow. External power is being used to accelerate the flow. This illustrates the basic mechanism of the *electromagnetic pump* and *plasma accelerator* (cf. Section 5–8).

The Hartmann velocity profile (5–22) is sketched in Fig. 5–2 for a single value of j_0 and various values of M. (Varying j_0 for a fixed value of M merely alters the value of U_0 (cf. (5–20)). The effect, therefore, is to vary the scale of the ordinate axis.) The curve corresponding to $M = 0$ is the hydrodynamic limit ($B_0 = 0$), that is, the familiar parabolic profile

$$U = \tfrac{3}{2}U_0(1 - z^2/d^2) \tag{5–22a}$$

Given U_0, increasing the applied field B_0 causes a flattening of the profile with an increased velocity gradient at the edges. When $M \gg 1$, (5–22) becomes

$$U = U_0\{1 - \exp{[-M(1 - |z|/d)]}\} \tag{5–22b}$$

which is approximately constant except for boundary layers of thickness $\sim(d/M)$ where the velocity gradient is $\sim(U_0 M/d)$.

The physical reason for this flattening of the velocity profile is easily understood when one looks at the current density profile, obtained from (5–11). On substituting (5–12), (5–19), and (5–21) in (5–11), $j_y(z)$ is given by

$$j_y = \frac{M(cP_0 + j_0 B_0)}{B_0 \sinh M} \cosh (Mz/d) - \frac{cP_0}{B_0} \qquad (5\text{–}23)$$

and this is shown in Fig. 5–3 for $j_0 = 0$. The negative current density in the centre of the duct interacts with B_0 to give a Lorentz force in the negative x

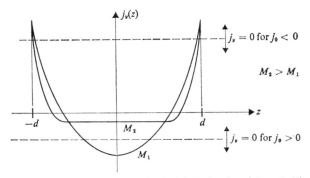

Fig. 5-3 Current profile for two values of M with P_0 fixed and $j_0 = 0$. The dotted lines indicate the z-axis for given non-zero values of j_0

direction and hence slows down the flow there. Meanwhile, at the edges the current is positive and acts, therefore, to accelerate the flow near the boundaries; hence the flattening of the parabolic profile. From (5–23) it can be seen that the effect of non-zero j_0 on Fig. 5–3 is to move the $j_y = 0$ axis as shown.

The induced magnetic field B_x is obtained by integrating (5–11). For this one needs a boundary condition on B_x and in general the existence of an external circuit may complicate matters. However, assuming symmetry about the central plane of the duct the boundary condition is $B_x(0) = 0$, so that substituting (5–23) in (5–11) and integrating gives

$$B_x = \frac{4\pi d(cP_0 + B_0 j_0) \sinh (Mz/d)}{cB_0 \sinh M} - \frac{4\pi P_0 z}{B_0} \qquad (5\text{–}24)$$

This is shown in Fig. 5–4. For $j_0 = 0$, the induced field B_x is also zero at the edges of the duct. However, this is not the case when a mean current flows. Since the direction of the magnetic field line at any point is given by $dx/dz = B_x/B_z = B_x/B_0$ one may obtain from Fig. 5–4 the distortion of the applied field caused by the flow. This is indicated in Fig. 5–5. When $j_0 = 0$,

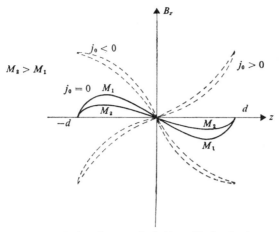

Fig. 5-4 Induced magnetic field profile for fixed P_0

the field lines are normal to the planes $z = \pm d$ at the boundaries. The distortion of the applied field for $j_0 \leqslant 0$ is as one would expect from the arguments concerning the stretching of field lines used in Chapter 4. The moving fluid pulls the lines of force in the direction of flow; the weaker the field, the

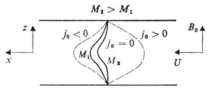

Fig. 5-5 Distortion of applied magnetic field lines; the stronger the field the less it is distorted by the flow. The dotted lines refer to non-zero j_0

more it is pulled out of shape. For $j_0 > 0$, a current is being driven through the fluid by an external power supply and the induced field due to this current results in a bending of the field lines in the opposite direction.

The other components of the magnetic field are given by (5–9) and (5–10b), and are constant. Finally, the pressure is given by (5–13) and (5–10c):

$$P = -P_0 x + P_1(z) = \text{const.} - \frac{B_x^2}{8\pi} - P_0 x$$

5–4 Couette flow

Another steady, one-dimensional flow which permits an exact analytic solution is *Couette flow*, depicted in Fig. 5–6. The problem is very similar to that of Hartmann flow, the main difference being that the initial flow is now due to the motion of the wall at $z = 2d$ relative to the stationary wall at $z = 0$

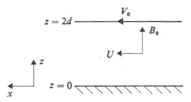

Fig. 5-6 Couette flow. The viscous drag due to the steady motion of the upper wall causes the initial flow

rather than a constant pressure gradient in the x direction. Making the same assumptions concerning the dimensions of the flow as were made for Hartmann flow, secondary flows and end-effects may be ignored. The flow velocity and applied magnetic field are given by (5–6a) and (5–6b) and the differential equations describing Couette flow are the same as those for Hartmann flow, excepting only that $\partial P/\partial x = 0$ in (5–10a).† Thus, with the boundary conditions $U(0) = 0$, $U(2d) = V_0$, it follows (cf. Problem 5–4) that

$$U(z) = \frac{V_0 \sinh (Mz/d)}{\sinh 2M} + \frac{2cE_0 \sinh (Mz/2d) \sinh [M(1 - z/2d)]}{B_0 \cosh M} \quad (5\text{–}25)$$

As for Hartmann flow, the value of the constant induced field E_0 depends on the mean current flow.

The remaining variables follow as before:

$$j_y(z) = \sigma(E_0 - B_0 U(z)/c) \quad (5\text{–}26)$$

$$B_x(z) = \frac{4\pi}{c} \int_0^z j_y(z) \, dz + B_x(0) \quad (5\text{–}27)$$

$$P(z) = \text{const.} - \frac{B_x^2(z)}{8\pi} \quad (5\text{–}28)$$

The velocity profile (5–25) for the short-circuit condition ($E_0 = 0$) is shown in Fig. 5–7. Note that the velocity gradient is increased at the moving wall ($z = 2d$) compared with the hydrodynamic flow ($M = 0$) and decreased at the stationary wall ($z = 0$).

Since the thickness of a boundary layer is usually very much less than its other dimensions, Couette flow is often used to analyse the effect of hydromagnetic forces on boundary layers. The stationary wall then corresponds to some real wall or body surface, while the moving wall represents the free

† Note that the only change in Cartesian axes compared with Section 5–3 is that the $z = 0$ plane is chosen as the lower wall instead of the mid-plane of the duct.

streamline on the opposite side of the boundary layer. The analysis of Sec-

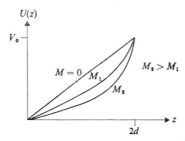

Fig. 5-7   Couette velocity profile for various values of Hartmann number and $E_0 = 0$

tions 5–3 and 5–4 is not, of course, restricted to rectangular ducts. It may be applied, for example, to flow between co-axial cylinders provided the distance between the cylinder surfaces is small compared with their radii and length.

5–5   Flow stability†

A number of experimental investigations of hydromagnetic flows have been carried out using liquid metals, particularly mercury. The earliest experiments investigated Hartmann flow for the open-circuit condition, $j_0 = 0$, and showed good agreement with the theory provided the flow remained laminar. Not surprisingly, agreement breaks down whenever the flow becomes turbulent. An interesting illustration of this appeared in the experiments of Hartmann and Lazarus,[35] who measured the pressure gradient P_0 required to maintain a specified mean flow rate U_0 as the applied field B_0 was increased. In Section 5–3 we noted that for $j_0 \leqslant 0$ the effect of the Lorentz force is always to act against the flow. For a constant U_0, therefore, the theory predicts that P_0 should increase monotonically with B_0; cf. (5–20). In fact, it was observed that for higher flow rates P_0 first *decreased* as B_0 was increased from zero and then increased with B_0 as predicted (cf. Fig. 5–8). The explanation of this behaviour lies in the turbulent nature of the initial ($B_0 = 0$) flow for sufficiently fast flow rates. As B_0 increases from zero the turbulence decreases. The effect of the increase in order of the flow dominates the hydromagnetic drag of the Lorentz force to begin with, so that the required P_0 decreases as B_0 increases. Eventually, as the turbulence disappears and the Lorentz force becomes large, P_0 increases with B_0.

   The ability of a magnetic field to inhibit turbulence in flows might be expected from the discussion of magnetostatics in Chapter 4. We saw there that a magnetic field imparts a certain rigidity to a conducting fluid by restricting motion across the lines of force—a phenomenon perhaps most easily pictured in terms of magnetic tubes of force as discussed in Section 4–3.

---

† Quantitative discussions of this topic are to be found in references 26 and 35.

Magnetic rigidity has been demonstrated experimentally by Lundquist, using mercury, and Lehnert, using liquid sodium. Quantitative investigations of the transition between Hartmann flow and turbulence have been carried out theoretically and experimentally by Lock and Murgatroyd,[35] respectively. Unfortunately, their results differ considerably even though the discrepancy may be partially explained in that Lock calculated the Reynolds number for the *onset of instability* while Murgatroyd measured the Reynolds number

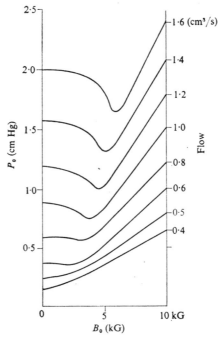

Fig. 5-8 Pressure gradient plotted against applied field for various mean flow rates of mercury across a magnetic field. The curves for the higher flow rates show an initial decrease of P_0 as B_0 is increased, an effect which is due to turbulence. As B_0 increases it inhibits the turbulence and then P_0 increases with B_0 as predicted by Hartmann theory (from Hartmann and Lazarus; *Kgl. Danske Vid. Sel. Mat-fys. Medd.*, **15**, 7, 1937)

required to *eliminate turbulence*. It is known that the latter is lower than the former. Lock's theory uses familiar techniques of perturbation theory as employed in Chapter 4 to investigate the stability of static configurations. A different approach by Lykoudis[35] postulates the existence of a laminar sublayer in a turbulent flow and plots the theoretical thickness of this sublayer for a varying magnetic field. In this way, Lykoudis calculated the Reynolds number below which turbulence cannot exist and found excellent agreement with Murgatroyd's experimental result.

Chandrasekhar[26] has investigated the stability of hydromagnetic Couette flows (for an applied field in the *y* direction, *not* the *z* direction as in Section

5–4). The theoretical predictions have been confirmed experimentally by Donnelly and Ozima.[26]

5–6 Parallel flows

In this and the following section, no specific flow problems are *solved* but the equations for two classes of hydromagnetic flows are transformed to equivalent hydrodynamic equations.[37] Thus, with appropriate specification of boundary conditions, large sections of hydrodynamic theory may be applied to these two classes of hydromagnetic flow problems.

The first class of problems to be considered is that of steady incompressible flow in which the magnetic field is everywhere parallel to the flow velocity, so that

$$\mathbf{B} = \lambda \mathbf{U} \qquad (5\text{–}29)$$

and

$$\frac{D\rho}{Dt} = (\mathbf{U}.\nabla)\rho = 0 \qquad (5\text{–}30)$$

Hence, (5–2) becomes

$$\nabla.\mathbf{U} = 0 \qquad (5\text{–}31)$$

so that substituting (5–29) in $\nabla.\mathbf{B} = 0$ gives

$$(\mathbf{U}.\nabla)\lambda = 0 \qquad (5\text{–}32)$$

as a condition on λ; namely, λ should be constant along streamlines (likewise along field lines by (5–29)). Of course, λ may be different for different streamlines (as may ρ, cf. (5–30) with (5–32)). Using (5–5) and (5–31), for steady flows the momentum equation, (5–1), may be written

$$\rho(\mathbf{U}.\nabla)\mathbf{U} = -\nabla P + \frac{(\mathbf{B}.\nabla)\mathbf{B}}{4\pi} - \nabla\left(\frac{B^2}{8\pi}\right) + \eta\nabla^2\mathbf{U} \qquad (5\text{–}33)$$

Now, introducing new variables

$$P^* = P + B^2/8\pi \qquad \rho^* = \rho - \lambda^2/4\pi \qquad (5\text{–}34)$$

and using (5–29) and (5–32), (5–33) reduces to

$$\rho^*(\mathbf{U}.\nabla)\mathbf{U} = -\nabla P^* + \eta\nabla^2\mathbf{U} \qquad (5\text{–}35)$$

while (5–30) enables us to write (5–32) as

$$(\mathbf{U}.\nabla)\rho^* = 0 \qquad (5\text{–}36)$$

Equations (5–31), (5–35), and (5–36) are the *hydrodynamic* equations describing a steady, incompressible flow with density ρ^* and pressure P^*.

P D—H

However, the analogy is not quite exact since the electromagnetic fields must satisfy Ohm's law and Maxwell's equations, (5–3) to (5–5). The electric field **E** and current density **j** may be eliminated, as in Section 4–2, to give for steady, parallel flows

$$\mathbf{V} \times \left(\frac{\mathbf{V} \times \mathbf{B}}{\sigma}\right) = 0 \qquad (5\text{–}37)$$

If this equation is satisfied identically then the analogy with hydrodynamics *is* exact. One obvious case of this is for perfectly conducting fluids. A second special case arises in potential flows with constant λ. Potential flows are those for which there exists some scalar function ϕ such that $\mathbf{U} = \mathbf{V}\phi$. Hence, using (5–29), $\mathbf{V} \times \mathbf{B} = 0$.

If the flow is inviscid, the analogy with hydrodynamics may be taken a step further. With $\eta = 0$, the dot product of **U** with (5–35) gives

$$\tfrac{1}{2}\rho^*(\mathbf{U}.\mathbf{V})U^2 + (\mathbf{U}.\mathbf{V})P^* = 0$$

or, using (5–36)

$$(\mathbf{U}.\mathbf{V})(\tfrac{1}{2}\rho^*U^2 + P^*) = 0$$

i.e., along a streamline

$$\tfrac{1}{2}\rho^*U^2 + P^* = \text{const.} = \tfrac{1}{2}\rho U^2 + P \qquad (5\text{–}38)$$

from (5–29) and (5–34). Equation (5–38) is *Bernoulli's equation.*

5–7 Transverse flows

Consider now flows in which **U** is perpendicular to $\mathbf{B} = (0, 0, B)$ and all variables are functions of x, y, and t only. As a further restriction consider only ideal fluids ($\sigma^{-1} = 0 = \eta$); however, the flows may be unsteady and compressible. The ideal hydromagnetic equations (3–83) to (3–87) for such two-dimensional, transverse flows may be written†

$$\frac{\partial \rho}{\partial t} + \mathbf{V}.(\rho\mathbf{U}) = 0 \qquad (5\text{–}39)$$

$$\rho\frac{D\mathbf{U}}{Dt} + \mathbf{V}(P + B^2/8\pi) = 0 \qquad (5\text{–}40)$$

$$\frac{D(P/\rho^\gamma)}{Dt} = 0 \qquad (5\text{–}41)$$

$$\frac{\partial B}{\partial t} + \mathbf{V}.(B\mathbf{U}) = 0 \qquad (5\text{–}42)$$

† Note that the magnetic field appears only as a scalar in (5–39) to (5–42).

Before transforming this set of equations into an equivalent set of hydro-dynamic equations it is desirable to say a little more about the energy equation, (5–41). This is the same equation in ideal hydromagnetics as in hydro-dynamics. Energy equations are sometimes written in terms of the tempera-ture T instead of P, and the set of equations is then completed by an *equation of state* relating P and T, for example $P = N\kappa T$; (cf. (3–44a, b)). Other forms of the energy equation are also used. Thus, carrying out the differentiation in (5–41) and using (5–39) one gets

$$\rho\frac{DI}{Dt} + P\mathbf{V}.\mathbf{U} = 0 \qquad (5\text{–}41a)$$

where the *internal energy* I is given in terms of P by[20]

$$I = \frac{P}{(\gamma - 1)\rho} \qquad (5\text{–}43)$$

Yet another form for the energy equation is in terms of the *entropy* S. This is defined by the thermodynamic equation

$$T\,dS = dI + P\,d(1/\rho) \qquad (5\text{–}44)$$

from which it follows, using (5–39) and (5–41a), that

$$\frac{DS}{Dt} = 0 \qquad (5\text{–}41b)$$

This is known as the *entropy equation* and states that entropy is constant along a streamline. If (5–41b) is used instead of (5–41), an equation of state relating P and S is required. In view of (5–44), this usually takes the form

$$P = f(\rho, S) \qquad (5\text{–}45)$$

In transforming the set of hydromagnetic equations, (5–39) to (5–42), into their hydrodynamic analogue it is convenient to use either (5–41a) (with the auxiliary equation (5–43)) or (5–41b) (with the auxiliary equation (5–45)) instead of the usual energy equation, (5–41). For clarity, the discussion is restricted to the latter possibility (cf. Problem 5–7).

The transformation is straightforward. The continuity equation, (5–39), and entropy equation, (5–41b), are already in hydrodynamic form and are not affected by the redefinition of pressure

$$P^* = P + B^2/8\pi \qquad (5\text{–}46)$$

by which (5–40) becomes the hydrodynamic momentum equation,

$$\rho\frac{D\mathbf{U}}{Dt} + \nabla P^* = 0 \qquad (5\text{–}47)$$

and the equation of state (5–45) becomes

$$P^* = f(\rho, S) + B^2/8\pi \qquad (5\text{–}48)$$

Equation (5–42) appears as a second continuity equation for the additional (magnetic flux density) variable B; or, using (5–39), as a second entropy equation

$$\frac{DS^*}{Dt} = 0 \qquad (5\text{–}49)$$

for the additional 'entropy' variable $S^* = B/\rho$. The appearance of this extra equation means that in the general case, as for parallel flows, the transformed set of hydromagnetic equations is not exactly analogous to a hydrodynamic system. The analogy *is* exact if the additional equation is satisfied identically. An obvious example is $S^* =$ constant. One then has the set of hydrodynamic equations (5–39), (5–41b), and (5–47), for the variables ρ, S, and $\mathbf{U}$ with the equation of state

$$P^* = f(\rho, S) + \rho^2 S^{*2}/8\pi$$

A second possibility which makes the analogy exact is $S =$ constant (isentropic flow). In this case, the set of equations comprises (5–39), (5–47), and (5–49) for the variables ρ, $\mathbf{U}$, and S^* with the equation of state

$$P^* = f(\rho) + \rho^2 S^{*2}/8\pi$$

It is then easy to see that a third possibility is $S^* = g(S)$, since (5–49) is then guaranteed by (5–41b). The set of hydrodynamic equations is the same as for the case $S^* =$ constant, but with the equation of state

$$P^* = f(\rho, S) + \rho^2 g^2(S)/8\pi$$

5–8 Plasma propulsion

The basic mechanism of a number of technological applications of hydromagnetic flows was illustrated by the discussion of Hartmann flow in Section 5–3. To complete this chapter, two of these applications are considered in more detail.

The principle of electromagnetic pumps and electromagnetic plasma accelerators is the same. Energy is supplied to the device in the form of an electric current at right angles to a magnetic field in such a way that the resulting

Lorentz force accelerates the flow of a conducting fluid. Pumps are particularly useful for the movement of corrosive substances, like liquid sodium in nuclear reactors (as are flowmeters for measuring the flux of such fluids). A plasma accelerator may simply be a device for accelerating ionized gases, for example in plasma experiments, or it may be a propulsion device in which plasma is accelerated and subsequently expelled in order to produce a thrust. Space research has given a great impetus to the development of plasma propulsion since it has important potential advantages over conventional propellants, especially for long-range missions lasting months or even years. A critical factor in the design of rocket accelerators is the exhaust velocity of the propellant. This depends on the initial temperature, T, and molecular weight, M, of the propellant as $(T/M)^{1/2}$. The molecular weight dependence makes hydrogen an obvious choice for propellant although from other considerations, such as storage, metal-derived plasmas may be preferable.

There are many kinds of plasma propulsion device. Some operate on short pulsed currents while others are continuous. Clearly, a small continuous current may produce a thrust equivalent to the average thrust provided by a

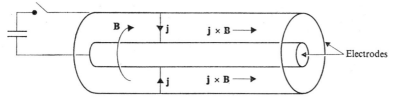

Fig. 5-9 Coaxial plasma gun

machine operating on a larger, pulsed current. The advantages of a continuously operating device are that less auxiliary equipment, such as capacitors, need be carried. On the other hand, large currents are able to produce the plasma by ionization and provide the necessary magnetic field, thereby dispensing with ionization equipment and magnets. Some devices are designed to use the *Hall current* rather than the direct current (see Section 5–9). Such accelerators are low-pressure devices, since it is under these conditions that the Hall current dominates the direct current; compare the fourth and last terms in (3–68).

One type of plasma accelerator, which was first developed by Marshall[38] for use in thermonuclear fusion research and has since been adapted and tested for rocket propulsion, is the *co-axial plasma gun*. Plasma is driven out of a hole at one end of the gun. Depending on the size of the hole, the dominant driving mechanism is either (a) the plasma pressure due to the pinch mechanism—that is, the operative Lorentz force is the product of the *axial* current and the induced azimuthal magnetic field (the device is essentially an open-ended linear pinch, cf. Section 4–3), or (b) the direct $\mathbf{j} \times \mathbf{B}$ force depicted in Fig. 5–9. This second force mechanism is not present in the ordinary linear pinch and is introduced in the gun by the co-axial electrodes which allow the passage of a radial current. As the plasma moves along the gun, this radial current moves with it, thus prolonging the action of the Lorentz force; this is a feature of so-called *rail plasma accelerators*. Mechanism (b) produces higher plasma velocities than (a) and is the dominant effect if the escape hole is large.

The limitations of plasma accelerators lie in low efficiency and thrust-to-weight ratios of much less than unity. This latter point means that a plasma-accelerated rocket must first be put into orbit by some other motive power.

5–9 MHD generators

The fundamental action of a magnetohydrodynamic (MHD) generator may be regarded, as indeed it appeared in Section 5–3, as complementary to the action of an accelerator. In the latter, current is supplied which accelerates the conducting fluid via the Lorentz force. On the other hand, current is extracted from a generator and results in a slowing down of the fluid flow. Both, of course, are direct applications of Faraday's laws, and it is perhaps more helpful to view the generator in this light; a conductor is moved across a magnetic field and hence a current is generated in it. This is also true of conventional generators in which the conductors are solid metals. The fundamental difference between conventional and MHD generators lies not in the basic action but in the conversion of heat into mechanical energy of the conductor. In a conventional generator, the heat is supplied to a compressible fluid which then drives a so-called prime mover (such as turbine blades) and this moves the conductor. In the MHD generator, the conductor (ionized gas) is itself heated and by a suitable arrangement (for instance, a convergent–divergent expansion nozzle) this heat energy is, in part, converted into flow energy. Thus the first potential advantage of an MHD generator is the elimination of the prime mover. This leads to a second advantage, namely that the potential working temperature is much higher (making possible higher efficiencies) since there are not the same mechanical stresses involved as with prime movers and solid conductors. Unfortunately, one must go to very much higher temperatures ($\sim$2,500°K) before a useful degree of ionization is achieved. At such high temperatures, many new problems arise, such as the choice of suitable refractory materials. Indeed, most of the technological difficulties in the design and construction of a commercially viable MHD generator stem directly or indirectly from the need for a high operating temperature.

A high degree of ionization is *not* required, 0·1 per cent being sufficient to give a gas a conductivity of the order of half its potential maximum. However, most common gases are not easily ionized even at such a high temperature as 2,500°K and a process known as *seeding* is usually adopted. This involves the addition of about 0·1 to 1 per cent of a substance (such as an alkali metal vapour) with low ionization potential.

The discussion of Hartmann flow in Section 5–3 illustrated the simplest of the MHD generator geometries, the *linear MHD generator* with *continuous electrodes* (Fig. 5–10). A variation of this is the linear generator with *segmented electrodes* shown in Fig. 5–11. The electrodes are segmented to prevent

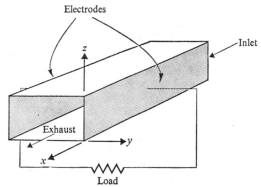

Fig. 5–10 Linear MHD generator with continuous electrodes

the flow of the Hall current. To discuss this it is necessary to modify the equations somewhat. In Section 5–2, we assumed that $\Omega_e \ll \nu_c$, in which case the Hall current is negligible compared with the Faraday current (compare

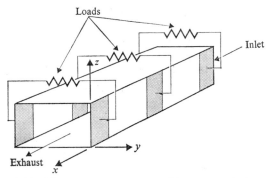

Fig. 5–11 Linear MHD generator with segmented electrodes

the fourth and last terms in (3–68)). We can attempt to find out the effect on Hartmann flow of relaxing this condition by reintroducing the Hall term into Ohm's law while ignoring any additional complications in the Navier–Stokes equation. Thus, from (3–61), we have

$$\mathbf{j} = \sigma\left[\mathbf{E} + \frac{\mathbf{U} \times \mathbf{B}}{c} - \frac{m^+}{\rho ec}(\mathbf{j} \times \mathbf{B})\right] \qquad (5\text{–}50)$$

From (5–8), we still have $j_z = 0$ and, if the induced field is neglected compared with the applied field, $\mathbf{B}_0$, (5–6a, b) may be substituted in (5–50) to give for the other components of the current

$$j_x = \frac{\sigma}{1+\alpha^2}\left[E_x - \alpha\left(E_y - \frac{UB_0}{c}\right)\right] \left.\vphantom{\begin{array}{c}1\\1\\1\\1\end{array}}\right\}$$

$$j_y = \frac{\sigma}{1+\alpha^2}\left[E_y - \frac{UB_0}{c} + \alpha E_x\right] \tag{5-51}$$

where $\alpha = \dfrac{\sigma B_0 m^+}{\rho e c} \approx \dfrac{\Omega_e}{\nu_c}$, using (3-60a, b).

In considering the geometry of a generator, if the electrodes are continuous, $E_x = 0$ and (5-51) gives

$$i_y = \frac{\sigma}{1+\alpha^2}\left(E_y - \frac{UB_0}{c}\right) \qquad j_x = -\alpha j_y \tag{5-52}$$

which shows that the effect of the Hall term is to reduce the Faraday current, j_y, by the factor $(1+\alpha^2)^{-1}$ and to give a Hall current, j_x, of magnitude αj_y. If, on the other hand, the electrodes are segmented, E_x need not be zero and the Hall current may be suppressed, that is $j_x = 0$, to retrieve the situation in which the total current flows across the channel. Then (5-51) gives

$$E_x = \alpha(E_y - UB_0/c)$$

$$j_y = \sigma(E_y - UB_0/c)$$

and the Faraday current is no longer reduced by $(1+\alpha^2)^{-1}$.

As in the case of accelerators, the Hall current may be used instead of the Faraday current. This has particular advantages in the regime $\alpha > 1$. In the

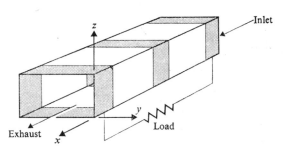

Fig. 5-12   Hall generator

Hall generator, the Faraday current is short-circuited and the load is connected between electrodes at the entrance and exit of the duct; see Fig. 5–12.

Variations of the linear and Hall generators are the *vortex* and *radial outflow generators*, respectively. In the vortex generator, shown in Fig. 5–13, the flow is azimuthal and the (Faraday) current radial. The radial outflow generator, Fig. 5–14, is a Hall generator. The Hall current, flowing radially,

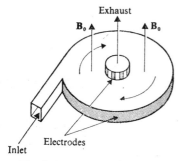

Fig. 5–13   Vortex generator

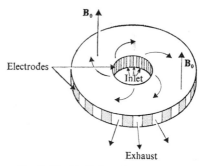

Fig. 5–14   Radial outflow generator

interacts with the field $\mathbf{B}_0$ to 'stir' the fluid so that the flow becomes a spiral. Both of these generators have the advantage of a longer interaction length, because of the spiral nature of the flow, for a given area of magnetic field.

Summary

1. *Hartmann flow*

$$U(z) = \frac{cM(cP_0 + B_0 j_0)}{\sigma B_0^2 \sinh M}[\cosh M - \cosh (Mz/d)]$$

External circuit		Application
(a) Short-circuit	$(j_0 = \min)$	electromagnetic brake
(b) Resistive	$(j_0 < 0)$	MHD generator
(c) Open-circuit	$(j_0 = 0)$	flowmeter
(d) Power supply	$(j_0 > 0)$	pump or accelerator

2. *Parallel flows*

Conditions: $\mathbf{B} = \lambda\mathbf{U}$, steady, incompressible.
Transformation: $P^* = P + B^2/8\pi, \quad \rho^* = \rho - \lambda^2/4\pi.$

Equation additional to hydrodynamic equations: $\nabla \times \left(\dfrac{\nabla \times \mathbf{B}}{\sigma}\right) = 0.$

Analogy exact for:
   (a) $\sigma = \infty.$
   (b) $\mathbf{U} = \nabla\phi, \quad \lambda = \text{const.}$

3. *Transverse flows*

Conditions: $\mathbf{U} \perp \mathbf{B} = (0, 0, B), \quad \eta = 0, \quad \sigma = \infty, \quad$ no z dependence.
Transformation: $P^* = P + B^2/8\pi, \quad S^* = B/\rho.$

Equation additional to hydrodynamic equations: $DS^*/Dt = 0$.
Analogy exact for
  (a) $S^* = $ const.
  (b) $S = $ const.
  (c) $S^* = g(S)$.

Problems

5–1 For the example of Hartmann flow in Section 5–3, use (5–3), (5–8a), (5–10b) and $\mathbf{\nabla . E} = 0$ to show $E_z = 0 = B_y$.

5–2 Find the average energy dissipation per unit volume in Hartmann flow due to

ohmic $(\mathbf{j.E})$ and viscous $\left[\eta \dfrac{\partial U_i}{\partial r_k}\left(\dfrac{\partial U_i}{\partial r_k} + \dfrac{\partial U_k}{\partial r_i}\right)\right]$ dissipation.

Compare their magnitudes for $E_0 \sim U_0 B_0/c$ in the cases $M \to 0$ and $M \to \infty$.

5–3 Find the proportion of the total flow that occurs outside the 'boundary layers' of thickness d/M in Hartmann flow and discuss the limits $M \to 1$ and $M \to \infty$.

5–4 Derive (5–25).

5–5 Find the flow velocity as a function of z for the case of combined Hartmann and Couette flow; i.e., solve equation (5–14) subject to the boundary conditions $U(\pm d) = \pm V_0$. Find the average velocity U_0 and sketch the velocity profile for large M when $U_0 \ll V_0$, $U_0 = V_0$, and $U_0 \gg V_0$.

5–6 Derive (5–41b).

5–7 Show that the transformation

$$I^* = I + \frac{B^2}{8\pi\rho}$$

together with (5–46) enables one to write the hydromagnetic equations (5–39), (5–40), (5–41a), (5–42), and (5–43) in hydrodynamic form, the additional equation being either (5–42) or (5–49).

6

Shock waves in plasmas

6–1 Introduction

Consider the propagation of a wave in a non-conducting fluid, produced, for example, by a small compressive displacement of a piston at one end of a long tube of the fluid initially at rest. Describing this situation by the linearized hydromagnetic equations, one finds that a pressure wave of constant shape travels down the tube with the speed of sound $c_s = (\gamma P / \rho)^{1/2}$. Such a wave is called a *compression* or *sound wave*.

Next, suppose the displacement of the piston is no longer small; to describe the motion one must now retain the non-linear terms in the equations. Then one finds that the profile of the wave changes shape as it moves down the tube, the slope of the wavefront gradually steepen-

Fig. 6–1 Profile of pressure wave for times $t_1 < t_2 < t_3$. Density, velocity, and temperature profiles also show a steepening of the wavefront

ing (see Fig. 6–1). The increasing slope means, of course, that the pressure gradient across the wavefront is increasing, as are the gradients of density, velocity, and temperature. Eventually (if the tube is long enough) these gradients become so large that the neglect of viscosity and heat conduction is no longer justified. Extending the fluid equations to include these effects, one finds the steepness of the wavefront is limited so that it reaches a steady shape known as the *permanent regime profile*. It is the steady profile pro-

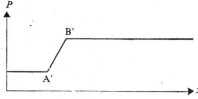

Fig. 6–2 Profile of an expansion wave

duced by the balance of non-linear and dissipative effects which corresponds to a *shock wave*.

The physical explanation of these modifications of the simple wave picture is as follows. The non-linear effects arise from two sources. Since the sound speed increases with pressure the *wave* velocity at A (Fig. 6–1) is greater than at B. Moreover, the *fluid* velocity varies across the wavefront, being greater at A than at B. This *convective effect* is actually the greater of the two and it is customary to refer to the non-linear effects collectively as convective effects. Thus both fluid and wave velocity are greater at A than at B, so that A tends to catch up with B, causing the wavefront to steepen. (By the same arguments the wavefront of an expansion wave, Fig. 6–2, would tend to flatten. For this reason expansion waves do not give rise to shocks; see also Section 6–2.) The physical explanation of dissipative effects is obvious from the manner of their introduction. These oppose any large gradients. For example, as the wavefront steepens, the temperature gradient across it increases. Therefore, the conduction of heat across the wavefront also increases until the steady state is reached. In practice, one always has this situation with convective and dissipative effects opposing each other. It is not possible, of course, to have significant non-linear effects without some dissipation mechanism being brought into play. The smaller the coefficients of dissipation the more nearly the shock wave approaches a discontinuity.

Fluid properties may change considerably across a shock wave and the shock is thus a steady transition region between the undisturbed (unshocked) fluid and the fluid through which the shock has passed. In Fig. 6–3, region 1

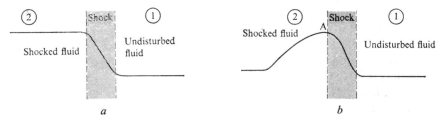

a   *b*

Fig. 6–3 Shock wave. (a) Idealized picture showing both undisturbed and shocked fluid as uniform states; (b) undisturbed fluid is in uniform state but following the shock is a relief or expansion wave so that region 2 is not uniform

is the unshocked fluid and is said to be *in front of* the shock; region 2 is the shocked fluid, said to be *behind* the shock. Usually one regards a shock as a transition region between two uniform states (cf. Fig. 6–3a) although in practice this is difficult to realize and the situation depicted in Fig. 6–3b is more likely. Here, the shocked fluid is not in a uniform state but is subject to a relief or expansion wave. This means that the state of the fluid behind the shock does not persist but changes with time after the shock has passed on. Nevertheless, it is very convenient to take both region 1 and region 2 as uniform and this will be assumed throughout the chapter unless otherwise stated. (The validity of this assumption depends on the time of relaxation from the state represented by the point A.) Since the establishment of a new equilibrium state in a non-conducting fluid can *only* be achieved by collisions, the width or thickness of the shock is of the order of a few mean free paths.

The theory of hydrodynamic shocks is reasonably well understood.[39] As usual, the hydromagnetic case is considerably more complicated. For a start, conducting fluids in a magnetic field can support two further modes of wave propagation. Returning to the piston analogy, transverse movements of the piston in a non-conducting fluid have no effect (beyond a boundary layer where viscous forces maintain a velocity gradient). Thus there is effectively only one (longitudinal) degree of freedom and, therefore, only one mode of propagation, with the velocity c_s. However, the transverse movement of a conducting piston in an infinitely conducting fluid carries any longitudinal component of a magnetic field with it, thereby producing a wave. Thus a conducting fluid in the presence of a magnetic field has three degrees of freedom (one longitudinal and two transverse) and hence three modes of wave propagation. There are three propagation speeds, therefore, generally known as *fast, intermediate,* and *slow* (cf. Chapter 8). Intermediate waves are purely transverse and do not steepen to form shock waves. The fast and slow modes in general contain both transverse and longitudinal components and the compression waves of these modes give rise to shocks.

However, a more serious and fundamental complication of hydromagnetic shocks concerns questions of shock structure. Joule heating—due to finite conductivity—provides an additional dissipative mechanism (see Section 6–5). Furthermore, in an ionized gas shock thicknesses less than the smallest collision mean free path may occur. The structure of these *collisionless shocks,* in which hydrodynamic dissipative processes play no part, is a complicated problem and is discussed briefly in Section 6–6.

In spite of the many difficulties associated with hydromagnetic shocks, two factors greatly facilitate discussion of the *effects* of a shock wave on a fluid. First, the shock transition region may usually be approximated by a discontinuity in fluid properties. Secondly, the macroscopic conservation equations and the Maxwell equations may be integrated across the shock to give a set of equations which are independent of shock structure and relate fluid properties on either side of the shock. This most useful and straight-forward aspect of hydromagnetic shock theory is developed in the following section and applied to special cases in Sections 6–3 and 6–4.

Some shock experiments are discussed in Section 6–7.

6–2 Hydromagnetic shock equations

For simplicity we restrict discussion to plane shocks moving in a direction normal to the plane of the shock. Let this be the x direction; then all variables

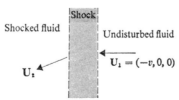

Fig. 6–4 Plane shock in frame of reference moving with the shock velocity. The undisturbed fluid flows into the front of the shock and the shocked fluid flows out from the back of the shock

are functions of x only inside the shock and are constant outside the shock (in regions 1 and 2). It is convenient to use a frame of reference moving with the shock; if in the laboratory frame the unshocked plasma (region 1) is at rest, then in the moving frame (Fig. 6–4) it enters the front of the shock with a velocity $U_1 = (-v, 0, 0)$, where v is the speed of the shock. Steady state conditions apply (that is, all variables are time-independent). With these conditions, the Maxwell equation for $\mathbf{j}$ is

$$\mathbf{j} = \frac{c}{4\pi}\left(0, -\frac{dB_z}{dx}, \frac{dB_y}{dx}\right)$$ (6–1)

The hydromagnetic equations do not apply in the shock region since dissipative processes take place; however they do apply on either side of the shock, i.e., in regions 1 and 2. Thus from Ohm's law

$$\mathbf{E}_1 = \frac{\mathbf{j}_1}{\sigma_1} - \frac{\mathbf{U}_1 \times \mathbf{B}_1}{c} \qquad \mathbf{E}_2 = \frac{\mathbf{j}_2}{\sigma_2} - \frac{\mathbf{U}_2 \times \mathbf{B}_2}{c}$$ (6–2)

Now from $\nabla \cdot \mathbf{B} = 0$ and $\nabla \times \mathbf{E} = -(1/c)\partial \mathbf{B}/\partial t = 0$ one finds

$$\frac{dB_x}{dx} = 0$$ (6–3a)

$$\frac{dE_y}{dx} = 0$$ (6–3b)

$$\frac{dE_z}{dx} = 0$$ (6–3c)

The other equations to be integrated across the shock are the general macroscopic equations derived in Chapter 3. The integration is most easily achieved by using the conservation form of these equations given in Problem 3–4. Since variables depend on x only one has

$$\frac{d(\rho U_x)}{dx} = 0$$ (6–4)

$$\frac{d\Pi_{xx}}{dx} = 0 \qquad \frac{d\Pi_{xy}}{dx} = 0 \qquad \frac{d\Pi_{xz}}{dx} = 0$$ (6–5)

$$\frac{dS_x}{dx} = 0$$ (6–6)

Integrating (6–3a, b, c) and using (6–1) and (6–2) gives, on observing that gradients are zero in regions 1 and 2,

$$[B_x]_1^2 = 0$$ (6–7)

$$[U_x B_y - U_y B_x]_1^2 = 0 \qquad (6\text{-}8)$$

$$[U_x B_z - U_z B_x]_1^2 = 0 \qquad (6\text{-}9)$$

where $[\phi]_1^2 = (\phi_2 - \phi_1)$ in the usual notation. The integration of (6–4) to (6–6) is trivial and on using the hydromagnetic approximations $P_{ij} = P\delta_{ij}$, $\mathbf{Q} = 0$, $E \ll B$, (which *apply* in regions 1 and 2) one finds (cf. Problem 6–1)

$$[\rho U_x]_1^2 = 0 \qquad (6\text{-}10)$$

$$[\rho U_x^2 + P + (B_y^2 + B_z^2)/8\pi]_1^2 = 0 \qquad (6\text{-}11)$$

$$[\rho U_x U_y - B_x B_y/4\pi]_1^2 = 0 \qquad (6\text{-}12)$$

$$[\rho U_x U_z - B_x B_z/4\pi]_1^2 = 0 \qquad (6\text{-}13)$$

$$[\rho U_x I + P U_x + \tfrac{1}{2}\rho U_x U^2 + U_x(B_y^2 + B_z^2)/4\pi - B_x(B_y U_y + B_z U_z)/4\pi]_1^2 = 0 \qquad (6\text{-}14)$$

where the internal energy $I = P/(\gamma - 1)\rho$ and we have replaced $\tfrac{3}{2}N\kappa T$ by ρI in S.

Equations (6–7) to 6–14) relate fluid variables on one side of the shock to those on the other, and these equations are sometimes called the *jump conditions* across the shock. Defining the unit vector $\hat{\mathbf{n}}$ in the direction of shock propagation, the jump conditions may be written in vector form:

$$[\rho \mathbf{U}.\hat{\mathbf{n}}]_1^2 = 0 \qquad (6\text{-}15)$$

$$[\rho \mathbf{U}(\mathbf{U}.\hat{\mathbf{n}}) + (P + B^2/8\pi)\hat{\mathbf{n}} - (\mathbf{B}.\hat{\mathbf{n}})\mathbf{B}/4\pi]_1^2 = 0 \qquad (6\text{-}16)$$

$$[\mathbf{U}.\hat{\mathbf{n}}\{(\rho I + \tfrac{1}{2}\rho U^2 + B^2/8\pi) + (P + B^2/8\pi)\} - (\mathbf{B}.\hat{\mathbf{n}})(\mathbf{B}.\mathbf{U})/4\pi]_1^2 = 0 \quad (6\text{-}17)$$

$$[\mathbf{B}.\hat{\mathbf{n}}]_1^2 = 0 \qquad [\hat{\mathbf{n}} \times (\mathbf{U} \times \mathbf{B})]_1^2 = 0 \qquad (6\text{-}18)$$

The first three equations represent the conservation of mass, momentum, and energy respectively for the flow of gas through the shock. The last pair of equations gives the jump conditions for the magnetic field.

Usually, one knows the state of the fluid before the passage of the shock wave and, if the shock speed is known or measured, this means that ρ_1, $\mathbf{U}_1$, P_1, $\mathbf{B}_1$ (and hence I_1) are given. Then, ρ_2, $\mathbf{U}_2$, P_2, $\mathbf{B}_2$ (and hence I_2) may be determined from (6–15) to (6–18). When $\mathbf{B} = 0$, (6–15) to (6–17) reduce to the corresponding hydrodynamic equations known as the *Rankine–Hugoniot equations*.

After considerable manipulation (cf. Problem 6–2), the velocity variables $\mathbf{U}_1$ and $\mathbf{U}_2$ may be eliminated from the energy equation, (6–17), and the result is

$$[I]_1^2 + \tfrac{1}{2}(P_1 + P_2)[1/\rho]_1^2 + \frac{1}{16\pi}\{[\mathbf{B}]_1^2\}^2[1/\rho]_1^2 = 0 \qquad (6\text{--}19)$$

The hydrodynamic equivalent of this equation,

$$[I]_1^2 + \tfrac{1}{2}(P_1 + P_2)[1/\rho]_1^2 = 0 \qquad (6\text{--}19a)$$

relates the pressure and density on either side of the shock and is known as the *Hugoniot relation*. It assumes the role played by the law $P\rho^{-\gamma} = \text{const.}$ in adiabatic changes of state. The adiabatic law does not hold for changes brought about by shocks since non-adiabatic processes (such as heat conduction) occur. Note that the *hydromagnetic Hugoniot* (6–19) reduces to (6–19a) not only when the magnetic field is zero but also when $\mathbf{B}_1$ and $\mathbf{B}_2$ are both parallel to the direction of shock propagation, since (6–18) then gives $\mathbf{B}_1 = \mathbf{B}_2$.

Before discussing particular solutions of the shock equations, the compressive nature of shocks (i.e., $P_2 > P_1$) will be proved, assuming the plasma is a *perfect gas*.† The proof follows as a consequence of the second law of thermodynamics—the law of increase of entropy. The entropy S of a perfect gas is given by (cf. Problem 6–3)

$$S = C_v \log (P/\rho^\gamma) + \text{const.} \qquad (6\text{--}20)$$

where C_v is the specific heat at constant volume. Thus,

$$\frac{dS_2}{d\rho_2} = \frac{C_v}{P_2}\frac{dP_2}{d\rho_2} - \frac{\gamma C_v}{\rho_2}$$

which, in terms of the ratios $r = \rho_2/\rho_1$, $R = P_2/P_1$, may be written

$$\frac{\rho_1}{C_v}\frac{dS_2}{d\rho_2} = \frac{1}{R}\frac{dR}{dr} - \frac{\gamma}{r} \qquad (6\text{--}21)$$

In this equation, we regard ρ_1 and P_1 as constants (the given values of ρ and P in region 1). A straightforward rearrangement of (6–19) gives

$$R = \frac{(\gamma + 1)r - (\gamma - 1) + (\gamma - 1)(r - 1)b^2}{(\gamma + 1) - (\gamma - 1)r} \qquad (6\text{--}19b)$$

where $b^2 = (\mathbf{B}_2 - \mathbf{B}_1)^2/8\pi P_1$. Now differentiating (6–19b) with respect to r

† In defining temperature by (3–38), the same assumption was made since this is essentially the perfect gas law.

and substituting in (6–21), we get

$$\frac{\rho_1}{C_v}\frac{dS_2}{d\rho_2}$$

$$= \frac{\gamma(\gamma+1)(\gamma-1)(r-1)^2+(\gamma-1)[\gamma(\gamma-1)r^2-2(\gamma+1)(\gamma-1)r+\gamma(\gamma+1)]b^2}{r[(\gamma+1)-(\gamma-1)r][(\gamma+1)r-(\gamma-1)+(\gamma-1)(r-1)b^2]}$$

$$(6\text{--}22)$$

The next step is to show that $dS_2/d\rho_2$ is positive, and we do this by proving that both numerator and denominator of the expression in (6–22) are positive. Since $\gamma > 1$, it is easy to verify that this statement is true for $r = 1$. Writing (6–19b) as

$$R = (A + Cb^2)/D$$

it follows for $r < 1$ that $C < 0$ and $D > 0$. Since the pressure ratio R must be positive, A must also be positive, and

$$r > (\gamma - 1)/(\gamma + 1) \qquad (6\text{--}23)$$

Likewise, if $r > 1$ then $C > 0$, $A > 0$ and, therefore, $R > 0$ implies $D > 0$,

i.e., $$r < (\gamma + 1)/(\gamma - 1) \qquad (6\text{--}24)$$

Combining (6–23) and (6–24),

$$\frac{\gamma - 1}{\gamma + 1} < r < \frac{\gamma + 1}{\gamma - 1} \qquad (6\text{--}25)$$

Thus, in the expression for R, the denominator D is positive. Therefore, $R > 0$ implies that the numerator $(A + Cb^2)$ is also positive. Since the denominator of the right-hand side of (6–22) is simply $rD(A + Cb^2)$, it is therefore positive.

Turning now to the numerator in (6–22), it is clear since $\gamma > 1$ that the first term is positive. The remaining term is quadratic in r and positive for $r = 1$. If this term is to be negative for some r it must pass through zero. However, equating it to zero one finds imaginary roots for r, which proves that the term is positive for all real r. The proof that $dS_2/d\rho_2 > 0$ is thus complete, showing that S_2 and ρ_2 increase or decrease together.

If $\rho_2 = \rho_1$, it follows from (6–19) that $I_2 = I_1$ and hence $P_2 = P_1$. Then, from (6–20), $S_2 = S_1$. This is the limiting case of no shock. Now since the second law of thermodynamics requires $S_2 \geqslant S_1$, and S_2 and ρ_2 change in the same sense, it follows that $\rho_2 \geqslant \rho_1$ and (6–25) must be replaced by

$$1 \leqslant r < (\gamma + 1)/(\gamma - 1) \qquad (6\text{--}26)$$

Then, from (6–19b), discounting $r = 1$ (no shock)

$$\frac{P_2}{P_1} = R > 1 + \frac{(r-1)(\gamma-1)b^2}{(\gamma+1)-(\gamma-1)r} \geqslant 1$$

That is, shocks are compressive, confirming the qualitative arguments used in the introductory section. Although this proof applies only to a perfect gas it seems that for all gases shocks are compressive. This may be proved quite generally for weak shocks[40] and with some further assumptions [41] for shocks of arbitrary strength.

6–3 Shock propagation parallel to the magnetic field

We now look for particular solutions of the shock equations, beginning with shocks propagating in the direction of the initial magnetic field. Thus $\mathbf{B_1} = B_1 \hat{\mathbf{n}}$, i.e.,

$$\mathbf{B_1} = (B_1, 0, 0) \tag{6–27}$$

In general, the magnetic field after the shock has passed may have both a normal and a tangential component; so, choosing the y-axis to coincide with its tangential component (if any), one has

$$\mathbf{B_2} = (B_x, B_y, 0) \tag{6–28}$$

Likewise, $\mathbf{U_2}$ may be written

$$\mathbf{U_2} = (U_x, U_y, U_z) \tag{6–29}$$

and $\qquad\qquad \mathbf{U_1} = (U_1, 0, 0) = (-v, 0, 0) \tag{6–30}$

Substituting (6–27) to (6–30) in (6–8) and (6–12) and eliminating U_y then gives

$$(U_x^2 - B_1^2/4\pi\rho_2)B_y = 0 \tag{6–31}$$

Thus, one possible solution for $\mathbf{B_2}$ is

$$\mathbf{B_2} = (B_x, 0, 0) = (B_1, 0, 0) = \mathbf{B_1} \tag{6–32}$$

by (6–7). Also, from (6–12) and (6–13), $U_y = U_z = 0$ so that

$$\mathbf{U_2} = (U_2, 0, 0) \tag{6–33}$$

Of the remaining equations, (6–9) is satisfied identically while (6–10), (6–11), and (6–14) become

$$\rho_1 U_1 = \rho_2 U_2 \tag{6–34}$$

$$\rho_1 U_1^2 + P_1 = \rho_2 U_2^2 + P_2 \tag{6–35}$$

$$\rho_1 U_1 I_1 + P_1 U_1 + \tfrac{1}{2}\rho_1 U_1^3 = \rho_2 U_2 I_2 + P_2 U_2 + \tfrac{1}{2}\rho_2 U_2^3 \tag{6–36a}$$

where the last equation may also be written, using (6–34) and the definition of I,

$$\frac{\gamma P_1}{(\gamma - 1)\rho_1} + \tfrac{1}{2}U_1^2 = \frac{\gamma P_2}{(\gamma - 1)\rho_2} + \tfrac{1}{2}U_2^2 \qquad (6\text{–}36b)$$

In this solution, the magnetic field plays no part. The field itself is not changed and it does not appear in (6–34) to (6–36) governing the changes in ρ, U, and P. This solution, therefore, corresponds to a hydrodynamic shock and it may be shown (cf. Problem 6–4) that one must have U_1 supersonic (relative to the sound speed in region 1) while U_2 is subsonic (relative to region 2).

Strong shocks are defined as those for which the pressure ratio $R \gg 1$. In this case, it follows (cf. Problem 6–5) that the *Mach number* in region 1, $M = (U_1/c_s(1)) \gg 1$, and the temperature ratio

$$\frac{T_2}{T_1} \approx \frac{2\gamma(\gamma - 1)}{(\gamma + 1)^2} M^2 \gg 1 \qquad (6\text{–}37)$$

Since M may be as large as 100, it is clear from (6–37)† that strong shocks may be used to generate high-temperature plasmas or to obtain a plasma from a neutral gas by creating a temperature, T_2, behind the shock sufficiently high to cause ionization. The conversion of flow energy into thermal energy in this situation can be demonstrated from (6–36b). In view of (6–37), the initial thermal energy may be neglected compared with the final thermal energy. Also (cf. Problem 6–6)

$$\frac{U_1}{U_2} = \frac{\rho_2}{\rho_1} \approx \frac{(\gamma + 1)}{(\gamma - 1)} \qquad (6\text{–}38)$$

This ratio is 4 for $\gamma = 5/3$ so (6–36b) may be approximated by

$$\frac{\gamma P_2}{(\gamma + 1)} = \tfrac{1}{2}\rho_1 U_1^2 \qquad (6\text{–}39)$$

That is, the flow energy is converted by the shock into thermal energy.

6–3–1 Switch-on-shocks

The hydrodynamic shock is not, however, the *only* possible solution for propagation parallel to the initial magnetic field. Suppose now that $B_y \neq 0$. In this case, (6–31) implies

$$U_x^2 = B_1^2/4\pi\rho_2 \qquad (6\text{–}40)$$

† This result is true for a perfect gas but is modified by various effects in real gases; see Section 6–7.

By (6–7), (6–27), and (6–28)

$$\mathbf{B}_2 = (B_1, B_y, 0)$$

Also, using (6–10) and (6–12)

$$U_y = B_1 B_y / 4\pi \rho_1 U_1 \tag{6-41}$$

From (6–9), $U_z = 0$ so that (6–13) is satisfied identically and (6–11) and (6–14) reduce to

$$\rho_1 U_1^2 + P_1 = \rho_2 U_x^2 + P_2 + B_y^2 / 8\pi \tag{6-42}$$

$$\frac{\gamma P_1}{(\gamma - 1)\rho_1} + \tfrac{1}{2}U_1^2 = \frac{\gamma P_2}{(\gamma - 1)\rho_2} + \frac{U_x^2 + U_y^2}{2} + \frac{B_y^2}{4\pi \rho_2} - \frac{B_1 B_y U_y}{4\pi \rho_2 U_x} \tag{6-43}$$

Substituting P_2 from (6–42) into (6–43) and eliminating velocities using (6–10), (6–40), and (6–41) gives

$$B_y^2 = 2(r - 1)B_1^2 \left[\frac{(\gamma + 1) - r(\gamma - 1)}{2} - \frac{4\pi \gamma P_1}{B_1^2} \right]$$

Hence real non-zero values of B_y are possible only for

$$\frac{c_s^2(1)}{v_A^2(1)} \equiv \frac{4\pi \gamma P_1}{B_1^2} < \frac{(\gamma + 1) - r(\gamma - 1)}{2} \equiv \alpha$$

where, by (6–26), $0 < \alpha \leqslant 1$. This solution is possible therefore only when the initial magnetic field is so large that the Alfvén speed is greater than the sound speed; for strong shocks $(\alpha \rightarrow 0)$ the condition becomes $v_A \gg c_s$. Note also from (6–10) and (6–40) that

$$U_1^2 = r \left(\frac{B_1^2}{4\pi \rho_1} \right)$$

so this solution occurs only for shocks with speeds greater than the local Alfvén speed.

These shocks, which produce a tangential component of magnetic field where none existed ahead of the shock, are called *switch-on shocks*. Similarly, one can have *switch-off shocks* where an initial tangential component is removed by the shock. A discussion of switch-off shocks[42] requires consideration of a general initial field with both normal and tangential components. They do *not* appear as a solution of the case $\mathbf{B}_1 = (0, B_1, 0)$ treated in the next section. Of course, switch-off shocks are not the only solution of the general case mentioned above. In general, if the initial field has both a normal and tangential component so too does the final field, but with a different magnitude for the tangential component. In other words, the field lines are refracted by the shock wave (Fig. 6–5). Because the tangential component of

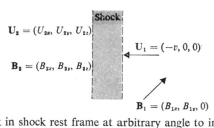

$$\mathbf{U_2} = (U_{2x}, U_{2y}, U_{2z})$$

$$\mathbf{B_2} = (B_{2x}, B_{2y}, B_{2z})$$

Shock

$$\mathbf{U_1} = (-v, 0, 0)$$

$$\mathbf{B_1} = (B_{1x}, B_{1y}, 0)$$

Fig. 6-5   Plane shock in shock rest frame at arbitrary angle to initial magnetic field. In general both **U** and **B** change direction on passing through shock wave

the magnetic field is not conserved, a shock in a fluid in general contains a current sheet.

6–4   Shock propagation perpendicular to the magnetic field

As a further illustration, consider a shock propagating at right angles to the field in the unshocked plasma. The y-axis may now be chosen such that

$$\mathbf{B}_1 = (0, B_1, 0) \tag{6-44}$$

and so, using (6–7), $\mathbf{B}_2$ may be written

$$\mathbf{B}_2 = (0, B_y, B_z) \tag{6-45}$$

Using (6–29) and (6–30), (6–10) gives

$$U_x = U_1/r \tag{6-46}$$

Then, substituting (6–44) to (6–46) in (6–9), (6–12), and (6–13) gives $B_z = U_y = U_z = 0$, so that (6–29) and (6–45) become

$$\mathbf{U}_2 = (U_1/r, 0, 0)$$
$$\mathbf{B}_2 = (0, B_y, 0) = (0, rB_1, 0)$$

by (6–8) and (6–46). Thus the magnetic field is constant in direction and increased in magnitude by the same ratio as the density. Finally, (6–11) and (6–14) are now

$$\rho_1 U_1^2 + P_1 + B_1^2/8\pi = \rho_2 U_2^2 + P_2 + B_2^2/8\pi$$

and

$$\frac{\gamma P_1}{(\gamma - 1)\rho_1} + \frac{U_1^2}{2} + \frac{B_1^2}{4\pi\rho_1} = \frac{\gamma P_2}{(\gamma - 1)\rho_2} + \frac{U_2^2}{2} + \frac{B_2^2}{4\pi\rho_2}$$

These may be written

$$\gamma M^2(1 - 1/r) = (R - 1) + Q(r^2 - 1) \tag{6-47}$$

and
$$\gamma M^2 \left(1 - \frac{1}{r^2}\right) = \frac{2\gamma}{(\gamma - 1)}\left(\frac{R}{r} - 1\right) + 4Q(r - 1)$$

respectively, where $Q = B_1^2/8\pi P_1$ is the ratio of magnetic to gas pressure ahead of the shock. Eliminating R and excluding the solution $r = R = 1$, which corresponds to no shock, one gets

$$Q(2 - \gamma)r^2 + [\gamma(Q + 1) + \tfrac{1}{2}\gamma(\gamma - 1)M^2]r - \tfrac{1}{2}\gamma(\gamma + 1)M^2 = 0$$

If r_1 and r_2 are the roots of this equation, then

$$r_1 r_2 = -\gamma(\gamma + 1)M^2/2(2 - \gamma)Q$$

and for $\gamma < 2$ one root is negative and, therefore, non-physical. Consequently, there is only one solution corresponding to a shock in this case. Since $r > 1$

$$\tfrac{1}{2}\gamma(\gamma + 1)M^2 > Q(2 - \gamma) + \gamma(Q + 1) + \tfrac{1}{2}\gamma(\gamma - 1)M^2$$

i.e.,
$$\gamma M^2 > 2Q + \gamma$$

i.e.,
$$U_1^2 > B_1^2/4\pi\rho_1 + \gamma P_1/\rho_1 = v_A^2 + c_s^2 = (c_s^*)^2$$

where the last equality defines c_s^*. Thus, for shocks to propagate perpendicular to a magnetic field the shock speed must be greater than c_s^*. The speed c_s^* assumes the role played by c_s in hydrodynamic shocks (cf. Problem 6–7); this is not a surprising result, since c_s^* is the speed of compressional waves propagating perpendicular to a magnetic field (cf. Section 8–2). Once again, we see that the effect of a magnetic field, for motion perpendicular to itself, is to increase the 'stiffness' of the conducting fluid.

The shock strength, R, is reduced by the introduction of the magnetic field (cf. Problem 6–8) since flow energy is now converted into magnetic energy as well as heat. However, since

$$B_2/B_1 = r < (\gamma + 1)/(\gamma - 1)$$

the increase in magnetic energy is limited while from (6–47)

$$\frac{P_2}{P_1} = R = 1 + \gamma M^2(1 - 1/r) - Q(r^2 - 1)$$

so that for large Mach number, relative to a fixed value of Q, the temperature ratio is approximately the same as for the hydrodynamic case (6–37).

6–5 Shock thickness

We now wish to consider the structure of a shock. Unfortunately, plasma shock structure (the variation of pressure, density, and so on within the shock region) is not well understood. This is particularly true for collisionless shock waves, which are discussed separately in the next section. Even for collision-dominated shock waves a quantitative calculation of shock structure involves considerable labour. The procedure is to solve the appropriate *transport equations* inside the shock region using either region 1 or region 2 as a set of boundary conditions. Just what form the appropriate transport equations take is not easily decided in general. Often the important dissipative mechanisms are viscosity, heat conductivity, and electrical conductivity (Joule heating). Of these, only Joule heating was retained in deriving the hydromagnetic equation (cf. (3–79)). The heat conduction term must be reintroduced into the energy equation; similarly, viscosity terms must be brought into the momentum and energy equations, as was done for the momentum equation in Chapter 4. Marshall[43] has such a set of equations to discuss shock structure in the limits of high and low electrical conductivity. However, there is considerable doubt as to whether such a one-fluid, hydromagnetic description is appropriate for a discussion of shock structure. In particular, it often happens that the electrons heat up first in a shock and then reach an equilibrium temperature with the ions after a longer period of time. Thus, a description involving separate ion and electron temperatures is usually necessary. Also, depending on the conditions, other dissipative mechanisms may be important—for instance, ionization if the unshocked gas has a zero or low degree of ionization.

However, leaving aside quantitative discussion of shock structure, one may obtain estimates for the thickness of a shock by a dimensional analysis of the transport equations[44] or, more simply, by order of magnitude arguments.[45] Since the conditions on either side of the shock are given by the initial conditions together with the solutions of the shock equations, the total rate of dissipation of energy is known and this must occur within the shock thickness, δ. If we know (or assume) that the dissipation is due principally to one particular mechanism, we can write an order of magnitude relationship. For example, suppose the appropriate dissipative process is viscosity. In the energy transport equation, the term involving viscosity is proportional to the square of the velocity gradient. Then, if other dissipative processes are negligible, the rate of dissipation of energy, $\Delta E/\Delta t$, is proportional to $\rho v (\Delta \mathbf{U}/\delta)^2$, where v is the kinematic viscosity. Since $\Delta t \sim \delta/U_1$, the order of magnitude relationship is

$$\frac{\Delta E}{\Delta t} \sim \frac{U_1 \, \Delta E}{\delta} \sim \rho v \left(\frac{\mathbf{U}_1 - \mathbf{U}_2}{\delta} \right)^2$$

i.e.,
$$\delta \sim \frac{\rho v (\mathbf{U}_1 - \mathbf{U}_2)^2}{U_1 \, \Delta E} \tag{6–48}$$

Applying this to the particular case of the strong hydrodynamic shock discussed in Section 6–3, the energy dissipated $\Delta E \approx \frac{1}{2}\rho_1 U_1^2$. Also, using (6–38), (6–48) implies

$$\delta \sim \nu/U_1 \tag{6-49}$$

or, in other words, the Reynolds number, $R = U_1\delta/\nu$, is of order unity. From kinetic theory (cf. Chapter 10), one can show approximately that $\nu \sim P\tau_c/\rho$, where τ_c is the ion–ion collision time. Using $P \sim \frac{1}{2}(P_1+P_2) \approx \frac{1}{2}P_2$, it follows from (6–38), (6–39), and (6–49) that

$$\delta \sim \tau_c \sqrt{\left(\frac{P_2}{\rho_2}\right)} \sim \tau_c \sqrt{\left(\frac{\kappa T_2}{m^+}\right)}$$

Thus the shock thickness is of the order of the ion collision mean free path, justifying the argument given in Section 6–1.

To give a further example, consider dissipation by Joule heating. Here, the dissipation of energy occurs at a rate proportional to j^2/σ. Since $\mathbf{j} = (c/4\pi)\nabla \times \mathbf{B}$, the order of magnitude relationship is

$$\frac{U_1 \Delta E}{\delta} \sim \frac{1}{\sigma}\left(\frac{c}{4\pi}\right)^2 \left(\frac{\mathbf{B}_1 - \mathbf{B}_2}{\delta}\right)^2$$

i.e.

$$\delta \sim \frac{1}{\sigma}\left(\frac{c}{4\pi}\right)^2 \frac{(\mathbf{B}_1 - \mathbf{B}_2)^2}{U_1 \Delta E} \tag{6-50}$$

This may be written in terms of the magnetic Reynolds number (cf. Eqn (4–18))

$$R_M = \frac{4\pi\sigma\delta U_1}{c^2} \sim \frac{(\mathbf{B}_1 - \mathbf{B}_2)^2}{4\pi \, \Delta E} \tag{6-51}$$

Now suppose this is applied to a strong shock propagating perpendicular to a magnetic field (cf. Section 6–4). With $\Delta E \approx \frac{1}{2}\rho_1 U_1^2$ and $B_1 \approx B_2/4$,

$$R_M \sim \frac{B_2^2/8\pi}{\frac{1}{2}\rho_1 U_1^2}$$

which gives the order of magnitude of the magnetic Reynolds number required if Joule heating is to be an adequate dissipative mechanism for this shock.

In general, more than one dissipative process may occur within a shock, and the shock may then be characterized by more than one thickness. For example, in Marshall's calculations[43] viscosity, thermal conductivity, and Joule heating are included. The results for the case of low electrical conductivity are shown in Figs. 6–6 and 6–7. If the initial magnetic field is low enough the shock displays two characteristic lengths (Fig. 6–6). Over one of these (such that $R_M \sim 1$) the magnetic field change occurs while the velocity and temperature show little change. This region is followed by a thinner one (such that $R \sim 1$) in which velocity and temperature change abruptly while the magnetic field is sensibly constant. For fields above a certain critical value, these abrupt changes disappear (Fig. 6–7); the shock is then characterized by a single thickness.

As mentioned earlier, the results from an analysis in which both ions and

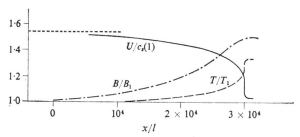

Fig. 6–6 Shock profile for case of low electrical conductivity, $r = 1.5$, and $Q = 0.25$. The curves show the ratio of B and T to their initial values and the ratio of $|U|$ to the initial sound speed; l is of the order of the mean free path in region 1 and the shock is moving to the left (after Marshall[43])

electrons have the same temperature are often of doubtful value. However, in a two-temperature plasma, dissipative mechanisms make things so complicated that an analytic approach to interpreting the results of an experiment is usually out of the question. Encouraging progress has been made using a

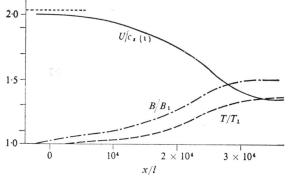

Fig. 6–7 Shock profile for case of low electrical conductivity, $r = 1.5$, and $Q = 1$. The shock is moving to the left (after Marshall[43])

numerical approach in which a number of physical effects is included and the consequences of these examined in detail using a computer. This method has been developed mainly by Hain and Roberts.[46] These 'numerical experiments' have been helpful in interpreting results from laboratory experiments in devices such as linear and theta pinches.

As an illustration, consider a numerical experiment on a pre-heat discharge in the partially ionized plasma (protons, electrons, and hydrogen atoms in the ground state) of a theta pinch.[47] In the description based on nine partial differential equations for the plasma and hydrogen atom densities and (radial) velocities, electron, ion, and atom temperatures, and magnetic field components B_z, B_θ, a great many effects were taken into account. These included ionization by electron collisions with hydrogen atoms, ion–atom charge exchange, adiabatic compression of the plasma and hydrogen gas separately, heat exchange by collisions between the three species, anisotropic

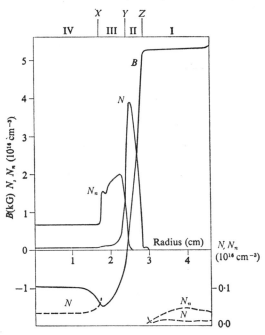

Fig. 6–8 Numerical calculation of plasma number density N, neutral number density N_n and magnetic field B at $t = 0.64\ \mu$s. When N, N_n are small, they are plotted (broken lines) on the right-hand scale (after Roberts[47])

plasma resistivity, and resistivity deriving from electron–neutral collisions. Thus, in addition to heating by adiabatic compression, heat exchange by collisions, and Joule heating, the electrons are cooled by ionization and excitation. The protons and hydrogen atoms are heated by viscosity and charge exchange.

The hydrogen gas was initially at a pressure of $100\ \mu$, being 5 per cent ionized and having a mean temperature of 0.2 eV in a discharge tube of radius 4.6 cm. Figures 6–8 and 6–9 show the magnetic field, plasma and neutral hydrogen number densities, ion and electron temperatures, at $t = 0.64\ \mu$s; Fig. 6–10 shows plasma and neutral gas velocities and the degree of ionization at the same time. The lines X, Y, Z denote the shock front, the point at which

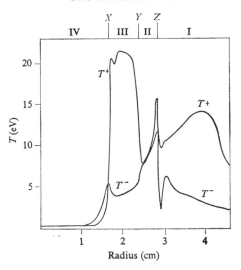

Fig. 6–9 Ion temperature T^+ and electron temperature T^- at $t = 0.64\ \mu s$ (after Roberts[47])

the magnetic field changes sign and the outer boundary of the main plasma. These lines divide the discharge into four regions. In region I, there is low-density partially ionized hydrogen; regions II and III are fully ionized, and hot, partially ionized, plasma, respectively, while IV is the 5 per cent ionized hydrogen gas through which the shock has yet to pass.

In I, the magnetic field is uniform and the plasma moves inwards with a steady velocity of 8.10^6 cm/s, equal to that of the lines of force. The hydrogen atoms are accelerated by charge exchange and this rise in gas temperature means that T^+ rises above T^-, since the ions are heated by viscosity.

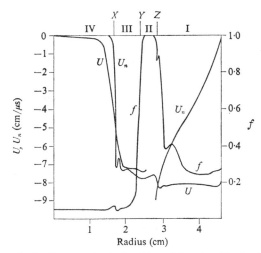

Fig. 6–10 Plasma velocity U, neutral velocity U_n and fractional ionization f at $t = 0.64\ \mu s$ (after Roberts[47])

In II, the dense plasma is pushed towards the axis by the magnetic field. The magnetic field now changes and steepens sharply, since the magneto-acoustic speed $(v_A^2 + c_s^2)^{1/2}$ is greatest at the outside (cf. Section 6–1); the steepening is balanced by diffusion of the magnetic field. The existence of a strong field gradient gives rise to Joule heating of electrons which, in turn, heat the ions by collisions; T^+ is smaller than T^- over most of the region.

The gas in region III is only partially ionized. As the hydrogen atom density rises, charge exchange between hot ions and neutral atoms provides gas pressure to drive the hydrodynamic shock, X, which heats the atoms; these, in turn, heat the ions by charge exchange.† Electron temperature, on the other hand, is fairly uniform throughout this region, the energy gain from Joule heating being offset by ionization loss.

The numerical data in Figs. 6–8 to 6–10 is valuable for comparison with the results of laboratory studies; it may, for example, suggest closer experimental examination of the behaviour of some parameter to uncover details not shown explicitly by the original diagnostics.

6–6 Collisionless shocks

The picture of a shock developed thus far is that of an abrupt, non-adiabatic change of fluid variables, accompanying the supersonic propagation of a disturbance, the sudden change in the state of the fluid being brought about by the conversion of ordered (flow) energy into random (thermal) motion. In a non-conducting fluid, this conversion can only take place via collisions. However, in ionized gases sudden changes accompanying supersonic disturbances have been observed to occur over distances much shorter than the mean free path. One example of this is the bow shock produced by the interaction of the solar wind with the earth's magnetic field (cf. Section 6–7). Satellite observations show that a sudden jump in the average magnetic field occurs over a distance $\lesssim 10^3$ km, the mean free path being $\sim 10^8$ km. This is perhaps the clearest example of a *collisionless shock wave*; more recently, laboratory experiments have confirmed their existence. For example, in the *Tarantula* experiment at Culham Laboratory (cf. Section 6–7) an annular imploding shock wave is produced with a thickness (~ 0.14 cm) which is appreciably less than the ion–ion and ion–neutral mean free paths. Thus viscosity may be discounted as an effective dissipative mechanism. Ion–electron collisions, while not so clearly ruled out, can be discounted since they are unlikely to provide an adequate explanation of the observed shock structure.

Given this experimental evidence for the existence of collisionless shocks, it is desirable to construct a theoretical model to explain them; in particular, a suitable mechanism for collisionless dissipation is required. Unfortunately, there is no generally accepted theory for collisionless shocks. The situation

† Neutral hydrogen temperatures have not been plotted but approximate T^+ over I and IV.

is complicated by a number of factors. First of all, in any given situation it is rarely clear, even qualitatively, which physical process provides the energy dissipation. Secondly, the absence of collisions allows the constituent particles to behave relatively independently so that a one-fluid hydromagnetic description is inadequate. Thirdly, all theories lead to a description in which initially an *adiabatic* change leads to a 'perturbed state' rather than a uniform one. The final uniform state is reached when the energy of these perturbations is fed into random particle motions by some damping process.

Roughly speaking, there are two approaches to the problem, depending on the nature of the perturbed state.[48] In one approach, more suitable for weak shocks, the energy to be dissipated is converted into plasma oscillations. This may be due to charge separation giving rise to strong electric fields which then dominate the motion and set up plasma oscillations (cf. the discussion in Section 2–4). Since no increase in random motion has occurred this is an adiabatic change. These oscillations are eventually damped—that is, energy is lost to random particle motions, so that a final uniform state with increased entropy is attained. The second approach, more likely to be valid for strong shocks, assumes that flow energy is initially converted into turbulent motion. This, too, is then gradually damped to give a final state with increased entropy.

In both cases, the shock thickness is defined by the initial change from undisturbed to perturbed plasma. The average values of the plasma variables in the perturbed state (averaged over plasma oscillations or turbulent motion) are significantly different from those in the undisturbed plasma but change little in the subsequent damping process. The shock thickness thus defined is found to be of the order of the *ion Larmor radius*, which is not too surprising since this is the next largest scale length to the mean free path. For the particular case of propagation perpendicular to the magnetic field, the shock thickness is found to be of the order of the geometric mean of the ion and electron Larmor radii. The thickness of the second region in which the damping takes place depends on the damping mechanism. This could, of course, be due to collisions.

6–7 Experiments

The simplest device for producing shock waves is an ordinary shock tube[49] in which a high-pressure region is initially separated from a low-pressure region by a diaphragm, Fig. 6–11. The pressure discontinuity propagates as a shock wave into the low-pressure region when the diaphragm is broken. These tubes may be used to produce high-temperature gases by employing high pressure ratios (strong shocks) as indicated in Section 6–3. The temperature ratio given by (6–37)

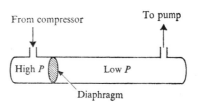

Fig. 6–11 Shock tube

holds only for a perfect gas. In a polyatomic gas, internal (rotational and vibrational) energy modes become increasingly important at high temperatures. Thus the heat required to raise the gas temperature by a given amount is greater at high temperatures than at low temperatures; in other words, the specific heat, C_v, increases with temperature. The reduction in temperature caused by variable specific heat is indicated in Fig. 6–12.

Diaphragm-type shock tubes are also used to produce weakly ionized plasmas which may then be used to study interaction with a magnetic field. Ionization is another effect which reduces the temperature ratio predicted by (6–37). The initial flow energy is not only converted into thermal energy but also strips electrons from their parent atoms.

Initial ionization rates are found which cannot be adequately explained on

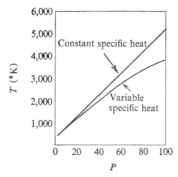

Fig. 6–12 Temperature plotted against pressure ratio for air with constant and variable specific heats (after Wright[49])

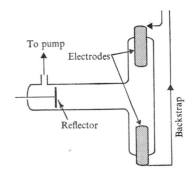

Fig. 6–13   T-tube

the basis of atom–atom collisions. A number of other factors probably contribute to the ionization, including ultraviolet radiation from the shocked gas. *Precursor effects* involving electrons may also be significant. Highly energetic electrons produced within the shock travel down the tube ahead of the shock front, causing pre-ionization.

Maximum Mach numbers of about 30 are obtainable in diaphragm-type shock tubes but much higher values (~100) are possible in *electromagnetic shock tubes*. One example of this type is the T-tube (Fig. 6–13) developed by Kolb.[50] Electrodes are inserted in the cross-piece of the T, the stem of the T serving as the expansion tube. When a current is discharged across the electrodes a shock wave is driven along the expansion tube. However, by making the return current run along a backstrap parallel and close to the discharge, the resulting Lorentz force increases the Mach number of the shock. (The effect is the same as the repulsion between parallel wires carrying opposite currents.) An interesting result found with this type of shock tube is that, under certain conditions, the temperature behind the shock wave is *greater* than that predicted in (6–37) by a factor of about two. The most likely explanation is pre-heating of the unshocked gas by radiation from the discharge and, to a lesser extent, from the shocked gas. This radiation also accounts for the high initial ionization rates observed. The heating can be further increased, yielding temperatures $\sim$100 eV, by joining two T-tubes in the shape of an H so that simultaneously discharged shock waves collide in the centre of the common expansion tube.

Other electromagnetic shock tubes also employ a Lorentz force as the driving mechanism of the shock. For example, the theta pinch mechanism described in Section 4–3 is used to produce imploding shock waves. Kolb[51] has combined this mechanism with the H-tube arrangement by placing magnetic field coils around the expansion tube. Thus the colliding shock waves

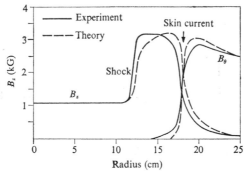

Fig. 6–14 Radial variation of axial and azimuthal fields at $t = 0.9$ μs in Tarantula experiment. The imploding shock wave travels ahead of the axial skin current which provides the pinch force (after Paul *et al.*[52])

from the two discharges are also subjected to an imploding shock, compressing the plasma towards the axis of the tube to give plasma densities as high as 10^{18}/cm³.

The Tarantula experiment,[52] designed to study collisionless shocks, uses the linear pinch mechanism (see Section 4–3). The production of high temperatures is facilitated by pre-ionization, since low conductivity allows the compressing magnetic field to leak into the gas ahead of the shock and work must then be expended against this field. In the Tarantula experiment, an initial ionization ($\sim$90 per cent) is created by electrical breakdown and heating by an oscillatory axial current. A much larger direct axial current is then suddenly applied to produce an imploding shock wave converging radially towards the axis. Results from this experiment (Figs. 6–14 to 6–16) show the radial variation of axial and azimuthal fields, a comparison of mean free paths with shock thickness, and the observed time variation of the electron temperature.

The earth's bow shock (Section 6–6), produced by the interaction of the solar wind with the magnetosphere, is a further example of collisionless shock behaviour. The solar wind is the flux of hydrogen plasma streaming outwards from the solar corona with a velocity $\sim 5 . 10^7$ cm/s and density ~ 10 particles/cm³, giving a flux density of $\sim 5 . 10^8$ particles/cm² s. When this ionized

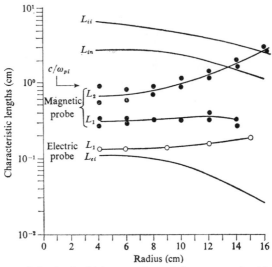

Fig. 6–15 Characteristic shock thicknesses L_1 and L_2 compared with mean free paths for ion–ion (L_{ii}), ion–neutral (L_{in}), and electron–ion (L_{ei}) collisions (after Paul et al.[52])

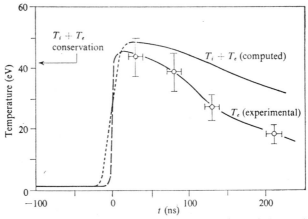

Fig. 6–16 Measured electron temperature T_e as a function of time. Plane geometry conservation equations predict a total temperature $T_i + T_e = 42$ eV while the computed curve shows the results of a computer calculation. The measured temperature agrees well with the theoretical predictions provided the electrons only are shock heated (after Paul et al.[52])

gas meets the magnetosphere it is unable to penetrate it, causing the *geomagnetic cavity* in the solar wind. The situation is analogous to high-speed aerodynamic flow around blunt bodies. Theoretical analyses based on this analogy predicted a transition region between the solar wind and the earth's

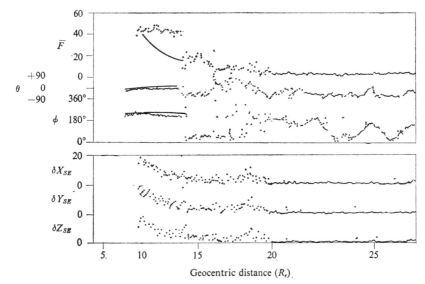

Fig. 6-17 Example of magnetic field data from satellite Imp 1, showing outward bound traversal of magnetopause at 13·6 R_e and shock wave at 19·7 R_e ($R_e = 1$ earth radius). The field is indicated by magnitude $\bar{F}$ and direction (θ, ϕ). Stabilization of the field after the satellite passed through the shock wave is seen clearly in the lower figure showing the variance of the X, Y, and Z components of the field (after Ness, Clearce, and Seek[53])

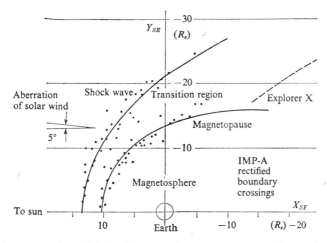

Fig. 6-18 Summary of rectified shock wave and magnetopause crossings and comparison with theoretical predictions. All boundary crossings have been rectified so as to reduce all distances to a geometry in which the solar wind is normal to the axis of the earth's magnetic dipole. Also the theoretical model has been rotated by 5°; this is comparable with the aberration effect of the earth's orbital motion (after Ness, Clearce, and Seek[53])

P D—K

magnetic field, bounded on the upstream side by a collisionless shock wave and on the other side by the magnetopause (the boundary of the magnetosphere). Satellite measurements[53] of the magnetic field showed clearly the transition region (large fluctuations indicating a high degree of turbulence) and its boundaries (sudden changes in magnitude and direction), Fig. 6–17. Figure 6–18 shows a comparison of predicted and observed location of the shock wave and magnetopause. Agreement is particularly good around the earth-sun axis (where the theoretical calculations are less sensitive to the parameters used).

Summary

1. *General conditions for plane shocks propagating in direction* $\hat{\mathbf{n}}$

   (a) *Jump conditions*

   $$\text{Mass } [\rho \mathbf{U}.\hat{\mathbf{n}}]_1^2 = 0$$
   $$\text{Momentum } [\rho \mathbf{U}(\mathbf{U}.\hat{\mathbf{n}}) + (P + B^2/8\pi)\hat{\mathbf{n}} - (\mathbf{B}.\hat{\mathbf{n}})\mathbf{B}/4\pi]_1^2 = 0$$
   $$\text{Energy } [\mathbf{U}.\hat{\mathbf{n}}\{(\rho I + \tfrac{1}{2}\rho U^2 + B^2/8\pi) + (P + B^2/8\pi)\}$$
   $$- (\mathbf{B}.\hat{\mathbf{n}})(\mathbf{B}.\mathbf{U})/4\pi]_1^2 = 0$$

   $$\left. \begin{array}{l} \text{Magnetic field } [\mathbf{B}.\hat{\mathbf{n}}]_1^2 = 0 \\ \qquad\qquad [\hat{\mathbf{n}} \times (\mathbf{U} \times \mathbf{B})]_1^2 = 0 \end{array} \right\} \text{ In general } \mathbf{B} \text{ is refracted by shock.}$$

   (b) *Hugoniot relation*

   $$[I]_1^2 + \tfrac{1}{2}(P_1 + P_2)[\rho^{-1}]_1^2 + \frac{1}{16\pi}[\rho^{-1}]_1^2\,([\mathbf{B}]_1^2)^2 = 0$$

   Shocks are *compressive*

   $$\frac{P_2}{P_1} > 1 \qquad 1 < \frac{\rho_2}{\rho_1} < \frac{\gamma + 1}{\gamma - 1}$$

2. *Propagation parallel to magnetic field*

   (a) *Hydrodynamic shock*: $\mathbf{B}_2 = \mathbf{B}_1$
   $$U_1 > c_s(1) \qquad U_2 < c_s(2)$$
   $$\frac{T_2}{T_1} \sim M^2 \quad \text{for } M \gg 1$$

   (b) *Switch-on shock*
   $$U_1 > v_A(1) > c_s(1)$$

3. *Propagation perpendicular to magnetic field*

   No refraction: $\mathbf{B}_2 = \dfrac{\rho_2}{\rho_1}\mathbf{B}_1$
   $$U_1 > (v_A^2(1) + c_s^2(1))^{1/2}$$

Shock strength *reduced* by introduction of magnetic field but

$$\frac{T_2}{T_1} \sim M^2 \quad \text{for } M \gg 1, \ B_1^2/8\pi P_1$$

4. *Shock thickness*

(a) *Collisional shocks*

$$\delta \sim \text{mean free path}$$

(b) *Collisionless shocks*

$$\delta \sim \left\{ \begin{matrix} r_L^+ \\ (r_L^+ \ r_L^-)^{1/2} \end{matrix} \right\} \ll \text{mean free path}$$

Problems

6-1 Integration of the conservation equations (as presented in Problem 3-4) across a plane shock gives

$$[\rho U . \hat{n}]_1^2 = 0 \qquad [\Pi . \hat{n}]_1^2 = 0 \qquad [S . \hat{n}]_1^2 = 0$$

Use the hydromagnetic approximations to reduce these equations to (6–15), (6–16), and (6–17).

6-2 Derive the hydromagnetic Hugoniot (6–19).
   Hint: let $\mathbf{U} = (u, v, w)$ and write (6–10) as

$$\rho_1 u_1 = \rho_2 u_2 = k \tag{6–10}$$

Then (6–12), (6–13) and (6–14) are

$$\left[v - \frac{B_x B_y}{4\pi k} \right]_1^2 = 0 \tag{6–12}$$

$$\left[w - \frac{B_x B_z}{4\pi k} \right]_1^2 = 0 \tag{6–13}$$

$$\left[I + \frac{P}{\rho} + \frac{k^2}{2\rho^2} + \frac{v^2}{2} + \frac{w^2}{2} + \frac{B^2 - B_x^2}{4\pi\rho} - \frac{B_x}{4\pi k}(B_y v + B_z w) \right]_1^2 = 0 \tag{6–14}$$

Square (6–12) and (6–13) and substitute for v^2 and w^2 in (6–14). This eliminates v and w entirely. Write (6–11) in the form

$$k^2 \left[\frac{1}{\rho} \right]_1^2 = -\left[P + \frac{B^2}{8\pi} \right]_1^2 \tag{6–11}$$

and multiply by $\frac{1}{2}(\rho_1^{-1} + \rho_2^{-1})$ to obtain

$$\frac{k^2}{2} \left[\frac{1}{\rho^2} \right]_1^2 = -\frac{1}{2}\left(\frac{1}{\rho_1} + \frac{1}{\rho_2} \right)\left[P + \frac{B^2}{8\pi} \right]_1^2 = \left[\frac{k^2}{2\rho^2} \right]_1^2$$

This is substituted in (6–14) leaving as the only term involving velocity

$$\left[\frac{1}{2}\frac{B_x^2}{(4\pi k)^2}(B^2 - B_x^2)\right]_1^2 = \frac{1}{2}\frac{B_x^2}{(4\pi k)^2}\left[B_y^2 + B_z^2\right]_1^2$$

$$= \frac{1}{2}\frac{B_x^2}{(4\pi k)^2}\{(B_{2y} - B_{1y})(B_{2y} + B_{1y}) + (B_{2z} - B_{1z})(B_{2z} + B_{1z})\}$$

Now using (6–12) and (6–13) in (6–8) and (6–9) one finds

$$\left[\frac{B_y}{\rho}\right]_1^2 = \frac{B_x^2}{4\pi k^2}\left[B_y\right]_1^2 \tag{6–8}$$

$$\left[\frac{B_z}{\rho}\right]_1^2 = \frac{B_x^2}{4\pi k^2}\left[B_z\right]_1^2 \tag{6–9}$$

Substitution of these equations in (6–14) then gives (6–19).

6-3 Derive (6–20) from (5-44) given that the specific heat at constant volume is defined by

$$C_v = \left(\frac{\partial I}{\partial T}\right)_v \qquad v = 1/\rho$$

6-4 Use (6–34) to rewrite (6–35) and (6–36b) as

$$\gamma M_1^2(1 - 1/r) = (R - 1)$$

$$\gamma M_1^2(1 - 1/r^2) = \frac{2\gamma}{\gamma - 1}\left(\frac{R}{r} - 1\right)$$

where $M_1^2 = \rho_1 U_1^2/\gamma P_1$. By solving for R, r and applying $R > 1$ show that U_1 is supersonic. Rewriting the solutions in terms of $M_2^2 = \rho_2 U_2^2/\gamma P_2$ show that U_2 is subsonic.

6-5 Deduce from your answer to the previous question that $R \gg 1$ implies $M_1 \gg 1$.

6-6 Verify (6–38) from your answer to Problem 6-4.

6-7 For the shock discussed in Section 6–4 it was shown that $U_1^2 > (c_*^*(1))^2$. By rewriting the equations in terms of $M_2^2 = \rho_2 U_2^2/\gamma P_2$ and $Q_2 = B_2^2/8\pi P_2$ show that

$$U_2^2 < (c_*^*(2))^2 = B_2^2/4\pi\rho_2 + \gamma P_2/\rho_2$$

6-8 For a shock propagating perpendicular to a magnetic field (Section 6–4) show that, for $\gamma < 2$, $r < r_0$ where r_0 is the solution for zero magnetic field. Hence, deduce from (6–47) that the shock strength R is reduced by the introduction of a magnetic field.

7

Waves in cold plasmas

7–1 Introduction
The subject of wave propagation is important to the development of plasma dynamics. One of the reasons why it occupies a key position is the link it provides between theory and experiment in a subject which, for a variety of reasons, lacked this basic essential of any branch of physics for too long. Another is because wave propagation, once understood, allows us to probe a plasma—for instance, to establish the electron density. The principal reason for the long time taken to reach the present state of development lies in the history of the subject, with its roots in astrophysics and ionospheric physics, gas discharges, electron tubes, and controlled thermonuclear research. For example, astrophysicists were interested primarily in violent hydromagnetic phenomena on the sun's surface rather than in wave propagation *per se*; likewise, workers in the CTR field were understandably obsessed in the beginning with the eventual goal of this project—fusion of light nuclei.

The most important developments came first in ionospheric physics, originating in the suggestion by Kennelly and Heaviside in 1902 that it might be possible to propagate radio waves across the Atlantic by reflecting them from the electrical layer (the Heaviside layer) in the atmosphere. The problem was examined theoretically at intervals (for instance, by Larmor in 1924) until resolved by Appleton and co-workers in Britain who, in 1925, simultaneously with Nichols and Schelleng in the United States, emphasized the need to take account of the geomagnetic field in discussing radio wave propagation. The full theory was developed in 1931 by Hartree and is known as the Appleton–Hartree magneto-ionic theory. About the same time, another important plasma mode was being studied, namely plasma oscillations. Penning in 1926 first suggested the existence of such oscillations to account for anomalously rapid scattering of electron beams by a plasma (i.e., over distances much shorter than would have been expected on grounds of a collision mechanism). These oscillations were studied exhaustively by Langmuir and a theory was worked out by Tonks and Langmuir in 1928. In addition to electron plasma oscillations, Langmuir also dealt with ion oscillations which, in the limit of long wavelength, are analogous to sound waves travelling through the ionized gas involving the motion of ions *and* electrons.

Historically, the next most important contribution to waves in plasmas came from studies in cosmic electrodynamics by Alfvén, who observed in 1942 that if one pictures magnetic field lines as elastic strings under tension there should exist magnetohydrodynamic waves analogous to the waves in ordinary elastic strings. These Alfvén waves have been discussed briefly in Chapter 4. Finally, one should mention experimental work in gas discharges and micro-wave electron-beam devices which flourished from 1946 onwards and is of particular importance in so far as bounded plasmas are concerned.

The variety of modes which propagate in a magnetized plasma is at first bewildering and for this reason it seems desirable to keep matters as simple as possible. In this chapter, therefore, only waves in cold plasmas are examined, i.e., plasma pressure is ignored and the set of cold plasma equations, (3–91) to (3–96), is appropriate. Neither boundary effects nor inhomogeneities in the plasma are admitted. Moreover, the analysis will be based on a linear perturbation theory. This might seem at first sight a curious way of going about getting an understanding of the subject with a view to correlating theory with experimental results. After all, waves propagating in a laboratory plasma will in general be non-linear disturbances in a bounded medium at the same time inhomogeneous and anisotropic, dissipative and dispersive. However, thanks to the ingenuity of experimentalists in providing plasmas which go a long way towards simulating even the gross idealizations of theoreticians, some common ground can be established. One other point worth stressing at this stage is the importance of recent experiments on wave motion in *solid-state plasmas* which approximate the cold plasma model very closely; further discussion of these comes in Section 7–8.

7–2 Some general wave concepts

Before developing a theory of waves in a cold plasma, we recall some elementary results.[16] For simplicity, first consider electromagnetic waves propagating in a vacuum. Then, the electrodynamic equations require that each component of $\mathbf{E}$, $\mathbf{B}$ satisfy the wave equation

$$\nabla^2 \psi = \frac{1}{c^2} \frac{\partial^2 \psi}{\partial t^2} \tag{7–1}$$

The plane-wave solution (a plane wave is one for which the wave disturbance is constant over all points of a plane normal to the direction of propagation of the wave) is

$$\psi(\mathbf{r}, t) = A \exp i(\mathbf{k}.\mathbf{r} - \omega t) + B \exp -i(\mathbf{k}.\mathbf{r} + \omega t) \tag{7–2}$$

where A, B are constants, $\mathbf{k}$ is the wave propagation vector, ω the frequency and, from (7–1), $\omega^2 = k^2 c^2$. Choosing the x-axis along the wave vector $\mathbf{k}$, the general plane-wave solution may be written, using the Fourier integral theorem,

$$\psi(x,t) = \int \{A_k \exp [ik(x - ct)] + B_k \exp [-ik(x + ct)]\} \, dk$$
$$= f(x - ct) + g(x + ct) \tag{7–3}$$

f and *g* being arbitrary functions representing waves propagating to right and left, respectively, each with velocity *c*. More generally, if one considers wave propagation in a medium characterized by a dielectric function ε and a magnetic permeability μ, this has the effect of replacing *c* by $v = c/\sqrt{(\mu\varepsilon)}$. In particular, (7–3) is still valid (with this substitution) provided that *v* is not a function of *k*. In that event, $v = v(k)$ is the velocity of each wave-number component. The velocity ω/k is known as the *phase velocity* of the wave (or, in the case of $v = v(k)$, the phase velocity of a wave-number component of the wave).

Considering the wave solutions† $\mathbf{E}(\mathbf{r}, t) = \mathbf{E}_0 \exp i(\mathbf{k}.\mathbf{r} - \omega t)$, $\mathbf{B}(\mathbf{r}, t) = \mathbf{B}_0 \exp i(\mathbf{k}.\mathbf{r} - \omega t)$ and using the vacuum divergence equations, one has the requirement

$$\mathbf{k}.\mathbf{E}_0 = 0 = \mathbf{k}.\mathbf{B}_0 \qquad (7\text{–}4)$$

Thus the electromagnetic wave has **E**, **B** orthogonal to **k** and is said to be *transverse* (in the sense of waves on a string—that is, the disturbance or displacement of the string is perpendicular to the direction in which it propagates).

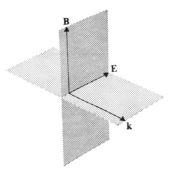

Fig. 7–1 Orientation of field vectors in a plane electromagnetic wave

7–2–1 *Wave polarization*

In general, to specify the electric field (or, equally, the magnetic field) in a plane wave one must write a superposition of two linearly independent solutions of the wave equation,

$$\mathbf{E}(x, t) = (E_y\hat{\mathbf{y}} + E_z\hat{\mathbf{z}}) \exp i(kx - \omega t) \qquad (7\text{–}5)$$

in which E_y and E_z are complex amplitudes. One may write

$$E_y = E_{y0}e^{i\alpha} \qquad E_z = E_{z0}e^{i\beta}$$

---

† Remember that we mean by this $\mathrm{Re}\{\mathbf{E}_0 \exp i(\mathbf{k}.\mathbf{r} - \omega t)\}$ and as long as everything is linear it is permissible to operate with complex quantities and omit the operator Re. Care is needed in discussing quadratic terms.

where E_{y0}, E_{z0} are real so that (7–5) becomes, if $\delta = \beta - \alpha$,

$$\mathbf{E}(x, t) = (E_{y0}\hat{\mathbf{y}} + E_{z0}e^{i\delta}\hat{\mathbf{z}}) \exp i(kx - \omega t + \alpha) \qquad (7\text{–}6)$$

The electric vector at each point in space rotates in a plane normal to $\hat{\mathbf{x}}$ and its tip, as time evolves, describes an ellipse. This is most easily seen for $\delta = \pm\pi/2$. Then,

$$\mathbf{E}(x, t) = (E_{y0}\hat{\mathbf{y}} \pm iE_{z0}\hat{\mathbf{z}}) \exp i(kx - \omega t + \alpha)$$

and taking the real part

$$\left.\begin{array}{l} E_y(x, t) = E_{y0} \cos (kx - \omega t + \alpha) \\ E_z(x, t) = \mp E_{z0} \sin (kx - \omega t + \alpha) \end{array}\right\} \qquad (7\text{–}7)$$

i.e.,

$$\frac{E_y^2}{E_{y0}^2} + \frac{E_z^2}{E_{z0}^2} = 1 \qquad (7\text{–}8)$$

The electromagnetic wave is said to be *elliptically polarized*.

In the special case of E_{y0} (or E_{z0}) $= 0$, the electric field is *linearly polarized* (or *plane polarized*), while if $E_{y0} = E_{z0}$ it is *circularly polarized*. To an observer looking at the approaching wave front, the $+$ sign in (7–7) gives a wave which is *left circularly polarized* (LCP), while the $-$ sign gives a *right circularly polarized* (RCP) wave.†

7–2–2   Energy flux

The time-averaged energy flux in a plane wave is given by the Poynting vector; with complex quantities, this means computing the real part of the complex Poynting vector;

$$\mathbf{S} = \frac{1}{2}\frac{c}{4\pi}\mathbf{E} \times \mathbf{H}^* \qquad (7\text{–}9)$$

Since $\mathbf{H} = \sqrt{(\varepsilon/\mu)}\hat{\mathbf{k}} \times \mathbf{E}$ and $\hat{\mathbf{k}}.\mathbf{E} = 0$, this gives

$$\mathbf{S} = \frac{c}{8\pi}\sqrt{\left(\frac{\varepsilon}{\mu}\right)} |\mathbf{E}_0|^2 \hat{\mathbf{k}} \qquad (7\text{–}10)$$

Moreover the time-averaged energy density of the wave is

$$W = \frac{1}{2}\cdot\frac{1}{8\pi}[\varepsilon\mathbf{E}.\mathbf{E}^* + \mathbf{B}.\mathbf{H}^*]$$

$$= \frac{\varepsilon}{8\pi} |\mathbf{E}_0|^2 \qquad (7\text{–}11)$$

† Alternatively, in terms of the complex amplitudes E_y, E_z, one sees from (7–6) with $\delta = +\pi/2(-\pi/2)$ that $iE_y/E_z = +1(-1)$ corresponds to right (left) circular polarization.

i.e.,
$$S = \frac{c}{\sqrt{(\varepsilon\mu)}} W\hat{\mathbf{k}} \qquad (7\text{–}12)$$

The energy flow is in the direction of propagation and its magnitude is just the product of the time-averaged energy density with the phase velocity of the wave.

7–2–3 Group velocity

So far we have only discussed monochromatic waves—those with a single frequency. However, in practice, even with such a monochromatic source as the laser, there is a small spread in frequency about a central frequency ω_0; equally well one might talk about a spread in wave number about k_0. Provided $\sqrt{(\varepsilon\mu)}$ is constant, then all modes composing this source propagate with the same phase velocity, ω_0/k_0. However, it often happens that the dielectric 'constant' of the medium is a function of ω and in this case the phase velocity of each mode is, in general, distinct. A medium, such that $\varepsilon = \varepsilon(\omega)$, in which only the wave profile of a monochromatic wave propagates unchanged is said to be *dispersive*.

Consider scalar waves propagating along Ox; using a Fourier representation†

$$E(x, t) = \frac{1}{\sqrt{(2\pi)}} \int_{-\infty}^{\infty} a(k) \exp\left[i(kx - \omega(k)t)\right] dk \qquad (7\text{–}13)$$

ω is a function of k, $\omega = \omega(k)$, which is known as the *dispersion relation*. Equation (7–13) and

$$a(k) = \frac{1}{\sqrt{(2\pi)}} \int_{-\infty}^{\infty} E(x, 0)e^{-ikx} dx$$

define a *wave packet*. Assume, for convenience, that $a(k)$ is peaked about some wave number k_0. The central question is this: given some definitely shaped wave packet at $t = 0$ (the *pulse shape*), what does it look like at some later time? Provided the medium is not too dispersive, $\omega(k)$ may be expanded about k_0:

$$\omega(k) = \omega_0 + \left(\frac{d\omega}{dk}\right)_{k=k_0}(k - k_0) + \ldots \qquad \omega_0 \equiv \omega(k_0)$$

† Equally one might write, using ω as the independent variable

$$E(x, t) = \frac{1}{\sqrt{(2\pi)}} \int_{-\infty}^{\infty} a(\omega) \exp\left[i(k(\omega)x - \omega t)\right] d\omega$$

in which case the dispersion relation (see below) is $k = k(\omega)$.

and so

$$E(x, t) \simeq \frac{1}{\sqrt{(2\pi)}} \int_{-\infty}^{\infty} a(k) \exp \{i[kx - \omega_0 t - (d\omega/dk)_{k_0}(k - k_0)t]\}\, dk$$

i.e.,
$$E(x, t) \simeq E(x - v_g t, 0) \exp [i(k_0 v_g - \omega_0)t] \qquad (7\text{-}14)$$

which represents a pulse travelling without distortion with a velocity $v_g = (d\omega/dk)_{k=k_0}$. This is the *group velocity*. The group velocity appears in this context as the propagation velocity of a wave packet; the concept is due to Hamilton (1839).

To relate v_g to the phase velocity $v_p = \omega/k$ is straightforward:

$$v_g = \frac{d\omega}{dk} = \frac{d}{dk}(kv_p) = v_p + k\frac{dv_p}{dk}$$

or, in terms of the wavelength,

$$v_g = v_p - \lambda\frac{dv_p}{d\lambda} \qquad (7\text{-}15)$$

Clearly, in the case in which the phase velocity does not depend on λ, v_g and v_p are identical, and the medium is *non-dispersive*. For $dv_p/d\lambda > 0$, $v_g < v_p$ which is the case of *normal* dispersion. In the opposite case, $dv_p/d\lambda < 0$, i.e., $v_g > v_p$, the dispersion is said to be *anomalous*.

As an example consider the dispersion relation for Alfvén waves in a plasma obtained in Section 4–7. In the dissipation-free case

$$\omega^2 = k^2 v_A^2$$

i.e., for this mode the plasma is non-dispersive. As an example of normal dispersion take electromagnetic wave propagation in a plasma; in Section 7–3

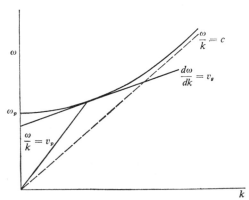

Fig. 7–2 Dispersion plot for electromagnetic wave propagating in an isotropic plasma

the dispersion relation for this mode in a collisionless, isotropic plasma is found to be

$$\omega^2 = \omega_p^2 + c^2 k^2$$

so that $d\omega/dk = kc^2/\omega$ and since $v_p = \omega/k > c$, $v_g < c$ (Fig. 7–2).

7–3 Waves in cold plasmas

Since the general theory of wave propagation in an anisotropic plasma is deferred until Section 7–6, we here only consider waves in the low-frequency region which propagate either in the direction of the applied magnetic field, $\mathbf{B_0}$, or orthogonal to it. The discussion will be based on the cold plasma equations, (3–91) to (3–96). Combining (3–95) and (3–96) gives

$$\nabla^2 \mathbf{E} - \nabla(\nabla.\mathbf{E}) = \frac{1}{c^2}\frac{\partial^2 \mathbf{E}}{\partial t^2} + \frac{4\pi}{c^2}\frac{\partial \mathbf{j}}{\partial t} \tag{7–16}$$

The momentum equation is

$$\rho\frac{D\mathbf{U}}{Dt} = q\mathbf{E} + \frac{1}{c}\mathbf{j} \times \mathbf{B} \tag{7–17}$$

and the generalized Ohm's law,

$$\frac{m^+m^-}{\rho e^2}\frac{\partial \mathbf{j}}{\partial t} = \mathbf{E} + \frac{\mathbf{U} \times \mathbf{B}}{c} - \frac{m^+}{\rho e c}\mathbf{j} \times \mathbf{B} \tag{7–18}$$

Only small-amplitude oscillations will be considered; that is, we shall examine the behaviour in time of initially small perturbations, dealing only with those which are stable. In general, if a plasma has a streaming motion there may exist a coupling between the streaming and a mode excited in the plasma which results in the growth in amplitude of this mode. Thus, for the purposes of this discussion, we shall assume zero streaming ($\mathbf{U_0} = 0$). Secondly, the basic non-linearity of (7–17) and (7–18) will be evaded by using linear perturbation analysis as we did in Section 4–5 when discussing the stability of the linear pinch. A plane wave form $\exp i(\mathbf{k}.\mathbf{r} - \omega t)$ will be taken for all perturbation quantities. The electrodynamic equation, (7–16), is without approximation, as is the momentum equation, (7–17) (within the $T = 0$ approximation). However, in (7–18) the extra approximation of neglecting terms of $O(m^-/m^+)$ has been made. Using a linear perturbation analysis, (7–17) and (7–18) become, since $\mathbf{E_0} = 0 = \mathbf{j_0}$,

$$\rho_0\frac{\partial \mathbf{U}}{\partial t} = \frac{1}{c}\mathbf{j} \times \mathbf{B_0} \tag{7–19}$$

(since $q\mathbf{E} = \mathbf{E}(\nabla.\mathbf{E})/4\pi$) and

$$\frac{m^+ m^-}{\rho_0 e^2} \frac{\partial \mathbf{j}}{\partial t} = \mathbf{E} + \frac{\mathbf{U} \times \mathbf{B}_0}{c} - \frac{m^+}{\rho_0 e c} \mathbf{j} \times \mathbf{B}_0 \qquad (7\text{--}20)$$

From (7–19) and (7–20), it follows immediately that

$$\frac{m^+ m^-}{\rho_0 e^2} \frac{\partial^2 \mathbf{j}}{\partial t^2} = \frac{\partial \mathbf{E}}{\partial t} + \frac{1}{c^2 \rho_0}(\mathbf{j} \times \mathbf{B}_0) \times \mathbf{B}_0 - \frac{m^+}{\rho_0 e c} \frac{\partial \mathbf{j}}{\partial t} \times \mathbf{B}_0$$

Multiplying across by $\rho_0 e^2/m^+ m^-$, substituting $(n^+ m^+ + n^- m^-)$ for ρ_0, replacing n^+ by n^- (since the difference is a first-order quantity $\nabla.\mathbf{E}/4\pi$), and dropping the m^-/m^+ term for consistency gives

$$\frac{\partial^2 \mathbf{j}}{\partial t^2} = \frac{n^- e^2}{m^-} \frac{\partial \mathbf{E}}{\partial t} + \frac{e^2}{m^+ m^- c^2}(\mathbf{j} \times \mathbf{B}_0) \times \mathbf{B}_0 - \frac{e}{m^- c} \frac{\partial \mathbf{j}}{\partial t} \times \mathbf{B}_0 \qquad (7\text{--}21)$$

Substituting the plane wave form for $\mathbf{E}$, $\mathbf{j}$ in (7–16) and (7–21) we find

$$k^2 \mathbf{E} - (\mathbf{k}.\mathbf{E})\mathbf{k} = \frac{\omega^2}{c^2}\mathbf{E} + i\frac{4\pi\omega}{c^2}\mathbf{j} \qquad (7\text{--}22)$$

$$\omega^2 \mathbf{j} = i\frac{\omega\omega_p^2}{4\pi}\mathbf{E} - \Omega_i\Omega_e[\mathbf{j} - (\mathbf{j}.\hat{\mathbf{b}}_0)\hat{\mathbf{b}}_0] + i\omega\Omega_e(\mathbf{j} \times \hat{\mathbf{b}}_0) \qquad (7\text{--}23)$$

in which $\hat{\mathbf{b}}_0 = \mathbf{B}_0/|\mathbf{B}_0|$.

The physics of wave propagation in a cold, anisotropic plasma in the linear approximation is contained in (7–22) and (7–23) with the exception of effects of $O(m^-/m^+)$. From these equations comes the *general dispersion relation*

$$F(\omega, \mathbf{k}) = 0 \qquad (7\text{--}24)$$

which relates the wave number $\mathbf{k}$ to the frequency ω. The appropriate dispersion relation is the key to any plasma mode since it contains all the information about the propagation of that mode. Before examining (7–24) in general, we consider some special cases of propagation either along or orthogonal to $\mathbf{B}_0$. With propagation along $\mathbf{B}_0$ (i.e., $\mathbf{k} \parallel \mathbf{B}_0$), the modes to be considered are the *Alfvén waves*—already encountered in Section 4–7—and *ion cyclotron waves*; for propagation across the field ($\mathbf{k} \perp \mathbf{B}_0$) the *magneto-acoustic* or *compressional Alfvén wave* will be studied. Note that when $\mathbf{B}_0 = 0$ the dispersion relation for transverse waves ($\mathbf{k} \perp \mathbf{E}$) may be written down at once from (7–22), (7–23):

$$\omega^2 = \omega_p^2 + k^2 c^2 \qquad (7\text{--}25)$$

which shows how the presence of the plasma alters the vacuum propagation of an electromagnetic wave for which $\omega^2 = k^2 c^2$. In particular, we see from

(7–25) that these transverse waves may propagate only at frequencies above ω_p; that is, the frequency range $0 < |\omega| < \omega_p$ is a stop-band for this mode.

For $\mathbf{k} \parallel \mathbf{E}$,[†] the mode is said to be *longitudinal* and in this case (7–22) and (7–23) give simply

$$\omega^2 = \omega_p^2 \qquad (7\text{–}26)$$

which are the *plasma oscillations* postulated by Penning in 1926 and studied by Langmuir, Tonks, and others thereafter. The oscillations occur only at the electron plasma frequency[‡] and do not propagate in the cold plasma limit.

7–3–1   Waves with k ∥ B₀: Alfvén waves

In Section 4–7 Alfvén waves appeared as a natural consequence of the tension $B_0^2/4\pi$ in magnetic lines of force. To re-derive the results of Section 4–7 within the framework of (7–22) and (7–23), write $\mathbf{E} = E_L \hat{\mathbf{k}} + E_T \hat{\mathbf{e}}$ where $\hat{\mathbf{e}}$ is a unit vector orthogonal to $\hat{\mathbf{k}}$. For the component E_L (7–22), (7–23) reduce to (7–26), so that the polarization of interest here is $\mathbf{E} = E_T \hat{\mathbf{e}}$. Dropping the subscript, (7–22) gives

$$(k^2 c^2 - \omega^2)\mathbf{E} = 4\pi i \omega \mathbf{j} \qquad (7\text{–}27)$$

and substituting (7–27) in (7–23), one finds

$$\left[\omega^2 - \frac{\omega^2 \omega_p^2}{\omega^2 - k^2 c^2} + \Omega_i \Omega_e \right] \mathbf{j} = i\omega \Omega_e \mathbf{j} \times \hat{\mathbf{b}}_0 \qquad (7\text{–}28)$$

The right-hand side of (7–28) expresses the essentially anisotropic nature of a plasma containing a magnetic field. Examining the size of the various terms in (7–28) we find, on dividing through by Ω_i^2,

$$\frac{\omega^2}{\Omega_i^2} : \frac{\omega^2}{\Omega_i^2}\left(\frac{\omega_p^2}{\omega^2 - k^2 c^2} \right) : \frac{\Omega_e}{\Omega_i} : \frac{\omega \Omega_e}{\Omega_i^2} \qquad (7\text{–}29)$$

Next suppose that we are only interested in very low frequencies, that is, $\omega \ll \Omega_i$. Then the fourth term in (7–29) is small compared with the third and the first is small compared with the fourth. Therefore, (7–28) becomes

$$\left[\frac{\omega^2 \omega_p^2}{\omega^2 - k^2 c^2} - \Omega_i \Omega_e \right] \mathbf{j} = 0 \qquad (7\text{–}30)$$

[†] For the general case in which $\mathbf{E}$ is neither parallel nor perpendicular to $\mathbf{k}$, recall that within the Coulomb gauge any $\mathbf{E}$ field can be written $\mathbf{E} = \mathbf{E}_L + \mathbf{E}_T$, $\mathbf{\nabla} . \mathbf{E}_T = 0$, $\mathbf{\nabla} \times \mathbf{E}_L = 0$ so that in such a situation there will exist two (uncoupled) plasma modes, one longitudinal, the other transverse.[23]

[‡] Had terms of $O(m^-/m^+)$ been retained, (7–26) would have read
$$\omega^2 = \omega_p^2(1 + m^-/m^+) = \omega_{pe}^2 + \omega_{pi}^2$$
using ω_{pe}, ω_{pi} to denote electron, ion plasma frequencies, respectively (cf. Section 7–6).

Equation (7–30) gives the dispersion relation for the mode under discussion in the frequency region $\omega \ll \Omega_i$,

i.e.,
$$\omega_p^2 - \Omega_i \Omega_e \left(1 - \frac{k^2 c^2}{\omega^2}\right) = 0$$

Using the definitions of ω_p, Ω_i, Ω_e

$$\frac{k^2 c^2}{\omega^2} = 1 + \frac{4\pi n^+ m^+ c^2}{B_0^2} \tag{7–31}$$

Recalling from Section 4–7 that $v_A^2 = \dfrac{B_0^2}{4\pi \rho_0} = \dfrac{B_0^2}{4\pi n^+ m^+}$, to $O\left(\dfrac{m^-}{m^+}\right)$ (7–31)

becomes simply

$$\frac{k^2 c^2}{\omega^2} = 1 + \frac{c^2}{v_A^2} \tag{7–32}$$

This is the dispersion relation for transverse waves propagating along $\mathbf{B}_0$ in a cold plasma with $\omega \ll \Omega_i$. In the limit of zero plasma density, (7–32) reduces to the familiar relation for electromagnetic waves propagating in a vacuum; if $c^2 \gg v_A^2$, then the result of Section 4–7 is retrieved,[†] that is,

$$\omega^2 = k^2 v_A^2 \tag{7–33}$$

The reason why (7–32) appears from this analysis rather than (7–33) stems from the fact that in getting the dispersion relation for Alfvén waves in Section 4–7, the ideal hydromagnetic equations were used in which the displacement current is dropped compared with $\mathbf{j}$ (cf. (3–66)). It is apparent from (7–32) that the Alfvén waves propagate without dispersion since $v_g = d\omega/dk = \omega/k$. Further, one may show that this mode is left circularly polarized (cf. Problem 7–3).

Another way of looking at Alfvén waves is to note from (7–32) that

$$\frac{\omega}{k} = \frac{c}{[1 + c^2/v_A^2]^{1/2}} = \frac{c}{\sqrt{\varepsilon_\perp}}$$

i.e., one may regard them as electromagnetic waves travelling in a medium of dielectric function

$$\varepsilon_\perp = 1 + \frac{c^2}{v_A^2} \tag{7–34}$$

[†] An interesting way of looking at this condition is to write it in the form
$$n^+ m^+ c^2 \gg B_0^2/4\pi$$
which requires that the 'rest energy' of the plasma be much greater than the magnetic energy.

This mode is known variously as the *shear Alfvén wave* or the *slow Alfvén wave*. Note, however, that this mode was labelled the intermediate wave in Section 6–1, where the observation was made that this purely transverse mode does not steepen to form a shock.

7–3–2 Waves with k ∥ B₀: ion cyclotron waves

Again consider a wave propagating along $\mathbf{B}_0$ with its $\mathbf{E}$ vector in the plane perpendicular to $\mathbf{B}_0$, and examine what happens close to the ion cyclotron frequency. From (7–29) it is clear that only the first of the four terms may now be dropped; thus (7–28) becomes

$$\left[\Omega_i\Omega_e - \frac{\omega^2\omega_p^2}{\omega^2 - k^2c^2}\right]\mathbf{j} = i\omega\Omega_e\,\mathbf{j} \times \hat{\mathbf{b}}_0 \tag{7–35}$$

so that the dispersion relation for this mode is

$$\left[\Omega_i\Omega_e - \frac{\omega^2\omega_p^2}{\omega^2 - k^2c^2}\right]^2 = \omega^2\Omega_e^2 \tag{7–36}$$

i.e.,
$$\left[1 - \frac{c^2/v_A^2}{(k^2c^2/\omega^2) - 1}\right]^2 = \frac{\omega^2}{\Omega_i^2} \tag{7–37}$$

Examining frequencies close to Ω_i requires, from (7–37), that $1 + c^2/v_A^2 \ll k^2c^2/\omega^2$; then to first order in small quantities,

$$\frac{k^2c^2}{\omega^2} = \frac{2c^2\Omega_i^2}{v_A^2(\Omega_i^2 - \omega^2)} = \frac{2\omega_{pi}^2}{\Omega_i^2 - \omega^2} \tag{7–38}$$

where ω_{pi} is the ion plasma frequency.

We see that as $\omega \to \Omega_i$ from below there is a marked change in the dispersion relation (cf. Fig. 7–3) for the mode propagating along $\mathbf{B}_0$, with $\mathbf{E}$ polarized in a plane perpendicular to $\mathbf{B}_0$, that is from (7–32) to (7–38). There is a *resonance* at $\omega = \Omega_i$ in the cold plasma limit. The waves propagating according to (7–38) in the region $\omega \lesssim \Omega_i$ are known as *ion cyclotron waves*, and were analysed by Stix.[54] The group velocity is given by

$$v_g = \frac{d\omega}{dk} = \frac{kv_A^2(\Omega_i^2 - \omega^2)^2}{2\omega\Omega_i^4} \tag{7–39}$$

and $v_g \to 0$ as the ion cyclotron resonance is approached.

To examine the polarization of this mode, one evaluates iE_x/E_y:

$$\frac{iE_x}{E_y} = \frac{ij_x}{j_y} = -\frac{\left[\Omega_i + \frac{\omega^2}{k^2c^2}\frac{\omega_p^2}{\Omega_e}\right]}{\omega}$$

$$= -\frac{\Omega_i}{\omega}\left[1 - \frac{\omega^2}{k^2c^2}\frac{c^2}{v_A^2}\right] \tag{7–40}$$

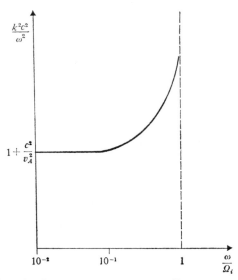

Fig. 7–3 Dispersion plot for transverse wave with $\mathbf{k} \parallel \mathbf{B}_0$ in the low-frequency region

Since the term in the bracket is positive and $\simeq 1$ we find that, in general, the electric field is elliptically polarized, degenerating to left circular polarization at ion cyclotron resonance. The electric field rotates in the same direction as the protons (cf. Fig. 2–1) and accelerates them.

One other point worth emphasizing is that these ion cyclotron waves are transverse, $\mathbf{k} \cdot \mathbf{E} = 0$; in other words they are not electrostatic. It so happens that when one admits finite temperature plasmas another mode appears in the vicinity of Ω_i but this wave is electrostatic in nature (cf. Section 8–2).

7-3-3 Waves with $k \perp B_0$: compressional Alfvén waves

For waves propagating orthogonally to $\mathbf{B}_0$, the electric field of the wave will, in general, be $\mathbf{E} = \mathbf{E}_\perp + \mathbf{E}_\parallel$ (Fig. 7–4). However, we can show (cf. Problem

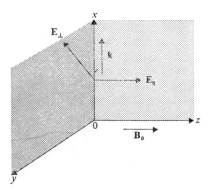

Fig. 7–4 Electric field components parallel and perpendicular to the magnetic field in a transverse wave propagating orthogonally to $\mathbf{B}_0$

7–7) that the component $E_\parallel$ is just a transverse wave obeying the dispersion relation (7–25); consequently, only $E = E_\perp$ need be considered. Choosing Ox along k and Oz along B_0, one has

$$\mathbf{E} = \mathbf{E}_\perp = E_x \hat{\mathbf{x}} + E_y \hat{\mathbf{y}} \qquad (7\text{–}41)$$

From (7–22),

$$\frac{\omega^2}{c^2} E_x + i\frac{4\pi\omega}{c^2} j_x = 0 \qquad (7\text{–}42a)$$

$$\frac{\omega^2}{c^2} E_y + i\frac{4\pi\omega}{c^2} j_y = k^2 E_y \qquad (7\text{–}42b)$$

and using these in (7–23) gives

$$[\omega^2 - \omega_p^2 + \Omega_i\Omega_e] j_x = i\omega\Omega_e j_y \qquad (7\text{–}43a)$$

$$\left[\omega^2 - \omega_p^2 \frac{\omega^2}{\omega^2 - k^2c^2} + \Omega_i\Omega_e\right] j_y = -i\omega\Omega_e j_x \qquad (7\text{–}43b)$$

From (7–43), the dispersion relation follows directly:

$$\begin{vmatrix} (\omega^2 - \omega_p^2 + \Omega_i\Omega_e) & -i\omega\Omega_e \\ -i\omega\Omega_e & -\left(\omega^2 - \omega_p^2 \dfrac{\omega^2}{\omega^2 - k^2c^2} + \Omega_i\Omega_e\right) \end{vmatrix} = 0 \quad (7\text{–}44a)$$

Using the ordering of (7–29), one has

$$(\Omega_i\Omega_e - \omega_p^2)\left(\omega_p^2 \frac{\omega^2}{\omega^2 - k^2c^2} - \Omega_i\Omega_e\right) = 0 \qquad (7\text{–}44b)$$

so that

$$\frac{k^2c^2}{\omega^2} = 1 + \frac{c^2}{v_A^2} \qquad (7\text{–}45)$$

which is the same dispersion relation as for shear Alfvén waves. However, there is an important physical distinction between the two modes. The shear Alfvén waves discussed in Section 7–3–1 are transverse in both the *electrodynamic* and the *mechanical* sense; the former since $\mathbf{k}.\mathbf{E} = 0$, the latter since $\mathbf{U} = \mathbf{j} \times \mathbf{B}_0/i\omega c\rho_0 \propto \mathbf{E} \times \mathbf{B}_0$ so that $\mathbf{k}.\mathbf{U} = 0$. The compressional Alfvén wave is transverse electrodynamically but now, since from (7–43b) and (7–44b) $|j_y| \gg |j_x|$, $\mathbf{k} \times \mathbf{U} \simeq 0$ so that mechanically this mode is approximately longitudinal. The fluid displacement is in the direction of propagation, as in a sound wave. For this reason it is sometimes known as the *magnetoacoustic wave* but a preferable label in the cold plasma limit is the *compressional Alfvén wave*† since it satisfies the same dispersion relation as the shear

† It is also known as the fast Alfvén wave (cf. Section 7–7).

Alfvén wave. Moreover, the magnetic field of the wave is parallel to $\mathbf{B}_0$ and since infinite conductivity (implied by the cold plasma limit) means that lines of force do not slip through the plasma, one may picture the motion as a sound wave but with magnetic pressure playing the role of gas kinetic pressure (cf. Fig. 7–5). Pressing this analogy further, one may substitute $B_0^2/8\pi$

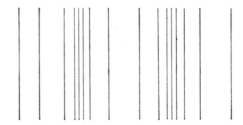

Fig. 7–5   Compression and rarefaction of magnetic field lines in a compressional Alfvén wave

for P in the expression for the sound speed $(\gamma P/\rho)^{1/2}$; then taking $\gamma = 2$ (cf. Problem 7–8) one finds that the 'sound speed' is given by v_A. Like a sound wave, the compressional Alfvén wave is non-dispersive. In the approximation of ignoring E_x compared with E_y (since $|j_x| \ll |j_y|$), it is clear that this wave is *linearly polarized*.

There is an important distinction between shear and compressional Alfvén waves as $\omega \rightarrow \Omega_i$. The character of the shear Alfvén mode alters drastically and degenerates to give an ion cyclotron wave. The compressional Alfvén wave, on the other hand, in general undergoes no radical change in character at this resonance. That this is so follows immediately from (7–44a), provided $\omega_p^2 \gg |\Omega_i\Omega_e|$ (i.e., $c^2 \gg v_A^2$); in consequence, compressional Alfvén waves propagate not only below $\omega = \Omega_i$ but also through the ion cyclotron resonance. For consistency with the condition $c^2 \gg v_A^2$, the compressional wave dispersion relation (7–45) should be written

$$\omega^2 = k^2 v_A^2 \tag{7–46}$$

For this mode there is no resonance until a higher frequency is reached. From (7–44a), the resonance frequencies for modes propagating across $\mathbf{B}_0$ are found simply by letting $kc/\omega \rightarrow \infty$, so that

$$\begin{vmatrix} \omega^2 - \omega_p^2 + \Omega_i\Omega_e & -i\omega\Omega_e \\ +i\omega\Omega_e & \omega^2 + \Omega_i\Omega_e \end{vmatrix} = 0$$

i.e., $2\omega^2 = (\omega_p^2 + \Omega_e^2 - 2\Omega_i\Omega_e)$
$$\pm [(\omega_p^2 + \Omega_e^2 - 2\Omega_i\Omega_e)^2 - 4(\Omega_i^2\Omega_e^2 - \Omega_i\Omega_e\omega_p^2)]^{1/2} \tag{7–47}$$

The minus sign gives a resonance at (cf. Problem 7–9)

$$\omega^2 = \frac{\Omega_i^2\Omega_e^2 - \omega_p^2\Omega_i\Omega_e}{\omega_p^2 + \Omega_e^2 - 2\Omega_i\Omega_e} \simeq -\Omega_i\Omega_e\frac{\omega_p^2}{(\omega_p^2 + \Omega_e^2)} \tag{7-48}$$

and the plus sign, one at

$$\omega^2 = \omega_p^2 + \Omega_e^2 \tag{7-49}$$

where terms of $O(m^-/m^+)$ have been dropped. Clearly, (7–48) may be written

$$\omega^2 = \alpha \mid \Omega_i\Omega_e \mid \qquad \frac{\omega_p^2}{\Omega^2} \lesssim \alpha \lesssim 1 \tag{7-50}$$

In the case of $\alpha \simeq 1$, the resonance occurs at $\omega^2 = \mid \Omega_i\Omega_e \mid$; the frequency $(\mid \Omega_i\Omega_e \mid)^{1/2}$ is known as the *lower hybrid frequency*. In general, the resonance occurs at some point between $\dfrac{\omega_p}{(\mid \Omega_i\Omega_e \mid)^{1/2}}\Omega_i$ and $(\mid \Omega_i\Omega_e \mid)^{1/2}$. The frequency determined by (7–49) is the *upper hybrid frequency*.

7–4 Experimental results for low-frequency waves
The theory of wave motion in cold plasmas thus far has centred on low-frequency modes propagating either along, or orthogonal to, $\mathbf{B}_0$. Before extending this development to other modes, it is worth asking what relevance this cold plasma model has for describing wave motion in laboratory plasmas which approximate to varying degrees the fiction of a cold, homogeneous, unbounded, fully ionized plasma. The cold plasma limitation means

$$\frac{8\pi P}{B^2} = \frac{\kappa T}{\tfrac{1}{2}m^+v_A^2} \ll 1$$

or, expressing T in electron volts, $2 . 10^{12}(T/v_A^2) \ll 1$. As long as this is satisfied, the consequences of ignoring P in the equations are not serious. A finite temperature also implies a finite collision frequency so that plasma waves are *attenuated* in laboratory experiments. Moreover, in practice the plane-wave dispersion relations often require modification to fit the experimental geometry; this, though often tedious, presents no problem in principle. Experiments carried out in conditions approximating a cold plasma have provided encouraging confirmation of theoretical predictions; illustrative examples of these are given in the following sub-sections.

7–4–1 *Shear Alfvén waves*
The first efforts to detect Alfvén waves were made with electrically conducting liquids by Lundquist (1951) and Lehnert (1954) using mercury and liquid sodium, respectively.[32, 33] Attenuation problems were so severe that conclusions were indefinite. Alfvén waves in a plasma were observed by Wilcox and his associates in a series of experiments from 1959 onwards.[55] In this work, the

plasma is contained in a copper cylinder with insulating end plates. The magnetic field is directed down the axis of the cylinder and hydromagnetic waves are induced by discharging a capacitor between the outside cylinder and a small co-axial electrode mounted in one of the end plates (Fig. 7–6). Discharging the capacitor causes a radial current to flow between the electrode and the cylinder. The interaction of this current with the axial magnetic field produces a torsional oscillation of the plasma. Since the high conductivity couples the plasma to the field effectively, these torsional waves propagate along B_0. The wave frequency is selected to conform to the restriction $\omega \ll \Omega_i$.

Small magnetic probes inserted into an access tube are used to detect the waves. The principal magnetic field component of the wave is azimuthal and

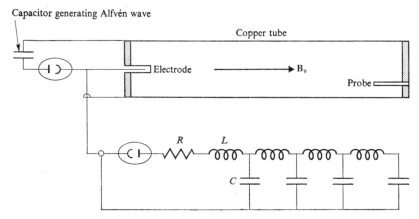

Fig. 7–6 Experiment designed to study Alfvén wave propagation (after Wilcox, Boley, and De Silva[55])

suitably oriented magnetic probes located at various axial positions can be used to determine the velocity and attenuation of the wave. The variation of phase velocity with axial magnetic field was measured in this way and yielded agreement with theory (Fig. 7–7). The proton density measured from Stark broadening was $5 \times 10^{15}/\text{cm}^3$, the plasma temperature being about 1 eV; to a good approximation the plasma is cold, since $\sqrt{(\kappa T/m^+)}$ $\sim 10^6$ cm/s whereas v_A is typically 10^7 cm/s.

The theory has been further verified in experiments by Jephcott and Stocker[56] using an argon plasma in which the ion density was $1 \pm 0.1 \times 10^{15}/$ cm³ and $T^- = 2.3 \pm 0.1$ eV from Langmuir probe measurements (cf. Appendix 3), while line broadening gave $T^+ = 1.7 \pm 0.2$ eV. The point in using argon was that it provided a more uniform plasma than the hydrogen plasma in the experiments of Wilcox. The wave was again excited by an oscillating discharge between an axial electrode and a ring electrode on the inner wall of

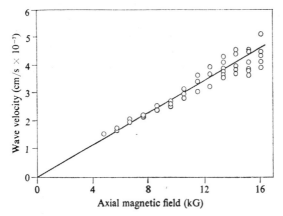

Fig. 7-7 Measurement of the dependence of the phase velocity of Alfvén waves on the axial magnetic field in a hydrogen plasma compared with theory (after Wilcox, Boley, and De Silva[55])

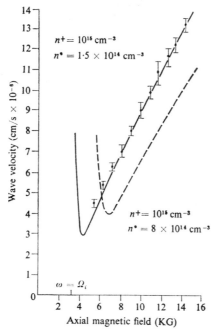

Fig. 7-8 Measurement of the dependence of the phase velocity of Alfvén waves on the magnetic field in an argon plasma. The curves were computed from a dispersion relation which took account of un-ionized gas in the plasma. The neutral argon density is then used as a parameter and it is seen that a value $n^0 = 1 \cdot 5 \times 10^{14}$ cm^{-3} gives good agreement with the measurements (after Jephcott and Stocker[56])

the discharge tube. Frequencies of 125 kc/s and 250 kc/s were used and the waves were essentially monochromatic. The tube was longer than that used by Wilcox and the wave was detected by magnetic probes at intervals of 40 cm along the tube. Figure 7–8 shows the comparison of theory and experiment. There is one fitted parameter, the neutral particle density, which is about one order of magnitude less than the ion density in this case.

7–4–2 Ion cyclotron waves

An early indication of the existence of ion cyclotron waves was found by Stix and Palladino.[57] Stix suggested that if the magnetic field were decreased some distance away from the wave exciter, then waves with $\omega < \Omega_i$ could be generated at the exciter and propagate into a region where the phase velocity vanishes. Strong absorption and plasma heating would be expected at the point $\omega = \Omega_i$ and this was later verified experimentally.

To illustrate the dispersion relation in the neighbourhood of the ion cyclotron frequency we consider an extension of the work on torsional Alfvén wave propagation by Wilcox and others[58] into a domain in which resonance

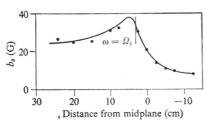

Fig. 7–9 Wave magnetic field measurements as a function of axial position at a constant radius in the plasma. The bar denotes the position at which ion cyclotron resonance occurs (after Boley et al.[58])

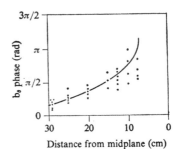

Fig. 7–10 Phase shift of wave magnetic field measured as a function of axial position; the curve is computed from theory (after Boley et al.[58])

effects are important. The torsional mode induced had $\omega \simeq \frac{1}{2}\Omega_i$ and propagated into a region of decreasing magnetic field until $\omega \simeq \Omega_i$ and the wave energy was dissipated. The wave field measurements showed that the fields increase as the wave propagates towards the region of lower magnetic field and then decrease to low values as ion cyclotron resonance is encountered (Fig. 7–9). The wave velocity may be determined from the axial dependence of the phase of the wave. As resonance is approached, the phase measurements become unreliable and the data depart from the solid curve (Fig. 7–10). Indications were that about 90 per cent of the wave energy is transferred from the wave to the plasma in the resonance region. Some results from a series of experiments at Princeton,[59] using a mirror machine designed to study plasma heating by the absorption of ion cyclotron waves, are shown in Figs. 7–11

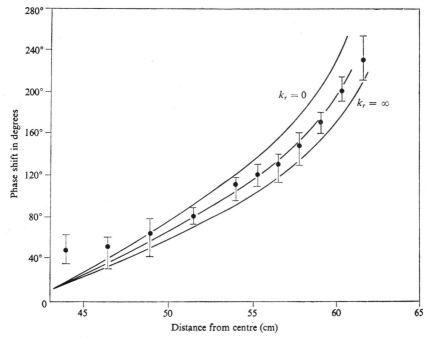

Fig. 7–11   Phase shift of ion cyclotron wave signal as a function of axial position (after Hooke *et al.*[59])

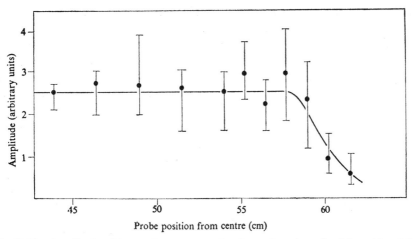

Fig. 7–12   Amplitude of ion cyclotron wave signal as a function of axial position (after Hooke *et al.*[59])

and 7–12. For waves in a cylindrical plasma of uniform density, the dispersion relation is

$$\frac{k_z^2 c^2}{\omega^2} = \frac{\omega_{pi}^2}{\Omega_i^2 - \omega^2} - \frac{k_r^2 c^2}{2\omega^2} + \left[\frac{\omega_{pi}^4}{(\Omega_i^2 - \omega^2)^2} + \left(\frac{k_r^2 c^2}{2\omega^2} \right)^2 \right]^{1/2}$$

k_z, k_r being longitudinal and radial wave numbers, respectively. This expression is valid provided $\omega_{pi}^2 \gg \omega^2$, Ω_i^2 and, in the limit $k_r \to 0$, reduces to (7–38), while for $k_r \to \infty$

$$\frac{k_z^2 c^2}{\omega^2} = \frac{\omega_{pi}^2}{\Omega_i^2 - \omega^2}$$

Thus the difference in k_z in these two limits is only $\sqrt{2}$.

Figure 7–11 shows the measured phase shift plotted against axial distance in the mirror machine; ion cyclotron resonance occurred at approximately 64 cm from the centre of the device. The curvature showing the decrease in wavelength is clear. Theoretical curves from the dispersion relation for $k_r = 0$, $k_r = \infty$, and for a value of k_r satisfying the plasma boundary conditions are plotted, showing good agreement with experimental results. Figure 7–12 represents the dramatic decrease in amplitude of the magnetic probe signal as ion cyclotron resonance is approached.

Other experiments on this device have investigated the heating of the plasma close to $\omega = \Omega_i$. Several hundred kilowatts of power have been transmitted to the plasma and ion temperatures of about 1 keV maintained in a local region for times of $\sim$100 μs.[60]

7–4–3 Compressional Alfvén waves

The first evidence for this mode was obtained in 1960 by Stix and Palladino.[61] More recent results due to Jephcott and Malein[62] and Swanson, Gould, and Hertel[63] are summarized here. The latter group studied the mode theoretically using a three-fluid cold plasma model. In their experimental work on a hydrogen plasma, resistivity, ion density, and ion–neutral collision frequency were determined from wave propagation measurements. This means that the experiment is not quite such a discriminating test of theory as one using *independently measured* parameters. Their experimentally determined dispersion relation is shown in Fig. 7–13.

Jephcott and Malein used a 22 cm diameter tube and worked with a much more uniform plasma. Variations of n^+, T^-, and j were less than 8 per cent from the axis out to $r = 10$ cm. All relevant plasma properties were measured *directly*. Ion–neutral collisions could be ignored since spectroscopic evidence suggested that neutrals were not present in significant amounts during wave propagation. The absolute value of n^+ on the axis was found using a helium–neon gas laser interferometer.

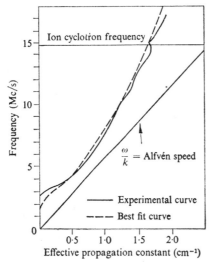

Fig. 7–13 Experimental investigation of compressional Alfvén wave propagation; plot of frequency against effective propagation constant (after Swanson, Gould, and Hertel[63])

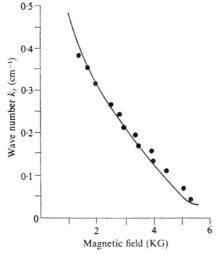

Fig. 7–14 Comparison of dispersion curve for compressional Alfvén wave with experiment; plot of wave number against magnetic field for $\omega = 2{\cdot}5 \times 10^5$ c/s (after Jephcott and Malein[62])

The elliptical polarization of the wave was demonstrated.† Dispersion curves (Fig. 7–14) agreed with theory over a frequency range $(1{\cdot}2{-}6{\cdot}0)\Omega_i$, $(\Omega_i \sim 10^6$ c/s).

† In Section 7–3–3 we saw that the polarization of the compressional Alfvén wave propagating *orthogonally* to $\mathbf{B}_0$ was linear. For *oblique* propagation, the polarization of this mode is elliptic.

7–5 Further wave concepts

Clearly, even the cold plasma model is a prolific source of waves, only a limited number of which have yet been considered. We could continue the discussion along the lines adopted so far, namely examining (7–22) and (7–23) and making approximations appropriate to the frequency range being considered. However, the multitude of modes and the need to get an overall picture of wave motion make it desirable to have an improvement on the simple classification according to frequency adopted so far. Such a scheme has been evolved by Allis, adapting ideas due to Clemmow and Mullaly (Section 7–7). Before discussing this classification, however, we must introduce some further wave concepts.

The notion of a *wave-normal surface* (or normal surface) appears in the study of crystal optics and is due to Fresnel, but it is equally relevant to wave propagation in any anisotropic medium, such as a plasma containing a magnetic field. This surface is the locus of the tip of the vector having the direc-

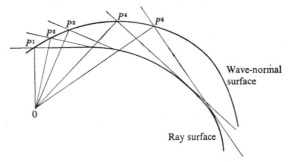

Fig. 7–15   Wave-normal and ray surfaces

tion of **k** and the magnitude of the wave velocity—that is, a polar plot of the phase velocity $\mathbf{v}_p = \omega \mathbf{k}/k^2$.

Consider an electromagnetic disturbance radiating outwards from a point. With this point as the origin, let plane wavefronts travel in all directions. After unit time, each of these wavefronts will have travelled a distance $OP^j = v_p^j$ in the direction of their propagation vectors (cf. Fig. 7–15). The set of points P^j form the wave-normal surface. One may also introduce the *ray surface*, defined as the envelope at unit time of wave planes which passed through O at $t = 0$ (Fig. 7–15).

The direction of energy flow may be shown to be along the group-velocity vector, a result which holds for anisotropic, dispersive, loss-free dielectrics. This direction is the *ray direction* and in anisotropic media is generally distinct from **k**, the propagation vector of the wave (cf. Fig. 7–16). In unit time, the plane wavefront travels a distance v_p in the direction **k** and v_r in the ray direction; $\mathbf{v}_r$ is the *ray velocity*. Note that though it has the same direction, it does *not* have the same magnitude as the group velocity.

One may further define an inverse wave normal surface as the polar plot

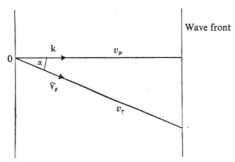

Fig. 7-16   Phase and ray velocity directions

of the vector $\mathbf{n} = \mathbf{k}c/\omega$, the equation of which is the dispersion relation to be derived in Section 7–6. This surface was defined by Sir W. R. Hamilton and is generally known as the *refractive index surface*. The importance of this sur-

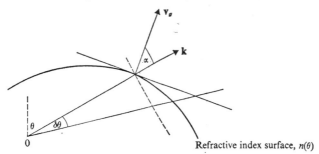

Fig. 7-17   Refractive index surface

face stems from the fact that the ray direction is normal to it (cf. Fig. 7–17 and Problem 7–15).

7–6   General theory of waves in cold plasmas

In Section 7–3, various low-frequency cold plasma modes were examined in the special situations in which the wave propagation vector $\mathbf{k}$ is oriented along, or orthogonal to, $\mathbf{B}_0$. The choice of these particular directions of propagation obscures the basically anisotropic nature of a cold plasma in a magnetic field. The previous discussion will now be generalized to admit propagation *oblique* to $\mathbf{B}_0$.

From electrodynamics, the relation between current and electric field in an anisotropic medium is the tensor relation[40]

$$i_k = \sum \sigma_{kl} E_l \equiv \sigma_{kl} E_l \qquad (7\text{--}51)$$

where σ_{kl} are the components of the *conductivity tensor*. From the generalized Ohm's law, one has

$$\frac{i\omega(\omega_{pe}^2 + \omega_{pi}^2)}{4\pi}\mathbf{E} = (\omega^2 + \Omega_i\Omega_e)\mathbf{j} - \Omega_i\Omega_e(\mathbf{j}\cdot\hat{\mathbf{b}}_0)\hat{\mathbf{b}}_0 - i\omega(\Omega_e + \Omega_i)\mathbf{j} \times \hat{\mathbf{b}}_0 \quad (7\text{--}52)$$

which differs from (7–23) in that terms of $O(m^-/m^+)$ have now been retained (cf. Problem 7–16). So doing restores a symmetry to the expressions derived in this section, the absence of which complicates the algebra and conceals the physical interpretation. The inverse tensor σ_{kl}^{-1}, that is, the resistivity tensor, follows immediately from (7–52)

$$\sigma_{kl}^{-1} = \frac{4\pi}{\omega(\omega_{pe}^2 + \omega_{pi}^2)} \begin{bmatrix} -i(\omega^2 + \Omega_i\Omega_e) & -\omega(\Omega_e + \Omega_i) & 0 \\ \omega(\Omega_e + \Omega_i) & -i(\omega^2 + \Omega_i\Omega_e) & 0 \\ 0 & 0 & -i\omega^2 \end{bmatrix} \quad (7\text{–}53)$$

The determinant of the matrix in (7–53) is

$$\Delta = i\omega^2[(\omega^2 + \Omega_i\Omega_e)^2 - \omega^2(\Omega_e + \Omega_i)^2]$$

$$= i\omega^2(\omega^2 - \Omega_e^2)(\omega^2 - \Omega_i^2) \quad (7\text{–}54)$$

so that inverting (7–53) to obtain the conductivity tensor gives

$$\sigma_{kl} = \frac{i\omega^3(\omega_{pe}^2 + \omega_{pi}^2)}{4\pi\Delta} \begin{bmatrix} i(\omega^2 + \Omega_i\Omega_e) & -\omega(\Omega_e + \Omega_i) & 0 \\ \omega(\Omega_e + \Omega_i) & i(\omega^2 + \Omega_i\Omega_e) & 0 \\ 0 & 0 & \dfrac{\Delta}{\omega^4} \end{bmatrix} \quad (7\text{–}55)$$

Then, from (7–22),

$$\mathbf{k} \times (\mathbf{k} \times \mathbf{E}) = -\frac{\omega^2}{c^2}\mathbf{E} - i\frac{4\pi\omega}{c^2}\mathbf{j}$$

and using (7–51)

$$[\mathbf{k} \times (\mathbf{k} \times \mathbf{E})]_i = -\frac{\omega^2}{c^2}\left[E_i + \frac{4\pi i}{\omega}\sigma_{il}E_l\right]$$

$$= -\frac{\omega^2}{c^2}\left[\delta_{il} + \frac{4\pi i}{\omega}\sigma_{il}\right]E_l \quad (7\text{–}56)$$

Defining ε_{il}, the *dielectric tensor*, by

$$\varepsilon_{il} = \delta_{il} + \frac{4\pi i}{\omega}\sigma_{il} \quad (7\text{–}57)$$

one may then write (7–56) in dyadic notation

$$\mathbf{k} \times (\mathbf{k} \times \mathbf{E}) + \frac{\omega^2}{c^2}\boldsymbol{\varepsilon}.\mathbf{E} = 0 \quad (7\text{–}58)$$

Writing out the dielectric tensor, one has

$$\varepsilon_{11} = \varepsilon_{22} = 1 - \frac{(\omega_{pe}^2 + \omega_{pi}^2)(\omega^2 + \Omega_i\Omega_e)}{(\omega^2 - \Omega_e^2)(\omega^2 - \Omega_i^2)}$$

$$= 1 - \frac{\omega_{pe}^2(\omega^2 - \Omega_i^2) + \omega_{pi}^2(\omega^2 - \Omega_e^2)}{(\omega^2 - \Omega_e^2)(\omega^2 - \Omega_i^2)}$$

using the identities $\omega_{pe}^2\Omega_i\Omega_e = -\omega_{pi}^2\Omega_e^2$, $\omega_{pi}^2\Omega_i\Omega_e = -\omega_{pe}^2\Omega_i^2$. Thus

$$\varepsilon_{11} = \varepsilon_{22} = S = 1 - \sum_{s=e,i} \frac{\omega_{ps}^2}{(\omega^2 - \Omega_s^2)} \tag{7-59}$$

Also,

$$\varepsilon_{12} = -\varepsilon_{21} = \frac{-i\omega(\omega_{pe}^2 + \omega_{pi}^2)(\Omega_e + \Omega_i)}{(\omega^2 - \Omega_e^2)(\omega^2 - \Omega_i^2)}$$

$$= \frac{-i\omega(\omega_{pe}^2\Omega_e + \omega_{pi}^2\Omega_i)}{(\omega^2 - \Omega_e^2)(\omega^2 - \Omega_i^2)}$$

Defining

$$R \equiv 1 - \sum_{s=e,i} \frac{\omega_{ps}^2}{\omega^2}\left(\frac{\omega}{\omega + \Omega_s}\right) \qquad L \equiv 1 - \sum_{s=e,i} \frac{\omega_{ps}^2}{\omega^2}\left(\frac{\omega}{\omega - \Omega_s}\right) \tag{7-60}$$

one finds that

$$\varepsilon_{11} = \varepsilon_{22} = \tfrac{1}{2}(R + L) \equiv S \tag{7-61}$$

and also

$$\varepsilon_{12} = -\varepsilon_{21} = -\tfrac{1}{2}i(R - L) \equiv -iD \tag{7-62}$$

Finally,

$$\varepsilon_{33} = 1 - \sum_{s=e,i} \frac{\omega_{ps}^2}{\omega^2} \equiv P \tag{7-63}$$

so that the (Hermitian) dielectric tensor is

$$\varepsilon_{kl} = \begin{bmatrix} S & -iD & 0 \\ iD & S & 0 \\ 0 & 0 & P \end{bmatrix} \tag{7-64}$$

One sees at once that ε_{33} is simply the dielectric function for the plasma in the direction of $\mathbf{B_0}$. To reveal the significance of the other elements, it is helpful to use a coordinate system such that the antisymmetric tensor $\boldsymbol{\epsilon}$ transforms into the diagonal tensor $\boldsymbol{\epsilon}' = \mathbf{U}.\boldsymbol{\epsilon}.\mathbf{U}^{-1}$ where $\mathbf{U}$ is the unitary matrix

$$\mathbf{U} = \frac{1}{\sqrt{2}}\begin{bmatrix} 1 & -i & 0 \\ 1 & i & 0 \\ 0 & 0 & \sqrt{2} \end{bmatrix} \tag{7-65}$$

and U^{-1} its inverse. Then (cf. Problem 7–17)

$$
\boldsymbol{\epsilon}' = \begin{bmatrix} R & 0 & 0 \\ 0 & L & 0 \\ 0 & 0 & P \end{bmatrix} \tag{7–66}
$$

The significance of $\boldsymbol{\epsilon}'$ is that the modes represented by R, L, P propagate independently of one another. It will become apparent that L, R stand for left and right circularly polarized waves; P is the (longitudinal) plasma mode.

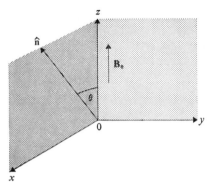

Fig. 7–18   Orientation of $\mathbf{n} = kc/\omega$ and $\mathbf{B_0}$

In (7–58), recalling that $kc/\omega = \mathbf{n}$ and choosing $\mathbf{n}$ in the plane Oxz (Fig. 7–18), so that $\mathbf{n} = (n \sin \theta, 0, n \cos \theta)$ one has

$$
\begin{bmatrix} S - n^2 \cos^2 \theta & -iD & n^2 \cos \theta \sin \theta \\ iD & S - n^2 & 0 \\ n^2 \cos \theta \sin \theta & 0 & P - n^2 \sin^2 \theta \end{bmatrix} \begin{bmatrix} E_x \\ E_y \\ E_z \end{bmatrix} = 0 \tag{7–67}
$$

From this, the *general dispersion relation* follows, containing all the information concerning wave propagation in an anisotropic, cold plasma, i.e.,

$$
An^4 - Bn^2 + C = 0 \tag{7–68}
$$

where
$$
\left. \begin{aligned}
A &= S \sin^2 \theta + P \cos^2 \theta \\
B &= RL \sin^2 \theta + PS(1 + \cos^2 \theta) \\
C &= PRL
\end{aligned} \right\} \tag{7–69}
$$

(cf. Problem 7–18).

Then,
$$
n^2 = \frac{B \pm [(RL - PS)^2 \sin^4 \theta + 4P^2D^2 \cos^2 \theta]^{1/2}}{2A} \tag{7–70}
$$

A more illuminating way of looking at the dispersion relation is to solve for θ as a function of the refractive index n giving

$$\tan^2 \theta = \frac{-P(n^2 - R)(n^2 - L)}{(Sn^2 - RL)(n^2 - P)} \tag{7-71}$$

From (7–71) the dispersion relations for propagation parallel, and perpendicular, to $\mathbf{B_0}$ are immediately apparent:

$$\theta = 0 \qquad\qquad P = 0 \qquad n^2 = R \qquad n^2 = L \tag{7-72a}$$

$$\theta = \pi/2 \qquad\qquad n^2 = RL/S \qquad n^2 = P \tag{7-72b}$$

From (7–67), the polarization of the oblique waves is determined by

$$\frac{iE_x}{E_y} = \frac{n^2 - S}{D} \tag{7-73}$$

from which, using (7–72a, b) we may retrieve the polarization of modes propagating parallel and perpendicular to $\mathbf{B_0}$.

At $\theta = 0$
$$\frac{iE_x}{E_y} = \frac{R - S}{D} \quad \text{or} \quad \frac{L - S}{D}$$
$$= +1 \quad \text{or} \quad -1 \tag{7-74}$$

giving *right circular polarization* $(+1, n^2 = R)$ and *left circular polarization* $(-1, n^2 = L)$.

At $\theta = \pi/2$
$$\frac{iE_x}{E_y} = \frac{(RL/S) - S}{D} \quad \text{or} \quad \frac{P - S}{D}$$
$$= -\frac{D}{S} \quad \text{or} \quad \frac{P - S}{D}$$

(since $S^2 - D^2 \equiv RL$) so that this mode is in general *elliptically polarized*.

7–6–1 Cut-offs and resonances

The terms *cut-off* and *resonance* have been used by Allis[64] to describe conditions under which $n^2 = 0$ (infinite phase velocity) and $n^2 \to \infty$ (zero phase velocity) respectively. In the cold plasma model, the cut-offs and resonances are sharp but in real plasmas (with finite electron temperature and density inhomogeneities) they are diffuse.

From (7–68) and (7–69), cut-off conditions are satisfied whenever one of the following holds:

$$P = 0 \qquad R = 0 \qquad L = 0 \tag{7-75}$$

provided A and B do not vanish simultaneously.

The resonance condition is satisfied by $A = 0$ so that, from (7–69),

$$\tan^2 \theta = -\frac{P}{S} \tag{7–76}$$

Thus at $\theta = 0$ resonance occurs for $S \to \infty$; when $P = 0$ all coefficients in (7–68) vanish and under these singular conditions whether or not one has resonance depends critically on how the double limit $\theta \to 0$, $P \to 0$ is taken. For $S \to \infty$, that is $(R + L) \to \infty$, it is obvious on inspection of (7–60) that $L \to \infty$ is the ion cyclotron resonance condition met in Section 7–3–2. Similarly, $R \to \infty$ corresponds to the *electron cyclotron resonance* condition.

At $\theta = \pi/2$, resonance occurs for $S = 0$ and contains the condition for the *upper* and *lower hybrid resonances* discussed in Section 7–3–3. The resonances at $\theta = 0$, $\pi/2$ have been called by Allis the *principal resonances*.

At a cut-off, the wavelength and phase velocity become infinite while at a resonance they vanish. In both cases, the group velocity vanishes. Physically the wave will, in general, be *reflected* at a cut-off and *absorbed* at a resonance.[54]

7–7 The CMA diagram

We can now discuss the Clemmow–Mullaly–Allis (CMA) diagram, which introduces some systematization into the classification of wave motion in anisotropic plasmas. The diagram is based on the notion of the wave-normal surface introduced in Section 7–5. From the dispersion relation in the form (7–71), we obtain the equation of the wave-normal surface. Defining

$$v_0^2 = \frac{c^2}{P} \qquad v_R^2 = \frac{c^2}{R} \qquad v_L^2 = \frac{c^2}{L} \tag{7–77}$$

one has for the phase velocity v

$$\frac{\sin^2 \theta}{(v_R^2 - v^2)(v_L^2 - v^2)} + \frac{2 \cos^2 \theta}{(v_0^2 - v^2)(v_R^2 + v_L^2 - 2v^2)} = 0$$

i.e.,

$$\frac{\cos^2 \theta}{v_0^2 - v^2} + \frac{1}{2} \frac{\sin^2 \theta}{v_R^2 - v^2} + \frac{1}{2} \frac{\sin^2 \theta}{v_L^2 - v^2} = 0 \tag{7–78}$$

Clearly, (7–78) is similar to the equation for the phase velocity surface in crystal optics:

$$\frac{\sin^2 \theta \cos^2 \phi}{v_1^2 - v^2} + \frac{\sin^2 \theta \sin^2 \phi}{v_2^2 - v^2} + \frac{\cos^2 \theta}{v_3^2 - v^2} = 0$$

v_1, v_2, v_3 being the principal velocities, θ, ϕ, the polar angles defining the direction of the wave vector. In an anisotropic plasma, unlike a crystal, there is cylindrical symmetry so that $\cos^2 \phi$ and $\sin^2 \phi$ are replaced by $1/2$. Also, with an anisotropic plasma, there can be arbitrarily large differences between v_0, v_R, v_L whereas the principal velocities in a crystal differ very little.

The essential point in the use of the wave-normal surface in classifying plasma modes is that, for considerable variations of plasma parameters, the basic *shapes* of the wave-normal surfaces show no change. Previously, the dispersion relations obtained were plotted in $(\omega, kc/\omega)$ diagrams. In the CMA diagram, a coordinate system is used involving plasma parameters (electron and ion density, magnitude of $\mathbf{B}_0$) as well as mode frequency. This space (known as *parameter space*) is divided into a number of volumes by surfaces representing the cut-offs and principal resonances discussed in Section 7–6, and within these volumes the topological type of the wave-normal surface does not change. Only on crossing one of these boundaries can a change occur.

For a two-component (electron–proton) plasma, the diagram is particularly simple as the parameter space is two-dimensional, divided by the cut-off and principal resonance lines. The abscissa represents $(\omega_{pe}^2 + \omega_{pi}^2)/\omega^2$, so that particle density increases to the right or, for fixed densities, frequency decreases. The ordinate represents $|\Omega_i\Omega_e|/\omega^2$ and the magnetic field, $\mathbf{B}_0$, increases upwards or, for fixed $\mathbf{B}_0$, frequency decreases. Thus for fixed

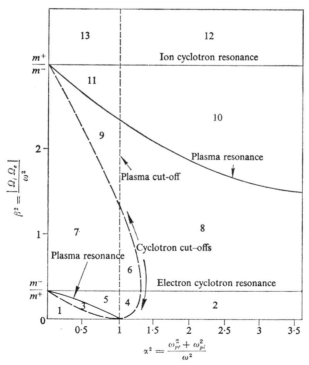

Fig. 7–19 Subdivision of parameter space (α^2, β^2) by principal resonance and cut-off curves into thirteen regions, in each of which the wave-normal surfaces are topologically distinct

plasma parameters, lowering the frequency corresponds to moving outward radially from the origin. To simplify the representation, the figures are drawn for an unrealistically low m^+/m^- ratio.†

Figure 7–19 shows parameter space divided into areas by the cut-off and principal resonance lines, namely

$$\textit{cut-off:} \qquad\qquad P = 0; \quad R = 0; \quad L = 0$$

$$\textit{principal resonance:} \; R = \infty; \quad L = \infty; \quad S = 0 \; (\theta = \pi/2)$$

There are 13 areas in all. The numbering in Fig. 7–19 follows the system used by Allis, namely alternatively left and right of the plasma cut-off and increasing with magnetic field. Regions whose characteristic number differs by 6 have similar properties. Note the degeneracy appearing at $\alpha^2 = (\omega_{pe}^2 + \omega_{pi}^2)/\omega^2 = 1, \beta^2 \equiv |\Omega_i \Omega_e|/\omega^2 = 0$; at this singular point the frequency $(\omega_{pe}^2 + \omega_{pi}^2)^{1/2}$ is a cut-off frequency, even though a plasma resonance line $(S = 0)$ passes through this point. The degeneracy is resolved by applying a magnetic field. Further, note that the two branches of the plasma resonance correspond to the upper hybrid resonance and that given by (7–48) which, in the limit of large α^2, occurs at the lower hybrid frequency and, for small α^2, at the ion cyclotron frequency.

Having finished dissecting parameter space, let us next turn to the wave-normal surfaces characterizing mode propagation within each of these compartments. From (7–70), it is clear that for a cold plasma of uniform density there are in general two values of n^2 and a wave-normal surface associated with each. Moreover, these surfaces do not intersect although in certain *singular* cases in which the discriminant of (7–70) vanishes, that is,

$$(RL - PS)^2 \sin^4 \theta + 4P^2 D^2 \cos^2 \theta = 0$$

they are coincident (cf. Problem 7–22). One wave-normal surface is contained within the other and the two modes are labelled fast (F) and slow (S). This labelling holds over the entire wave-normal surface.

Other labels which may be applied to the wave-normal surface derive from the dispersion relations for propagation at $\theta = 0, \pi/2$. For $\theta = 0$, we saw in Section 7–6 (Eqn (7–74)) that waves propagating along $\mathbf{B}_0$ are either right (R) or left (L) circularly polarized. For propagation at $\theta = \pi/2$, from (7–72b), either $n^2 = P$ or $n^2 = RL/S$. The mode satisfying the dispersion relation which does not involve $\mathbf{B}_0$ is labelled *ordinary* (O); that satisfying $n^2 = RL/S$, namely a $\mathbf{B}_0$-dependent dispersion relation, is labelled *extraordinary* (X).‡

---

† Unrealistic, that is, for a gaseous plasma though not necessarily so for a semiconductor plasma, in which this corresponds to the ratio of hole to electron mass.

‡ A note of caution must be sounded over the terms ordinary and extraordinary. The sense in which they are used here comes from ionospheric physics (cf. K. G. Budden, *Lectures on Magneto-ionic Theory*, p. 31. Gordon & Breach, New York and London, 1964). Unfortunately, in hydromagnetics the terms ordinary and extraordinary are taken to have the meaning ascribed to them in crystal optics. One wave has $n^2 = S$, *regardless* of the direction of propagation (the crystal behaves isotropically); this wave obeys Snell's law and is labelled ordinary. For the other mode, $n^2 = n^2(S, P, \theta)$; this is the extraordinary wave.

Polar plots of the phase velocity may be drawn in Fig. 7–19. The phase velocity surface is generated by rotating each polar plot about a vertical axis (parallel to $\mathbf{B_0}$) passing through the point $v = 0$. Inside each of the areas 1–13 of Fig. 7–19 the topological type of each polar diagram does not change.[54]

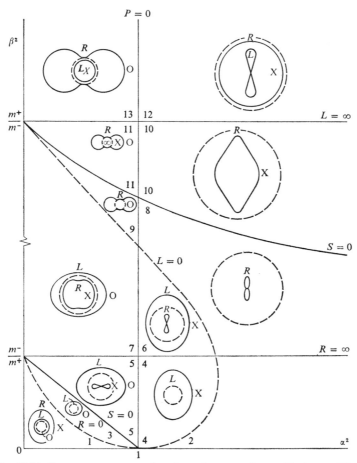

Fig. 7–20 CMA diagram; wave-normal surfaces for waves in a cold plasma. The figures are not plotted to scale but the dotted circle represents the velocity of light in each region (after Allis, Buchsbaum, and Bers[64])

The wave-normal surfaces are sketched in Fig. 7–20. Let us examine briefly the low-frequency region of the CMA diagram. Since lowering the frequency corresponds to moving outwards radially from the origin, the region of interest is 12, in which $\omega < \Omega_i$, $\omega^2 < \omega_{pe}^2 + \omega_{pi}^2$, that is, the region of Alfvén wave propagation. In Section 7–3 we considered Alfvén wave propagation at $\theta = 0$, $\pi/2$; for the general case (cf. Problem 7–23) (7–68) factorizes to give

$$n^2 \equiv \frac{k^2 c^2}{\omega^2} = 1 + \frac{c^2}{v_A^2} \qquad (7\text{–}79)\dagger$$

and, provided θ is not too close to $\pi/2$,

$$n^2 \cos^2 \theta \equiv \frac{k^2 c^2 \cos^2 \theta}{\omega^2} = 1 + \frac{c^2}{v_A^2} \qquad (7\text{–}80)$$

In other words, (7–79) represents the fast mode, (7–80) the slow mode; the former has an isotropic phase velocity whereas the latter cannot propagate at angles greater than some critical angle. Hence the polar diagram for the F mode will be circular while that for the S mode will be the lemniscate shown in Fig. 7–21. The F mode is labelled X since its dispersion relation involves B_0; since the S mode does not propagate at $\theta = \pi/2$, the question of an O or X label does not arise. Both propagate at $\theta = 0$ so that polarization labels may be attached: the fast mode is RCP, and the slow mode, LCP.

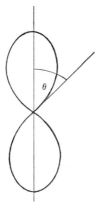

Fig. 7–21 Wave-normal diagram for helicons (whistlers)

In Section 7–3–2, we examined the change in behaviour of the slow mode near to ion cyclotron resonance while in Section 7–3–3 we saw that the fast mode did not change radically in character on passing through this resonance. This is reflected in the CMA diagram, in which FRX persists in region 10 on crossing the ion cyclotron resonance, whereas SL disappears. Note, however, that the FRX wave-normal surface undergoes a reshaping.

7–7–1 *Appleton–Hartree dispersion relation*

Before examining some other regions in the diagram, we obtain a dispersion relation valid for frequencies high enough to let us neglect ion motions. Rearranging (7–68) by adding An^2 to each side,

$$n^2 = \frac{An^2 - C}{An^2 + A - B} \qquad (7\text{–}81)$$

and replacing n^2 by (7–70) on the right-hand side (cf. Problem 7–24),

$$n^2 = 1 - \frac{2(A - B + C)}{2A - B \pm (B^2 - 4AC)^{1/2}} \qquad (7\text{–}82)$$

---

† Observe that (7–79) was obtained in Section 7–3–3 in the special case of propagation orthogonal to the magnetic field; however, it is now established independent of direction.

Now, provided $\Omega_i/\omega \equiv \beta_+ \ll 1$, the ion motions may be neglected and (7–82) simplifies to give (cf. Problem 7–25)

$$n^2 = 1 - \frac{2\alpha^2(1 - \alpha^2)}{2(1 - \alpha^2) - \beta_-^2 \sin^2 \theta \pm \Gamma} \tag{7–83}$$

with
$$\Gamma = [\beta_-^4 \sin^4 \theta + 4\beta_-^2(1 - \alpha^2)^2 \cos^2 \theta]^{1/2} \tag{7–84}$$

and $\beta_- \equiv |\Omega_e|/\omega$, $\alpha^2 \simeq \omega_{pe}^2/\omega^2$. This limiting form of the dispersion relation is a special case (found by neglecting inter-particle collisions) of the *Appleton–Hartree formula* first derived in connection with radio-wave propagation in the ionosphere.[65]

The Appleton–Hartree result (7–83) will apply unreservedly in regions (1)–(5) of the CMA diagram and equally will *not* be valid for (9)–(13).†
For regions (6)–(8), it will generally be valid.

An interesting method of factorizing (7–83) has been devised by Booker[65] based on approximating the discriminant Γ in (7–84) when one or other of its two components may be neglected: i.e.,

$$\beta_-^4 \sin^4 \theta \ll 4\beta_-^2(1 - \alpha^2)^2 \cos^2 \theta, \quad Q_\| \tag{7–85}$$

$$\beta_-^4 \sin^4 \theta \gg 4\beta_-^2(1 - \alpha^2)^2 \cos^2 \theta, \quad Q_\perp \tag{7–86}$$

where $Q_\|$ and $Q_\perp$ mean quasi-parallel and quasi-perpendicular to $\mathbf{B}_0$, respectively.‡ Then, (7–83) gives

$$Q_\| : \quad n_{L,R}^2 \simeq 1 - \frac{2\alpha^2(1 - \alpha^2)}{2(1 - \alpha^2) - \beta_-^2 \sin^2 \theta \pm 2\beta_-(1 - \alpha^2)\cos \theta} \tag{7–87}$$

$$Q_\perp : \quad n_O^2 \simeq \frac{1 - \alpha^2}{1 - \alpha^2 \cos^2 \theta} \tag{7–88}$$

$$n_X^2 \simeq \frac{(1 - \alpha^2)^2 - \beta_-^2 \sin^2 \theta}{(1 - \alpha^2) - \beta_-^2 \sin^2 \theta} \tag{7–89}$$

where subscripts O and X denote ordinary and extraordinary. The dispersion relation for the $Q_\|$ components simplifies further if, in addition, one has

$$\beta_-^2 \sin^2 \theta \ll |2(1 - \alpha^2)|$$

when
$$n_{L,R}^2 \simeq 1 - \frac{\alpha^2}{1 \pm \beta_- \cos \theta} \tag{7–90}$$

---

† One should recall that for a real plasma, region (9) is an exceedingly small area although for the artificial m^-/m^+ ratio chosen, this does not show up in the diagram. The same is true of region (11).

‡ Booker used quasi-longitudinal and quasi-transverse, but we have used longitudinal and transverse in a different sense. Allis, Buchsbaum, and Bers[64] use quasi-circular (QC) and quasi-plane (QP), which refer to the polarization.

We may now look briefly at some of the remaining regions of the CMA diagram.

Region 8: *Whistlers and helicons.* In the transition from region (12) to region (10) FRX, while preserved, was deformed from a spherical surface to one approximating an ellipsoid and SL disappeared entirely on crossing the ion cyclotron resonance. On crossing from (10) to (8), a more drastic reshaping of this mode takes place. If waves propagating in directions close to $\mathbf{B}_0$ have frequencies in the range $\Omega_i \ll \omega < \Omega_e$, and if in addition $\omega \, | \, \Omega_e \, | < \omega_{pe}^2$, one has from (7–90)

$$n_R^2 \simeq \frac{\alpha^2}{\beta_- \cos \theta} \qquad (7\text{--}91)$$

while $n_L^2 < 0$. This is the appropriate dispersion relation for most of region (8). To demonstrate the change in shape of the wave-normal surface from that in region (10), consider (7–89) under the conditions for which (7–91) is valid; then,

$$n_X^2 \simeq -\frac{\alpha^4 - \beta_-^2 \, \sin^2 \theta}{\alpha^2 + \beta_-^2 \, \sin^2 \theta} < 0$$

Thus the polar diagram for this mode is now a lemniscate as in Fig. 7–21. The waves obeying (7–91) are known as *whistlers* or *helicons*; the former derives from ionospheric physics, being the name given by Barkhausen in 1919 to electromagnetic waves in the audio-frequency range which are produced by lightning flashes and for which the geomagnetic field acts as a wave guide. These waves may undergo reflection at the geomagnetic poles; in any case their long path gives rise to frequency dispersion which produces the characteristic whistle with steadily falling pitch heard at the receiver. The other name for these waves, *helicons*, comes from solid state physics and is due to the fact that the tip of the wave $\mathbf{B}$ vector traces a helix. Some experimental work on helicons is discussed in Section 7–8.

Observe that if one continues downwards in the CMA diagram, the next region, (2), is empty. This represents a stop-band for wave propagation.

Regions (1), (3), (5), (7): *electromagnetic waves.* Next, consider electromagnetic wave propagation with $\omega > \omega_{pe}$; for regions (1), (3), (5), $\omega > | \, \Omega_e \, |$, while for region (7), $\omega_{pe} < \omega < | \, \Omega_e \, |$. For propagation at $\theta = 0$, from (7–87)

$$n_{L,R}^2 \simeq 1 - \frac{\omega_{pe}^2}{\omega(\omega \pm | \, \Omega_e \, |)} \qquad (7\text{--}92)$$

while for propagation at $\theta = \pi/2$,

$$n_O^2 = 1 - \frac{\omega_{pe}^2}{\omega^2} \qquad n_X^2 = \frac{(\omega^2 - \omega_{pe}^2)^2 - \omega^2 \Omega_e^2}{\omega^2(\omega^2 - \omega_{pe}^2 - \Omega_e^2)} \qquad (7\text{--}93)$$

These dispersion relations are displayed in Fig. 7–22; the shaded areas represent stop-bands for propagation. Beginning with the highest frequencies, namely those above the line $R = 0$ for which

$$\omega = \omega_2 = -\frac{\Omega_e}{2} + \left(\frac{\Omega_e^2}{4} + \omega_{pe}^2\right)^{1/2}$$

it is clear from Fig. 7–22 that both an FRX and an SLO mode can propagate (region (1) in the CMA diagram).

Crossing the cyclotron cut-off $R = 0$, the FRX mode meets a stop-band but the other wave is unaffected (region (3)). Continuing downwards in frequency, one passes out of the stop-band at the upper-hybrid resonance ($S = 0$), i.e.,

$$\omega = \omega_{uh} = (\Omega_e^2 + \omega_{pe}^2)^{1/2}$$

Thus once again propagation occurs in two modes for $\theta = \pi/2$ but only the LO mode propagates at $\theta = 0$; moreover, this is now labelled FLO and the other SX (region (5)). The next change in dispersion characteristics appears at the electron cyclotron resonance $R = \infty$ at which a right circularly polarized wave reappears so that in region (7) of the CMA diagram one finds both FLO and SRX modes.

Region (6): *electromagnetic waves.* Moving still further down the frequency axis in Fig. 7–22 the plasma cut-off $P = 0$ is met, at which point important changes occur. (In the CMA diagram we have now moved sideways into region (6).) The O wave now disappears as it must, since no propagation is possible below the plasma frequency in an isotropic plasma. Further, the phase velocity of the extraordinary wave now increases above the velocity of light (represented by the dotted lines in Fig. 7–22) so that in region (6) one now has an FLX mode with an SR mode.

Finally, continuing downwards one meets the cyclotron cut-off, $L = 0$, which removes the FLX mode. Thus in region (8) one finds only the right circularly polarized wave propagating in directions close to $\mathbf{B}_0$, namely the whistler mode, discussed earlier.

One final point worth emphasizing is the similarity between wave-normal surfaces in regions (3), (4), (5), (6), and (7) in the high-frequency part of the CMA diagram and regions (9), (10), (11), (12), and (13) in the low-frequency part (those whose region numbers differ by 6). The main difference is that

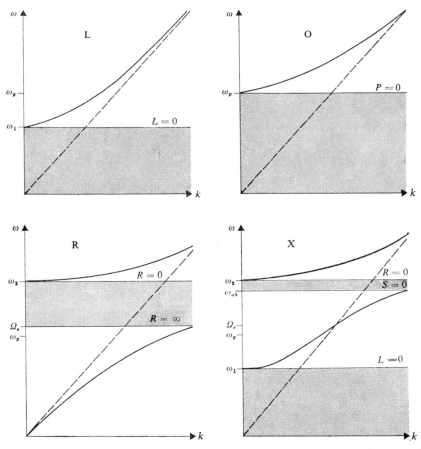

Fig. 7–22   Dispersion plots of waves propagating along and orthogonally to the magnetic field in a cold plasma. The shaded parts of the diagram denote stop-bands for wave propagation (after Bekefi[88])

the R, L labels are interchanged. This similarity is nothing but the mirroring of electron behaviour at high frequencies by the ions at low frequencies, the switching of R and L being due to the fact that electrons and ions gyrate in different senses round the magnetic field.

7–8   Further experimental work

Previously, we discussed some experimental work in the low-frequency regime; in this section we examine experiments on helicon waves. While this mode was first observed in the form of ionospheric whistlers, comparison with theory is difficult because we lack precise information on ionospheric electron density. Laboratory experiments on helicon wave propagation have now been carried out.

In work by Lehane and Thonemann,[66] a xenon plasma with particle densities between $3.10^{12}/cm^3$ and $2.10^{13}/cm^3$ was produced in magnetic fields between 70 to 500 G. Helicon waves were generated by discharging a capacitor through either a double loop around the discharge tube (producing the $m = 0$ mode) or a single-turn loop inside the tube having its axis perpendicular to the tube axis—that is, to $\mathbf{B}_0$—producing the $m = 1$ mode. The amplitude and phase of the signal transmitted down the discharge tube were measured using Langmuir and magnetic probes.

The experimentally determined dispersion curves are shown in Fig. 7–23 together with the theoretical dispersion relation for plane waves in an infinite medium; the measured wavelengths fall below these theoretical values. However, when these results are fitted to curves drawn from a modified

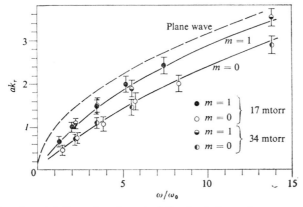

Fig. 7–23 Experimental dispersion characteristics for helicons in a xenon plasma compared with theoretical plots; a is the tube radius and ω_0 the frequency of a helicon of wave number a^{-1} in an unbounded plasma (after Lehane and Thonemann[66])

theory (applying to a uniform, cylindrical plasma bounded by insulating walls), using the plasma density which gives the best simultaneous fit for both modes, good agreement is found.

It is interesting to note that helicons were previously studied in solid-state plasmas, being observed by Bowers[67] at audio-frequencies ($\sim$30 c/s) in sodium and by Libchaber and Veilex[68] in indium antimonide at 10 Gc/s. In Bowers' experiment, a slab of sodium was cooled to liquid helium temperatures, so increasing the conductivity of the metal to $\sim$7,000 times that at laboratory temperature. To an excellent approximation this is a cold plasma; this feature, together with uniformity in density, is the great advantage of a solid-state plasma over a gaseous one.

Unlike the experiment of Lehane and Thonemann in which travelling waves were used, the helicons produced in the sodium plasma were standing waves. The set-up is shown in Fig. 7–24; a slab of sodium is set in a strong magnetic field $\mathbf{B}_0$ and two coils are placed as shown, one of which excites the helicons, the other acting as a detector. The voltage in the pickup coil is monitored as the driving frequency is varied and found to peak only at the characteristic frequencies for which the system resonates (i.e., when the length of the slab in the direction of the magnetic field is an integral multiple of half the helicon

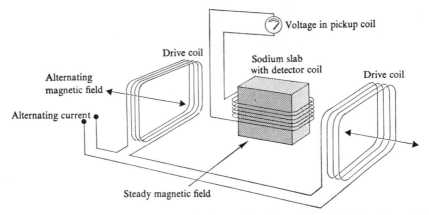

Fig. 7-24   Experiment to study standing helicon waves in a sodium slab (after Bowers[67])

wavelength). The frequency at which the peaks occur is a function of magnetic field strength, the dimensions of the slab of sodium and the carrier density. The first two quantities are known precisely and one can compute the number of carriers by assuming there is one conduction electron per atom of sodium.

From (7–91), one has

$$\omega = \frac{cB_0 k^2 \cos\theta}{4\pi n_0 e} \tag{7-94}$$

where n_0 ($\sim 10^{22}/cm^3$) is the density of conduction electrons in the metal. The wave vector in the sodium slab is

$$\mathbf{k}_{lmn} = \pi\left(\frac{l\hat{\mathbf{x}}}{X} + \frac{m\hat{\mathbf{y}}}{Y} + \frac{n\hat{\mathbf{z}}}{Z}\right)$$

where l, m, n are integers. Thus, in (7–94), since $\cos\theta = \hat{\mathbf{k}}.\hat{\mathbf{b}}_0$

$$\omega_{lmn} = \frac{\pi cB_0}{4n_0 e}\cdot\frac{n^2}{Z^2}\left(1 + \frac{l^2 Z^2}{n^2 X^2} + \frac{m^2 Z^2}{n^2 Y^2}\right)^{1/2} \tag{7-95}$$

i.e.,

$$\omega_{00n} = \frac{\pi cB_0}{4n_0 e Z^2}n^2 \tag{7-96}$$

In other words, the ratio of the resonant frequencies is as the square of the harmonic number. For technical reasons,† only odd-integer modes are detected, so that the first four should appear at frequencies

$$\omega_{001} : \omega_{003} : \omega_{005} : \omega_{007} :: 1 : 9 : 25 : 49$$

One can check from the experimental curve in Fig. 7–25 that there is excellent agreement with theory.

Bowers' experiment concerned *standing* waves. In 1964 Harding and Thonemann[69] reported an experiment in which helicon waves were propagated parallel to the axis of a long cylinder of indium at liquid helium temperatures, in an attempt to check the theoretical dispersion relation by measuring the

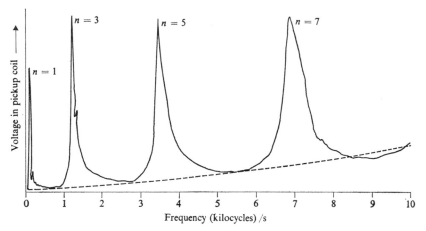

Fig. 7–25 Experimental results for standing helicon waves in a sodium slab (after Bowers[67])

wavelength directly (as opposed to inferring it from the dimensions of a sample as described above).

The indium rod was 9·4 cm long and 0·8 cm in diameter and an arrangement of coils at one end allowed various helicon modes to be generated. Harding and Thonemann studied the $m = 0$, $m = \pm 1$ modes, covering a frequency range from 1 c/s to 100 c/s, the corresponding wavelength range being 4·5 cm to 0·4 cm. Their results for $m = 0$, $m = 1$ helicons are shown in Figs. 7–26 and 7–27, together with the theoretical dispersion relation. Agreement with the cold plasma theory is excellent, using a carrier density estimated on the basis of one hole per atom of indium. This assumption gives a theoretical value for the *Hall coefficient*, $R = (n_0 e c)^{-1}$, that is very close to that derived from the measurements.

In fact, the helicon wave velocity offers a direct means of *measuring* the

† The sodium slab is wholly contained within the pickup coil, so that for even modes the electric fields cancel when summed over the total coil area.

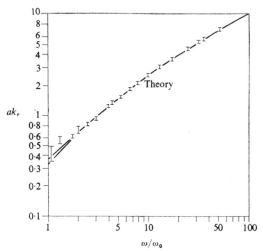

Fig. 7–26 Experimental dispersion characteristics for $m = 0$ helicons in an indium plasma compared with theory (after Harding and Thonemann[69])

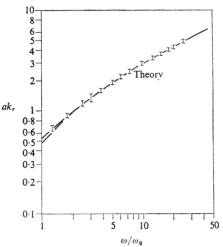

Fig. 7–27 Experimental dispersion characteristics for $m = 1$ helicons in an indium plasma compared with theory (after Harding and Thonemann[69])

carrier density of the metal or semiconductor. A further interesting application of helicons is in studying the Fermi surface in a solid.

Summary

1. *Wave polarization*

$$E_y(x, t) = E_{y0} \cos(kx - \omega t + \alpha)$$
$$E_z(x, t) = \pm E_{z0} \sin(kx - \omega t + \alpha)$$

i.e., $\qquad \dfrac{E_y^2}{E_{y0}^2} + \dfrac{E_z^2}{E_{z0}^2} = 1$: elliptically polarized wave

$$E_{y0} = 0 \text{ (or } E_{z0} = 0) : \text{linearly (plane) polarized wave}$$

$$E_{y0} = E_{z0} : \text{circularly polarized wave}$$

To observer looking at approaching wave front the $+$ sign gives an LCP wave and the $-$ sign an RCP wave.

2. *Wave velocity*

Phase velocity $v_p = \omega/k \qquad$ group velocity $v_g = \dfrac{d\omega}{dk}$

$$v_g = v_p + k\frac{dv_p}{dk} = v_p - \lambda\frac{dv_p}{d\lambda}$$

$$\frac{dv_p}{d\lambda}\begin{cases} > 0 : \text{normal dispersion} \\ = 0 : \text{no dispersion} \\ < 0 : \text{anomalous dispersion} \end{cases}$$

3. *Cold plasma modes: isotropic plasma*
   (i) Electron plasma oscillation: $\omega^2 = \omega_{pe}^2$
   (ii) Electromagnetic wave: $\omega^2 = \omega_{pe}^2 + k^2 c^2$

4. *Cold plasma modes: anisotropic plasma,* $\mathbf{k} \parallel \mathbf{B}_0$
   (i) Alfvén wave (LCP): $\dfrac{k^2 c^2}{\omega^2} = 1 + \dfrac{c^2}{v_A^2} \qquad \omega \ll \Omega_i$

   (ii) Ion cyclotron wave (LCP): $\dfrac{k^2 c^2}{\omega^2} \simeq \dfrac{2c^2 \Omega_i^2}{v_A^2(\Omega_i^2 - \omega^2)} \qquad \omega \lesssim \Omega_i$

   (iii) Whistler (helicon) mode (RCP):

   $$\frac{k^2 c^2}{\omega^2} \simeq \frac{\omega_{pe}^2}{\omega |\Omega_e|} \qquad \Omega_i \ll \omega < \Omega_e \qquad \omega |\Omega_e| < \omega_{pe}^2$$

   (iv) Electron plasma oscillation: $\omega^2 = \omega_{pe}^2$
   (v) Electron cyclotron wave (RCP):

   $$\frac{k^2 c^2}{\omega^2} \simeq 1 - \frac{\omega_{pe}^2}{\omega(\omega - |\Omega_e|)} \qquad \omega_{pe}^2 < \omega^2 \qquad \omega_{pe}^2 \ll c^2 k^2$$

   (vi) High-frequency electromagnetic waves (RCP, LCP):

   $$\frac{k^2 c^2}{\omega^2} \simeq 1 \mp \frac{\omega_{pe}^2}{\omega^2}\left[1 - \frac{\Omega_e}{(\omega_{pe}^2 + k^2 c^2)^{1/2}}\right] \qquad \Omega_e^2 \ll \omega_{pe}^2 + k^2 c^2 \qquad |\Omega_e| \ll \omega$$

5. *Cold plasma modes: anisotropic plasma,* $\mathbf{k} \perp \mathbf{B_0}$

(i) Compressional Alfvén wave (linear polarization):

$$\frac{k^2 c^2}{\omega^2} = 1 + \frac{c^2}{v_A^2} \qquad \omega \ll \Omega_i$$

(ii) Electromagnetic wave (linear polarization, $\mathbf{E} \parallel \mathbf{B_0}$):

$$\frac{k^2 c^2}{\omega^2} = 1 - \frac{\omega_{pe}^2}{\omega^2} \text{ (ordinary)}$$

(iii) Coupled longitudinal–transverse mode (elliptic polarization in plane $\perp \mathbf{B_0}$)

$$\frac{k^2 c^2}{\omega^2} = \frac{(\omega^2 - \omega_{pe}^2)^2 - \omega^2 \Omega_e^2}{\omega^2 (\omega^2 - \omega_{pe}^2 - \Omega_e^2)} \text{ (extraordinary)}$$

(iv) Resonances for propagation $\perp \mathbf{B_0}$:

$$\text{Lower hybrid resonance, } \omega^2 \simeq -\Omega_i \Omega_e \left(\frac{\omega_{pe}^2}{\omega_{pe}^2 + \Omega_e^2} \right)$$

$$\text{Upper hybrid resonance, } \omega^2 \simeq \omega_{pe}^2 + \Omega_e^2$$

6. *General dispersion relation*

$$\tan^2 \theta = -\frac{P(n^2 - R)(n^2 - L)}{(Sn^2 - RL)(n^2 - P)}$$

$$P = 1 - \sum_{s=e,i} \frac{\omega_{ps}^2}{\omega^2} \qquad R, L = 1 - \sum_{s=e,i} \frac{\omega_{ps}^2}{\omega^2} \left(\frac{\omega}{\omega \pm \Omega_s} \right) \qquad S = \tfrac{1}{2}(R + L)$$

$\theta = 0:$ $\qquad\qquad\quad P = 0 \qquad n^2 = R \qquad n^2 = L$

$\theta = \pi/2:$ $\qquad n^2 = \dfrac{RL}{S}$ (extraordinary) $\qquad n^2 = P$ (ordinary)

Cut-off ($n^2 = 0$) when $P = 0$, $R = 0$, $L = 0$.

Resonance ($n^2 \to \infty$) when $\tan^2 \theta = -\dfrac{P}{S}$.

Generally, waves are reflected at a cut-off and absorbed at a resonance.

Problems

7–1 Obtain expressions for the phase and group velocities of modes satisfying the following dispersion relations:

(i) $\omega^2 = k^2 c^2 \left(1 + \dfrac{c^2}{v_A^2} \right)^{-1}$ $\qquad\qquad$ Alfvén waves

(ii) $\omega = -k^2 c^2 \Omega_e / \omega_p^2$ $\qquad\qquad$ helicon waves
(iii) $\omega^2 - \omega \omega_p^2 (\omega + \Omega_e)^{-1} = k^2 c^2$ $\qquad$ electron cyclotron waves $\omega \to |\Omega_e|$

7–2 In obtaining the dispersion relation for Alfvén waves, (7–32), an ordering of terms was introduced in Eqn (7–28). Interpret this ordering physically.

7-3 Show that Alfvén waves propagating in the direction of the magnetic field are circularly polarized.

Hint: to examine Alfvén wave polarization it is essential to retain terms of order ω/Ω_i compared with 1. The dispersion relation is then

$$\frac{k^2c^2}{\omega^2} = 1 - \frac{\omega_p^2}{\Omega_e(\Omega_i \mp \omega)}$$

and it is straightforward to show that $iE_x/E_y = \mp 1$. The shear or slow Alfvén wave considered in Section 7-3-1 corresponds to $iE_x/E_y = -1$, i.e., it displays left circular polarization.

7-4 Compute Alfvén wave speeds in the following plasmas:

  (i) interstellar space $n \sim 10/\text{cm}^3, B \sim 10^{-3}$ G
  (ii) solar corona $n \sim 10^6/\text{cm}^3, B \sim 10$ G
  (iii) ionosphere $n \sim 10^5/\text{cm}^3, B \sim 1$ G
  (iv) laboratory discharge $n \sim 10^{14}/\text{cm}^3, B \sim 10^4$ G
  (v) solid-state plasma $n \sim 10^{22}/\text{cm}^3, B \sim 10^3$ G

The positive charges are protons in (i), (ii), and (iv), nitrogen ions in (iii) and in (v) are holes having a mass (say) $3m^-$.

7-5 Interpret physically what takes place as ion cyclotron resonance is approached, assuming that *some* collisions do occur. What would happen in the absence of any dissipative mechanism?

7-6 Verify the statement made in Section 7-3-2 that in ion cyclotron waves the electric field rotates in the same direction as the ions, accelerating them.

7-7 Consider a wave propagating orthogonally to a magnetic field $\mathbf{B}_0$ with its electric vector aligned along $\mathbf{B}_0$. Show that this wave satisfies the dispersion relation $\omega^2 = \omega_p^2 + k^2c^2$. Comment on this result physically.

7-8 In comparing compressional Alfvén waves in a plasma with acoustic waves in an ordinary fluid, the magnetic pressure was substituted for the kinetic pressure and γ taken to be 2. Explain this choice of γ.

7-9 Obtain (7-48).

7-10 Using the particle orbit equations consider the motion of electrons and ions in a plane electric field at the lower hybrid resonance frequency, $(|\Omega_i\Omega_e|)^{1/2}$. In particular confirm that the electron and ion displacements along $\mathbf{E}$ are both equal and in phase at this resonance.

7-11 Consider ion cyclotron waves propagating in a cylindrical plasma of uniform density. Show that the dispersion relation is

$$\frac{k_z^2c^2}{\omega^2} = \frac{\omega_{pi}^2}{\Omega_i^2 - \omega^2} - \frac{k_r^2c^2}{2\omega^2} + \left[\frac{\omega_{pi}^4}{(\Omega_i^2 - \omega^2)^2} + \left(\frac{k_r^2c^2}{2\omega^2}\right)^2\right]^{1/2}$$

where k_z, k_r are longitudinal and radial wave numbers respectively.

7-12 The right circularly polarized Alfvén wave propagating along the magnetic field is not affected by the ion cyclotron resonance and persists for frequencies $\omega > \Omega_i$. From (7–22), (7–23) show that this wave obeys the dispersion relation $\omega \simeq -k^2c^2\Omega_e/\omega_p^2$ for $\omega \gg \Omega_i$; it is known as a helicon wave or a whistler (cf. Section 7–7–1).

7–13 Consider the mode propagating with $\mathbf{k}$ perpendicular to $\mathbf{B}_0$; if the motion of the ions is neglected show that the dispersion relation for this mode is

$$\frac{k^2c^2}{\omega^2} = \frac{(\omega^2 - \omega_p^2)^2 - \omega^2\Omega_e^2}{\omega^2(\omega^2 - \omega_{pe}^2 - \Omega_e^2)}$$

Observe that this displays the resonance at the upper hybrid frequency and anticipates some results from Section 7–7–1 on the Appleton–Hartree dispersion relation.

7–14 Using the same model as in the last problem, show that transverse waves propagating along $\mathbf{B}_0$ obey the dispersion relation

$$\frac{k^2c^2}{\omega^2} = 1 - \frac{\omega_p^2}{\omega(\omega \pm |\Omega_e|)}$$

Comment on the resonance at the electron cyclotron frequency for the mode with right circular polarization.

7–15 Show that the ray direction is normal to the refractive index surface.

7–16 Verify that the generalized Ohm's law has the form (7–52) when terms of order m^-/m^+ are retained.

Hint: it is unnecessary to return to the analysis given in Chapter 3; the result follows directly by symmetrizing (7–23) with respect to the ions and electrons.

7–17 Verify that the dielectric tensor (7–64) may be reduced to the diagonal form (7–66).

Hint: note that $U^{-1} = (U^*)^T$ where $*$ denotes complex conjugate values and T stands for the transpose.

7–18 Reduce (7–67) to give the dispersion relation in the form $An^4 - Bn^2 + C = 0$ with A, B, C given by (7–69).

7–19 Show that at $\theta = \pi/2$, the resonance condition $S = 0$ reproduces the hybrid resonance conditions (7–48) and (7–49).

7–20 Reproduce Fig. 7–19 showing the division of parameter space into 13 regions for a more realistic choice of the mass ratio m^+/m^-.

7–21 Show that the points of intersection of the plasma cut-off line $P = 0$ with the plasma resonance and cyclotron cut-off lines ($S = 0$, $L = 0$ respectively) in a two-component cold plasma occur at

$$\alpha^2 = 1, \quad \beta_+^2 = 1 - \frac{m^-}{m^+} + \left(\frac{m^-}{m^+}\right)^2$$

$$\alpha^2 = 1, \quad \beta_+ = 1 - \frac{m^-}{m^+}$$

7-22 The wave-normal surfaces corresponding to fast and slow modes coincide when the discriminant in (7–70) vanishes, i.e.,

$$(RL - PS)^2 \sin^4 \theta + 4P^2 D^2 \cos^2 \theta = 0$$

Determine the modes for which such coincidence is possible for propagation along, and orthogonal to, $\mathbf{B}_0$.

7-23 Verify that in the frequency region $\omega < \Omega_i$, $\omega^2 < \omega_{pe}^2 + \omega_{pi}^2$, (7–68) factors to give the dispersion relations for the fast and slow Alfvén waves

$$\frac{k^2 c^2}{\omega^2} = 1 + \frac{c^2}{v_A^2}$$

$$\frac{k^2 c^2 \cos^2 \theta}{\omega^2} = 1 + \frac{c^2}{v_A^2}$$

provided θ is not too close to $\pi/2$.

7-24 Establish (7–82).

7-25 From (7–82), neglecting ion motions, establish the Appleton–Hartree dispersion relation (7–83).

7-26 Show that the tip of the wave magnetic field vector in a helicon wave traces out a helix.

7-27 Determine the stop band limit frequencies which appear in Fig. 7–22.

8

Waves in warm plasmas

8–1 Introduction

The cold plasma approximation used in Chapter 7 showed clearly the existence of a large number of waves in a cold, anisotropic, loss-free plasma. In general, one does not have to insist that the plasma temperature is really zero for the theory developed to be applicable. As pointed out in Chapter 3, this approximation will be viable, provided the phase velocity of the wave under consideration is much larger than $(\kappa T^+/m^+)^{1/2}$, $(\kappa T^-/m^-)^{1/2}$; in other words, it might even be useful in describing waves with a sufficiently high phase velocity in quite hot plasmas. However, under plasma resonance conditions, $v_p = 0$, so that in a warm plasma the theory developed in Chapter 7 breaks down completely near a resonance. Moreover, the existence of plasma temperature means, in general, that the pressure gradient terms appearing both in the equation of motion (3–54) and the generalized Ohm's law (3–61) must now be taken into consideration. Although daunting—in view of the considerable complexity of cold plasma waves—one has to recognize that admitting plasma temperature opens up the possibility of further modes. In this chapter we shall only look at a few special cases (although some plots of the general dispersion relation will be given in Section 8–7). The reason (quite apart from wanting to avoid some heavy algebra) is that although using pressure terms is a step towards realism, nevertheless the hydromagnetic equations give an incomplete picture of wave motion in a dissipation-free plasma. To describe warm plasma behaviour fully we need kinetic theory but, since this often involves rather lengthy calculations, a hydromagnetic approach may sometimes be expedient. Typically, this alternative fails when, for example, some significant fraction of the plasma electrons have thermal velocities $\sim v_p$ so that they interact rather strongly with the wave. The physical consequences of such an interaction are lost to a hydromagnetic analysis because of averaging over individual particle velocities (cf. Chapter 3).

A good example of the shortcomings of the hydromagnetic approach is the description of electron plasma oscillations. In the cold-plasma limit, these are simply oscillations at $(\omega_{pe}^2 + \omega_{pi}^2)^{1/2}$, i.e., they do not propagate. In finite temperature plasmas, on the other hand, the dispersion relation is $\omega^2 = \omega_{pe}^2 + \omega_{pi}^2 + k^2(V_e^2 + V_i^2)$ where $V_{i,e} = (\gamma^\pm \kappa T^\pm/m^\pm)^{1/2}$; moreover, this result is obtained (for sufficiently small k) regardless of whether one uses the hydromagnetic equations or kinetic theory (the Vlasov equation).

However, from a kinetic theory treatment, additional information is retrieved that is lost to hydromagnetic theory; in particular, one finds the electron plasma oscillations in an equilibrium plasma are damped even though inter-particle collisions are negligible. This phenomenon, known as Landau damping, will be discussed at length in Chapter 10 and is typical of the omissions from a hydromagnetic description.

In Section 8–2 the one-fluid hydromagnetic equations of Section 3–5 are used to discuss low-frequency (magnetodynamic) modes in a warm plasma; the assumption of high plasma conductivity is still made so that the $\mathbf{j}/\sigma$ term in the generalized Ohm's law may be omitted. In addition, the pressure is taken to be scalar rather than the pressure tensor required for the complete description of an anisotropic plasma; the consequences of this simplification are that shear effects (e.g., shear waves) are lost. In general, the possibility of distinct electron and ion temperatures exists which means that the one-fluid description is no longer useful. One must then resort to the two-fluid equations presented in Section 3–4; this model is used in Sections 8–3 and 8–4 to discuss purely longitudinal modes (electron plasma and ion plasma oscillations and ion acoustic waves) in warm plasmas. These modes exist in magnetic field-free plasmas or in special situations in anisotropic plasmas in which the wave vector is aligned along $\mathbf{B}_0$.

A brief discussion of Landau damping is given in Section 8–5 and followed by a description of experimental work on ion acoustic waves and electron plasma oscillations. The general dispersion relation is presented schematically in dispersion plots in Section 8–7. In Section 7–6, the cold plasma dispersion relation was of order 3 in ω^2 giving three modes; now the general dispersion relation for waves in a warm anisotropic plasma (ignoring electron–ion collisions) gives six roots for ω^2 as one would expect, since, on a two-fluid picture, there are six degrees of freedom.

Once again the plasma being considered is unbounded and spatially uniform.

8–2 Magnetodynamic waves

As in Chapter 7, we begin by examining the lowest frequency modes which can propagate on the basis of the one-fluid equations. From the Maxwell equations one has, ignoring displacement current (i.e., taking $k^2 c^2/\omega^2 \gg 1$),

$$\mathbf{\nabla} \times \mathbf{\nabla} \times \mathbf{E} = -\frac{4\pi}{c^2} \frac{\partial \mathbf{j}}{\partial t} \tag{8-1}$$

From (3–54) and (3–61), the equation of motion and the generalized Ohm's law become (on the assumption of scalar pressure and high plasma conductivity)

$$\rho \frac{D\mathbf{U}}{Dt} = -\mathbf{\nabla}P + q\mathbf{E} + \frac{1}{c}(\mathbf{j} \times \mathbf{B}) \tag{8-2}$$

$$\frac{m^+m^-}{\rho e^2}\frac{\partial \mathbf{j}}{\partial t} = \mathbf{E} + \frac{\mathbf{U} \times \mathbf{B}}{c} - \frac{m^+}{\rho e c}(\mathbf{j} \times \mathbf{B}) + \frac{m^+}{2\rho e}\nabla P \qquad (8\text{-}3)$$

In the low-frequency limit $\omega \ll \Omega_i$, both the term on the left-hand side of (8-3) together with the Hall field, $(m^+/\rho e c)(\mathbf{j} \times \mathbf{B})$, are dropped, as in Section 7-3. Thus from (8-1) to (8-3), on linearizing and taking a plane wave form for first-order quantities, we get

$$k^2\mathbf{E} - \mathbf{k}(\mathbf{k}.\mathbf{E}) = \frac{4\pi i\omega}{c^2}\mathbf{j} \qquad (8\text{-}4)$$

$$-i\omega\rho_0\mathbf{U} = -ik P + \frac{\mathbf{j} \times \mathbf{B}_0}{c} \qquad (8\text{-}5)$$

$$\mathbf{E} + \frac{\mathbf{U} \times \mathbf{B}_0}{c} + \frac{im^+\mathbf{k}}{2\rho_0 e}P = 0 \qquad (8\text{-}6)$$

The cold plasma dispersion relation followed from the three analogous equations; clearly, a further relation is now required on account of the pressure. This follows from (3-56) which, on linearization, gives the adiabatic law

$$P\rho^{-\gamma} = \text{constant} \qquad (8\text{-}7)$$

From (8-7), one has

$$\frac{\nabla P}{P_0} = \gamma\frac{\nabla\rho}{\rho_0} \qquad (8\text{-}8)$$

together with the linearized continuity equation

$$\frac{\partial \rho}{\partial t} = -\rho_0\nabla.\mathbf{U} \qquad (8\text{-}9)$$

i.e., for a plane wave $\rho = (\mathbf{k}.\mathbf{U}/\omega)\rho_0$ so that

$$\nabla\rho = ik\frac{(\mathbf{k}.\mathbf{U})}{\omega}\rho_0 \qquad (8\text{-}10)$$

Consequently, substituting for $\nabla\rho$ in (8-8), gives

$$ikP = i\gamma P_0 k\frac{(\mathbf{k}.\mathbf{U})}{\omega} \qquad (8\text{-}11)$$

Introducing a sound speed c_s defined by

$$c_s^2 = \frac{\gamma P_0}{\rho_0}$$

one finds from (8–5),

$$U = \left(\frac{k \cdot U}{\omega^2}\right)c_s^2 k + i\frac{j \times B_0}{c\omega\rho_0} \qquad (8\text{–}12)$$

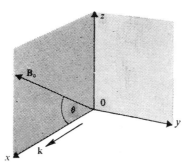

Fig. 8-1 Orientation of **k** and **B$_0$**

Choose a coordinate system having **k** along Ox with **B$_0$** lying in the Oxz plane (cf. Fig. 8–1). Then, if $\hat{k} \cdot \hat{b}_0 = \cos\theta$, (8–12) gives

$$\left.\begin{aligned}
U_x &= \frac{iB_0}{c\omega\rho_0}\frac{j_y \sin\theta}{1 - k^2 c_s^2/\omega^2} \\
U_y &= \frac{iB_0}{c\omega\rho_0}(j_z \cos\theta - j_x \sin\theta) \\
U_z &= \frac{-iB_0}{c\omega\rho_0}(j_y \cos\theta)
\end{aligned}\right\} \qquad (8\text{–}13)$$

From (8–4),

$$\left.\begin{aligned}
j_x &= 0 \\
E_y &= \frac{4\pi i\omega}{c^2 k^2}j_y \\
E_z &= \frac{4\pi i\omega}{c^2 k^2}j_z
\end{aligned}\right\} \qquad (8\text{–}14)$$

so that substituting in the generalized Ohm's law (8–6) for E_y, E_z, and U_x, U_y, U_z gives

$$\left[\frac{4\pi\omega}{c^2 k^2} - \frac{B_0^2}{c^2\omega\rho_0}\left(\cos^2\theta + \frac{\sin^2\theta}{1 - k^2 c_s^2/\omega^2}\right)\right]j_y = 0 \qquad (8\text{–}15a)$$

$$\left[\frac{4\pi\omega}{c^2 k^2} - \frac{B_0^2}{c^2\omega\rho_0}\cos^2\theta\right]j_z = 0 \qquad (8\text{–}15b)$$

From (8–15a), if v_A denotes the Alfvén velocity,

$$\frac{\omega^4}{k^4} - \frac{\omega^2}{k^2}\left(c_s^2 + v_A^2\right) + c_s^2 v_A^2 \cos^2 \theta = 0 \qquad (8\text{–}16a)$$

while (8–15b) gives

$$\frac{\omega^2}{k^2} = v_A^2 \cos^2 \theta \qquad (8\text{–}16b)$$

Thus, the dispersion relation factorizes and gives in all three roots for ω^2/k^2 as opposed to two modes (shear and compressional Alfvén waves) in the low-frequency cold plasma case. All three modes in (8–16) are non-dispersive. Observe that the mode in (8–16b) is independent of the sound speed and is known as the *oblique Alfvén wave*. The other two modes involve both the magnetic and the plasma kinetic pressure and are known as *magneto-acoustic waves*. From (8–16a),

$$v_p^2 \equiv \frac{\omega^2}{k^2} = \frac{1}{2}\left(c_s^2 + v_A^2\right)\left[1 \pm \left(1 - \frac{4c_s^2 v_A^2 \cos^2\theta}{(c_s^2 + v_A^2)^2}\right)^{1/2}\right] \qquad (8\text{–}17)$$

For situations in which

$$\frac{c_s v_A \cos\theta}{(c_s^2 + v_A^2)} \ll 1$$

i.e., when either $\cos\theta \ll 1$ or $c_s \ll v_A$ or $v_A \ll c_s$, (8–17) gives

$$\frac{\omega^2}{k^2} \simeq c_s^2 + v_A^2 \qquad (8\text{–}18a)$$

or

$$\frac{\omega^2}{k^2} \simeq \left(\frac{c_s^2 v_A^2}{c_s^2 + v_A^2}\right) \cos^2 \theta \qquad (8\text{–}18b)$$

The phase velocity given by (8–18a) is then much larger than that in (8–18b) and the two modes are distinguished as *fast* and *slow* magneto-acoustic waves. Moreover, since the phase velocity from (8–16b) is intermediate between those from (8–18), it is common to find the three modes described as *fast* (8–18a), *intermediate* (8–16b), and *slow* (8–18b). This was done, for example, in Section 6–1 where it was pointed out that the transverse intermediate waves do not steepen to form shocks as opposed to the fast and slow modes which contain both transverse and longitudinal components. However, this labelling scheme is somewhat undesirable in view of the use of the fast and slow labels for Alfvén waves in Chapter 7. In the limit $c_s^2 \to 0$, the fast mode does indeed identify with the compressional (fast) Alfvén wave but it is the intermediate mode which corresponds to the shear (slow) Alfvén wave; in this limit the slow wave (8–18b) disappears.

Observe too that in the case of propagation orthogonal to $\mathbf{B}_0$, the fast wave alone persists, while for propagation along $\mathbf{B}_0$ one finds from (8–16) $\omega^2 = k^2 c_s^2, \omega^2 = k^2 v_A^2$ (twice), that is, the modes decouple giving a longitudinal (acoustic) mode and two transverse (compressional and shear Alfvén) modes.

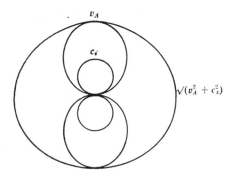

Fig. 8–2 Phase velocity surfaces for magnetodynamic waves; $v_A > c_s$

In other words, the physical effect of a *transverse* component of the magnetic field is to couple a purely longitudinal mode with one which is transverse, giving magneto-acoustic waves containing both components.

The three low-frequency modes are classed as *magnetodynamic waves*; their phase velocity surfaces are sketched in Fig. 8–2.

8–2–1 Behaviour close to Ω_i

We now examine what happens as the mode frequency is increased and approaches the ion cyclotron frequency. In the cold plasma approximation, the compressional Alfvén wave was not greatly affected on passing through $\omega = \Omega_i$ but the shear Alfvén wave disappeared completely giving way to an ion cyclotron wave close to the resonance. To find the behaviour in a warm plasma as the frequency is increased is straightforward; one must retain the Hall term in the generalized Ohm's law, (8–3), (cf. Section 7–3–2) so that (8–15a, b) are replaced by (cf. Problem 8–2)

$$\left[\frac{\omega^4}{k^4} - \frac{\omega^2}{k^2}(c_s^2 + v_A^2) + c_s^2 v_A^2 \cos^2 \theta\right]\left[\frac{\omega^2}{k^2} - v_A^2 \cos^2 \theta\right]$$
$$- \left(\frac{\omega v_A^2}{\Omega_i}\right)^2\left[\frac{\omega^2}{k^2} - c_s^2\right] \cos^2 \theta = 0 \quad (8\text{–}19)$$

At a resonance, $n^2 \equiv c^2 k^2/\omega^2 \to \infty$ so that from (8–19) it appears that a resonance occurs in a warm plasma at

$$\omega^2 = \Omega_i^2 \cos^2 \theta \quad (8\text{–}20)$$

In the neighbourhood of this resonance, the wave will be labelled the *second ion cyclotron wave* to distinguish it from the *first* ion cyclotron wave identified in the cold plasma limit (cf. Section 7–3–2). The behaviour of the other two modes in the neighbourhood of this resonance is determined by (cf. Problem 8–3).

$$\frac{\omega^4}{k^4} - \frac{\omega^2}{k^2}[c_s^2 + v_A^2(1 + \cos^2\theta)] + v_A^2 \cos^2\theta \left[2c_s^2 + v_A^2\left(1 - \frac{\omega^2}{\Omega_i^2}\right)\right] \simeq 0 \qquad (8\text{–}21)$$

i.e.,
$$\frac{\omega^2}{k^2} \simeq \frac{c_s^2 + v_A^2(1 + \cos^2\theta)}{2} \pm \frac{1}{2}\left\{[c_s^2 + v_A^2(1 + \cos^2\theta)]^2 \right.$$
$$\left. - 4v_A^2 \cos^2\theta\left[2c_s^2 + v_A^2\left(1 - \frac{\omega^2}{\Omega_i^2}\right)\right]\right\}^{1/2} \qquad (8\text{–}22)$$

A simple way of identifying these modes is to pass to the limit $c_s^2 \to 0$; then,

$$\left(\frac{\omega^2}{k^2}\right)_+ \to v_A^2(1 + \cos^2\theta) \qquad \left(\frac{\omega^2}{k^2}\right)_- \to \frac{v_A^2 \cos^2\theta}{1 + \cos^2\theta}\left(\frac{\Omega_i^2 - \omega^2}{\Omega_i^2}\right) \qquad (8\text{–}23)$$

which are the dispersion relations for the compressional Alfvén wave in the neighbourhood of $\omega = \Omega_i$ and the first ion cyclotron wave, respectively. In other words, the roots of (8–22) are to be identified with the fast magneto-acoustic (+) and oblique Alfvén (−) waves in a warm plasma.

Thus, as the mode frequency is increased from the region $\omega \ll \Omega_i$, the mode corresponding to the slow magneto-acoustic wave in this limit becomes the second ion cyclotron wave in the neighbourhood of $\omega = \Omega_i \cos\theta$ (cf. Problem 8–4).

With a further increase in frequency into the region $\omega \simeq \Omega_i$ the oblique Alfvén wave of the low-frequency limit shows *pseudo-resonant* behaviour (Fig. 8–14). Thus, another important effect of admitting a finite plasma temperature is that the true cold plasma resonance at $\omega = \Omega_i$ for the first ion cyclotron wave is destroyed. The warm plasma counterpart of this wave, corresponding to the minus sign in (8–22), persists at frequencies $\omega > \Omega_i$. The fast magneto-acoustic wave—like its counterpart in the cold plasma approximation—is scarcely affected by either the resonance at $\omega = \Omega_i \cos\theta$ or the pseudo-resonance at $\omega = \Omega_i$.

8–2–2 Behaviour at $\omega > \Omega_i$

To examine the behaviour of the two modes persisting above the ion cyclotron resonance at $\omega = \Omega_i \cos\theta$, consider the dispersion relation (8–19) in the form

$$\left(\frac{\omega^2}{k^2} - c_s^2\right)\left[\left(\frac{\omega^2}{k^2} - v_A^2 \cos^2\theta\right)^2 - \left(\frac{\omega}{\Omega_i}v_A^2 \cos\theta\right)^2\right]$$
$$- \frac{\omega^2}{k^2}v_A^2\left(\frac{\omega^2}{k^2} - v_A^2 \cos^2\theta\right)(1 - \cos^2\theta) = 0 \qquad (8\text{–}24)$$

Provided the last term is small enough to be ignored, one finds

$$\omega^2 = k^2 c_s^2 \tag{8-25a}$$

$$\frac{\omega^2}{k^2} = v_A^2 \cos\theta\left(\cos\theta \pm \frac{\omega}{\Omega_i}\right) \tag{8-25b}$$

Above the resonance at $\Omega_i \cos\theta$, only the mode corresponding to the plus sign in (8–25b) persists; thus for $\omega > \Omega_i$ the dispersion relation (8–25b) becomes

$$\frac{\omega^2}{k^2} \simeq \frac{\omega}{\Omega_i} v_A^2 \cos\theta \tag{8-26}$$

It is straightforward to reduce (8–26) to (7–91) which is the dispersion relation for helicons. One must next check the implications of ignoring the final term on the left-hand side of (8–24). This is the subject of Problem 8–6; one finds that the dispersion relations (8–25) are valid if $c_s^2 \gg v_A^2$ (provided $v_A^2 \omega \cos\theta \not\simeq c_s^2 \Omega_i$) and again if $v_A^2 \gg c_s^2$ together with $\cos^2\theta \simeq 1$ (provided $v_A^2 \omega \not\simeq c_s^2 \Omega_i$). Thus the oblique Alfvén wave in the magnetodynamic limit survives the resonance at $\omega = \Omega_i \cos\theta$ and transforms via the first ion cyclotron wave at $\omega = \Omega_i$ into either a helicon ($c_s^2 \gg v_A^2$) or an acoustic wave ($v_A^2 \gg c_s^2$, $\cos^2\theta \simeq 1$). The fast magneto-acoustic wave (8–18a) propagates above $\omega = \Omega_i$ effectively unchanged if $c_s^2 \gg v_A^2$; if, however, $v_A^2 \gg c_s^2$, $\cos^2\theta \simeq 1$, then this mode transforms into a helicon for $\omega > \Omega_i$.

8–3 Longitudinal waves in warm plasmas

In getting results from the one-fluid hydromagnetic equations we assumed that both species had the same temperature and that electron and ion velocities were approximately equal. In all three magnetodynamic waves, electron and ion velocities are in fact equal, so that using the two-fluid equations introduced in Section 3–4 leads to just the results found in Section 8–2. In this section, we examine the propagation of longitudinal waves in a warm plasma ($\mathbf{B}_0$ is now aligned along Ox in Fig. 8–1) and the restrictions of the one-fluid approach are relaxed. Linearizing (3–26) and (3–49), we find

$$\frac{\partial n^+}{\partial t} + n_0^+ \nabla.\mathbf{u}^+ = 0 \tag{8-27$^+$}$$

$$\frac{\partial n^-}{\partial t} + n_0^- \nabla.\mathbf{u}^- = 0 \tag{8-27$^-$}$$

$$n_0^+ m^+ \frac{\partial \mathbf{u}^+}{\partial t} + \nabla p^+ - n_0^+ e\left(\mathbf{E} + \frac{\mathbf{u}^+ \times \mathbf{B}_0}{c}\right) = 0 \tag{8-28$^+$}$$

$$n_0^- m^- \frac{\partial \mathbf{u}^-}{\partial t} + \nabla p^- + n_0^- e\left(\mathbf{E} + \frac{\mathbf{u}^- \times \mathbf{B}_0}{c}\right) = 0 \qquad (8\text{–}28)^-$$

The energy equations reduce to adiabatic laws for each species so that by analogy with (8–11)

$$i k p^+ = i \gamma^+ p_0^+ \mathbf{k}\left(\frac{\mathbf{k} \cdot \mathbf{u}^+}{\omega}\right) \qquad (8\text{–}29)^+$$

$$i k p^- = i \gamma^- p_0^- \mathbf{k}\left(\frac{\mathbf{k} \cdot \mathbf{u}^-}{\omega}\right) \qquad (8\text{–}29)^-$$

For longitudinal waves, $\mathbf{u}^\pm \times \mathbf{B}_0 = 0$ so that $(8\text{–}28)^\pm$ give, using $(8\text{–}29)^\pm$ and $n_0^+ = n_0^- \equiv n_0$,

$$-i\omega \mathbf{u}^+ + i\frac{\gamma^+ p_0^+}{n_0 m^+}\mathbf{k}\left(\frac{\mathbf{k} \cdot \mathbf{u}^+}{\omega}\right) - \frac{e}{m^+}\mathbf{E} = 0 \qquad (8\text{–}30)^+$$

$$-i\omega \mathbf{u}^- + i\frac{\gamma^- p_0^-}{n_0 m^-}\mathbf{k}\left(\frac{\mathbf{k} \cdot \mathbf{u}^-}{\omega}\right) + \frac{e}{m^-}\mathbf{E} = 0 \qquad (8\text{–}30)^-$$

The electric field may be eliminated from $(8\text{–}30)^\pm$ by substituting in $\nabla \cdot \mathbf{E} = 4\pi q$ for $q = e(n^+ - n^-)$; using $(8\text{–}27)^\pm$ and the fact that $\mathbf{u}^\pm$ and $\mathbf{k}$ are parallel to $\mathbf{B}_0$

$$\mathbf{E} = -i\frac{4\pi e n_0}{\omega k^2}[\mathbf{k} \cdot (\mathbf{u}^+ - \mathbf{u}^-)]\mathbf{k} \qquad (8\text{–}31)$$

Then, from $(8\text{–}30)^\pm$,

$$\mathbf{u}^+\left(1 - \frac{k^2}{\omega^2}\frac{\gamma^+ p_0^+}{n_0 m^+} - \frac{\omega_{pi}^2}{\omega^2}\right) + \frac{\omega_{pi}^2}{\omega^2}\mathbf{u}^- = 0 \qquad (8\text{–}32)^+$$

$$\mathbf{u}^-\left(1 - \frac{k^2}{\omega^2}\frac{\gamma^- p_0^-}{n_0 m^-} - \frac{\omega_{pe}^2}{\omega^2}\right) + \frac{\omega_{pe}^2}{\omega^2}\mathbf{u}^+ = 0 \qquad (8\text{–}32)^-$$

From (8–32), the dispersion relation for longitudinal waves follows immediately:

$$\left(\omega^2 - k^2 \frac{\gamma^+ p_0^+}{n_0 m^+} - \omega_{pi}^2\right)\left(\omega^2 - k^2 \frac{\gamma^- p_0^-}{n_0 m^-} - \omega_{pe}^2\right) - \omega_{pe}^2 \omega_{pi}^2 = 0 \qquad (8\text{–}33)$$

By analogy with the definition of the sound speed in the one-fluid analysis of Section 8–2, one may introduce ion and electron thermal speeds defined by

$$V_i^2 = \frac{\gamma^+ p_0^+}{n_0 m^+} = \gamma^+ \frac{\kappa T^+}{m^+} \qquad (8\text{–}34)^+$$

$$V_e^2 = \frac{\gamma^- p_0^-}{n_0 m^-} = \gamma^- \frac{\kappa T^-}{m^-} \tag{8-34}^-$$

and so the dispersion relation now reads

$$(\omega^2 - k^2 V_i^2 - \omega_{pi}^2)(\omega^2 - k^2 V_e^2 - \omega_{pe}^2) - \omega_{pe}^2 \omega_{pi}^2 = 0 \tag{8-35}$$

The roots of this dispersion relation separate into high- and low-frequency regions; one has

$$\omega^2 = \tfrac{1}{2}[\omega_{pe}^2 + \omega_{pi}^2 + k^2 V_e^2 + k^2 V_i^2].$$
$$\left[1 \pm \left\{1 - \frac{4(k^4 V_i^2 V_e^2 + k^2 V_i^2 \omega_{pe}^2 + k^2 V_e^2 \omega_{pi}^2)}{(\omega_{pe}^2 + \omega_{pi}^2 + k^2 V_e^2 + k^2 V_i^2)^2}\right\}^{1/2}\right] \tag{8-36}$$

Consider the second term in the discriminant; this is always small compared with one except under the rather unusual conditions $V_e \lesssim V_i$ (cf. Problem 8–7). In most physical situations $V_e \gg V_i$ so that one has

$$\omega^2 \simeq \omega_{pe}^2 + \omega_{pi}^2 + k^2(V_e^2 + V_i^2) \tag{8-37}$$

or
$$\omega^2 \simeq k^2 \frac{(k^2 V_i^2 V_e^2 + V_i^2 \omega_{pe}^2 + V_e^2 \omega_{pi}^2)}{[\omega_{pe}^2 + \omega_{pi}^2 + k^2(V_e^2 + V_i^2)]} \tag{8-38}$$

In most cases the ion terms in (8–37) may be dropped to a good approximation giving the dispersion relation for *electron plasma waves*

$$\omega^2 = \omega_{pe}^2 + k^2 V_e^2 \tag{8-39}$$

Thus admitting an electron temperature introduces an important change in the cold plasma result (7–26); the electron plasma oscillations may now propagate, their group velocity being given by

$$v_g = \frac{d\omega}{dk} = \frac{k V^2}{\omega} \tag{8-40}$$

Observe that $v_g \to 0$ in the limit of long wavelengths.

The dispersion relation (8–39) is very different from that for longitudinal sound waves in a neutral gas, $\omega^2 = k^2 c_s^2$, a distinction, of course, which simply mirrors the difference between the basic interactions in each case. In a neutral gas, the interaction is very short range (typically a few angstroms for hydrogen) and there is little or no coherence. In a plasma, on the other hand, long-range interactions represented by the self-consistent fields provide a coherence absent in a neutral gas and are the origin of the ω_{pe}^2 term in (8–39).

8–4 Ion acoustic waves and ion plasma oscillations

Next, consider the low-frequency dispersion relation (8–38) for the case $V_i \ll V_e$. In the limit of long wavelengths (i.e., $k^2 \ll \omega_{pe}^2/V_e^2$) this becomes

$$\omega^2 = k^2 \frac{(\omega_{pe}^2 V_i^2 + \omega_{pi}^2 V_e^2)}{\omega_{pe}^2}$$

or

$$\omega^2 \simeq k^2 \left(\gamma^+ \frac{\kappa T^+}{m^+} + \gamma^- \frac{\kappa T^-}{m^+} \right) \tag{8–41}$$

This is the dispersion relation for an acoustic wave with a 'sound speed' $[(\gamma^+\kappa T^+ + \gamma^-\kappa T^-)/m^+]^{1/2}$. However, there is a fundamental distinction between the mode described by (8–41) and a sound wave in a neutral gas which propagates only on account of inter-molecular collisions. In a collisionless plasma, on the other hand, the potential energy required to drive the wave is electrostatic in origin and is due to the difference in amplitudes of the electron and ion oscillations.

For shorter wavelengths satisfying $\omega_{pe}^2/V_e^2 \ll k^2 \sim \omega_{pi}^2/V_i^2$, for the case $V_i \ll V_e$ (8–38) becomes

$$\omega^2 = \omega_{pi}^2 + k^2 V_i^2 \tag{8–42}$$

i.e., (8–42) only holds if $T^- \gg T^+$.

In the limit of vanishing ion temperature, (8–42) corresponds to ion plasma oscillations in which the cold ion oscillate at the ion plasma frequency while the electrons (since $k^2 V_e^2 \gg \omega_{pe}^2$) do *not* behave collectively but are dominated by their random motion. Thus for electron plasma waves $\mathbf{u}^- \gg \mathbf{u}^+$ and the role of the ions is simply to provide an essentially immobile positive background for the co-operative electron motion, while for ion plasma oscillations $\mathbf{u}^+ \gg \mathbf{u}^-$ and the electrons perform the same service for the ions; however, because of their high thermal speeds they play a dynamic, rather than a static, neutralizing role.

Clearly, neither the dispersion relation for electron plasma waves (8–39) nor that for ion plasma oscillations could be obtained from the one-fluid hydromagnetic equations which require $\mathbf{u}^+ \simeq \mathbf{u}^-$ to be a valid description. This condition is violated by both the high- ($\mathbf{u}^+ \ll \mathbf{u}^-$) and low- ($\mathbf{u}^+ \gg \mathbf{u}^-$) frequency modes. However, one-fluid theory *does* give for waves propagating along $\mathbf{B}_0$, from (8–5),

$$\omega \rho_0 \mathbf{U} = \gamma P_0 \mathbf{k} \left(\frac{\mathbf{k} \cdot \mathbf{U}}{\omega} \right)$$

i.e.,

$$\omega^2 = k^2 \left(\frac{\gamma P_0}{\rho_0} \right) \tag{8–43}$$

which resembles the ion acoustic wave dispersion relation (8–41). The inference is that, in ion acoustic waves, electron and ion velocities are comparable (cf. Problem 8–8).

In the limit of short wavelengths, (8–38) becomes

$$\omega^2 \simeq k^2 \frac{V_i^2 V_e^2}{V_e^2 + V_i^2} \simeq k^2 V_i^2 \quad \text{(for } V_i^2 \ll V_e^2) \tag{8-44}$$

In the same limit the high-frequency branch of the dispersion relation, (8–39), reduces to $\omega^2 = k^2 V_e^2$, which has the form of an electron acoustic wave. However, this is a physically unattainable limit as both high- and low-frequency modes are severely attenuated (cf. Section 8–5). This is Landau damping, which we find despite neglect of inter-particle collisions since it originates from a resonance between the phase velocity of the modes and the thermal speeds of *some* of the plasma particles. Since only a special group of particles is involved, this effect is lost in a hydromagnetic description of wave motion as the fluid equations were obtained in Sections 3–3 to 3–4 by *averaging* over the distribution function in velocity space.

8–5 Landau damping of longitudinal plasma waves

We defer analysis of the Landau damping of longitudinal waves in a plasma until Chapter 10 since it forms part of a kinetic theoretic treatment of plasma oscillations. However a brief qualitative introduction is necessarily included at this stage; for example, we shall find that some of the dispersion relations in the previous section have no physical significance since the mode they describe is so severely attenuated that no propagation is possible.

Collisionless damping originates in the strong interaction between a longitudinal plasma wave and those particles in the plasma having velocities close to the phase velocity of the wave. Such particles will tend to extract energy from the wave when they are initially randomly phased with respect to it. Wave energy is absorbed if there are more particles moving slightly slower than the phase velocity of the wave than are moving slightly faster. On the other hand, when the reverse holds, energy is transferred from the particles into the wave, which grows in amplitude. For equilibrium plasmas with a Maxwellian distribution, longitudinal waves are damped; this phenomenon is known as *Landau damping*.

The mode frequency ω now has a real (ω_r) and imaginary (ω_i) part. A kinetic theory analysis gives a complex dispersion relation, the real part of which is (in one dimension, so that $\gamma = 3$)

$$\omega_r^2 = \omega_{pe}^2 + k^2 V_e^2 = \omega_{pe}^2 \left(1 + \frac{3k^2}{k_D^2} \right) \qquad \frac{k^2}{k_D^2} \ll 1 \tag{8-45}$$

k_D being the reciprocal of the Debye length, λ_D, while the imaginary part gives

$$\omega_i \simeq \sqrt{\left(\frac{\pi}{8}\right)} \omega_{pe} \left(\frac{k_D}{k}\right)^3 \exp\left[-\frac{k_D^2}{2k^2} - \frac{3}{2}\right] \qquad \frac{k^2}{k_D^2} \ll 1$$

or $$\omega_i \simeq 0.14\omega_{pe}\left(\frac{k_D}{k}\right)^3 \exp\left[-\frac{k_D^2}{2k^2}\right] \qquad \frac{k^2}{k_D^2} \ll 1 \qquad (8\text{-}46)$$

Asymptotic expressions for ω_r, ω_i exist in the opposite limit but then damping is so severe that the mode cannot propagate.

Thus kinetic theory provides not only an expression for the damping but also a condition for the validity of the dispersion relation, namely $k\lambda_D \ll 1$, which did not appear from the derivation of (8-39). However, with hindsight, one may convince oneself that this restriction is implicit in the dispersion relations derived from the hydromagnetic analysis. In Section 3-6 we saw that a hydrodynamic description of plasma behaviour could be meaningful when inter-particle collisions were negligible, since the particles constituting the fluid element could be held together by the self-consistent fields. This led to the set of cold plasma equations. However, if some thermal energy is retained a fluid element will only persist if the forces producing coherence (the self-consistent fields) dominate the disrupting influence of the random thermal motion. In other words, a fluid description will be consistent only if $\omega_{pe}^2 \gg k^2 V_e^2$, that is, $k^2 \ll k_D^2$.

Similar arguments in the low-frequency region require that $\omega_{pi}^2 \gg k^2 V_i^2$ for the dispersion relation for ion plasma oscillations to be valid. Thus for (8-42) one needs both

$$k^2 V_e^2 \gg \omega_{pe}^2 \qquad \omega_{pi}^2 \gg k^2 V^2$$

which may be combined to give

$$\frac{\omega_{pe}^2}{V_e^2} \ll k^2 \ll \frac{\omega_{pi}^2}{V_i^2} \qquad (8\text{-}47)$$

as a necessary condition for ion plasma oscillations to exist *without* suffering severe Landau damping. Thus ion plasma oscillations may propagate only for wavelengths *less* than the Debye length but *greater* than $(T^+/T^-)^{1/2}$ times this length.

In plasmas for which $T^- \lesssim T^+$ ion acoustic waves propagate with a velocity of the order of the ion thermal velocity. In consequence they are subject to Landau damping by the ions as well as by the electrons. If, however, $T^- \gg T^+$ then from (8-41)

$$\frac{\omega^2}{k^2} \simeq \gamma^- \frac{\kappa T^-}{m^+} \gg V^2 \qquad (8\text{-}48)$$

and in these circumstances the Landau damping of the ion acoustic waves is reduced. Thus for ion acoustic waves to propagate without suffering ion Landau damping one requires $T^- \gg T^+$.

8–6 Experimental results for waves in warm plasmas

We shall confine discussion of experimental work on waves in warm plasmas to:

(i) Experiments on the ion acoustic waves described by (8–41).

(ii) An experiment on electron plasma waves including a measurement of Landau damping.

Because ion plasma oscillations propagate with wavelengths much *shorter* than a Debye length their detection is extremely difficult.

8–6–1 *Ion acoustic waves*

Experimental dispersion plots for this mode were obtained in 1961 by Little[70] and by Hatta and Sato.[71] In both experiments, a weakly ionized plasma was used, the degree of ionization being between 10^{-5} and 10^{-2}. Hatta and Sato excited the ion waves at the anode of a dark discharge in argon and used a

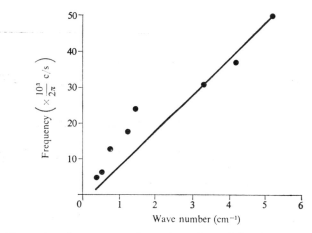

Fig. 8–3 Comparison between experimental results and the plane wave dispersion plot for ion acoustic waves in an argon plasma (after Hatta and Sato[71])

Langmuir probe to detect the plasma density fluctuations due to the wave. Because of the low degree of ionization, a hydromagnetic treatment including ion–neutral collisions was required. The plane wave dispersion curve (Fig. 8–3) shows good agreement with the measurements at higher frequencies; in this region $\omega \gg \nu_{in}$, the ion–neutral collision frequency, so that the dispersion relation (8–41) is then adequate. Below about $0\cdot5\,.\,10^4$ c/s there are marked discrepancies between theory and experiment.

Little and his collaborators used the plasma in a low-pressure mercury arc discharge. The principal drawback in this arrangement is the presence of sizeable fluctuations in density. The frequency range covered was from cut-off at 25 kc/s up to 50 kc/s and wavelengths from 30 cm down to 3 cm were measured. Some of their results are shown in Fig. 8–4; the theoretical curve was deduced from direct measurements of parameters for the steady state plasma, assuming the plasma column is of uniform density. In these experiments, $T^- \gg T^+$, which means that the damping of ion acoustic waves is decreased; moreover, this is a current-carrying plasma with an electron drift relative to the ions and under these conditions ion acoustic waves can actually

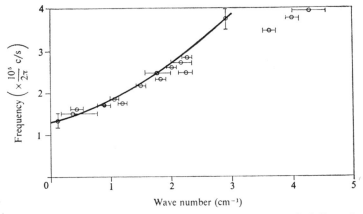

Fig. 8-4 Comparison between experimental results and the theoretical dispersion curve for ion acoustic waves in a mercury plasma (after Little[70])

grow in amplitude.[72] These effects have been eliminated in a series of experiments carried out at Princeton.[73, 74]

In this work, very stable plasmas in a device known as a Q (for quiescent) machine are used. An alkali metal plasma is created by contact ionization. The result is an almost completely ionized plasma very close to thermal equilibrium with the ionizing tungsten plates ($\sim$2,300°K). To a good approximation, the ions are collision-free. Magnetic fields of 6 or 12 kG mean that $\Omega_i \sim 100$ Mc/s which is very much higher than the mode frequencies studied. The mode examined in this experiment is that obeying the dispersion relation

$$\frac{\omega}{k} = c_s \cos \theta \qquad (8\text{–}49)$$

The apparatus is shown schematically in Fig. 8–5. Caesium or potassium atoms from an oven strike a hot tungsten plate giving a drifting plasma approximately 3 cm in diameter and 90 cm long with a density $10^{10}/cm^3$ to $10^{12}/cm^3$. A second hot plate contains the plasma axially and magnetic fields

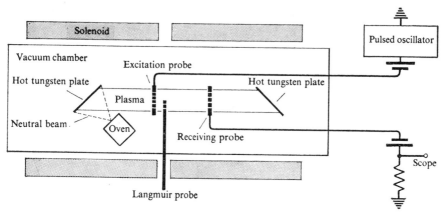

Fig. 8–5 Apparatus designed to study the properties of a quiescent plasma (after Wong, Motley, and D'Angelo[74])

confine it radially; grids perpendicular to the axis modulate the ion density and detect the transmitted modulation.

For propagation along the direction of $\mathbf{B}_0$, one has from (8–41)

$$\frac{\omega}{k} = \left(\frac{\gamma^+ \kappa T^+ + \gamma^- \kappa T^-}{m^+}\right)^{1/2} = 2{\cdot}4\sqrt{\frac{\kappa T^+}{m^+}}$$

for $T^+ = T^-$ and $\gamma^+ = \gamma^- = 3$ (for one-dimensional oscillations). Thus for $T^+ \sim 2{,}300°\mathrm{K}$ with potassium and caesium plasmas,

$$\left(\frac{\omega}{k}\right)_{\mathrm{K}} \simeq 1{\cdot}7 \times 10^5 \,\mathrm{cm/s} \qquad \left(\frac{\omega}{k}\right)_{\mathrm{Cs}} \simeq 0{\cdot}9 \times 10^5 \,\mathrm{cm/s}$$

Phase-shift measurements determined the phase velocities of the waves and confirmed that these were independent of frequency and dependent on ion mass. However, the measured phase velocities of $2{\cdot}5 \times 10^5 \,\mathrm{cm/s}$ in potassium and $1{\cdot}3 \times 10^5 \,\mathrm{cm/s}$ in caesium are appreciably higher than the values predicted by theory.

In later work Wong, Motley, and D'Angelo[74] observed that, since the plasma is produced at one end of the column, recombination losses induce a drift of plasma away from the producing plate. Accordingly, they measured the phase velocities of waves moving upstream and downstream and found these to be different:

$$\left(\frac{\omega}{k}\right)_{\mathrm{K}}^{\mathrm{up}} = 1{\cdot}3 \times 10^5 \,\mathrm{cm/s} \qquad \left(\frac{\omega}{k}\right)_{\mathrm{Cs}}^{\mathrm{up}} = 0{\cdot}9 \times 10^5 \,\mathrm{cm/s}$$

$$\left(\frac{\omega}{k}\right)_{\mathrm{K}}^{\mathrm{down}} = 2{\cdot}5 \times 10^5 \,\mathrm{cm/s} \qquad \left(\frac{\omega}{k}\right)_{\mathrm{Cs}}^{\mathrm{down}} = 1{\cdot}3 \times 10^5 \,\mathrm{cm/s}$$

Further, they calculated the phase velocity, allowing for a drift velocity V_0 (negligible compared with *electron* thermal velocity), and found

$$\frac{\omega}{k_r} = 2\cdot05\sqrt{\frac{\kappa T^+}{m^+}} \pm V_0 + \frac{0\cdot72\sqrt{\frac{\kappa T^+}{m^+}}}{2\cdot05 \pm V_0 \bigg/ \sqrt{\frac{\kappa T^+}{m^+}}} \qquad (8\text{--}50)$$

$(k = k_r + ik_i).$

Using (8–50), the theoretical and experimental phase velocities in potassium are found to be within 5 per cent of one another while in caesium the predicted velocities are ~12 per cent lower than those measured. The results are shown in Figs. 8–6 and 8–7.

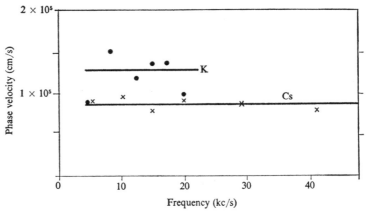

Fig. 8–6 Phase velocities of the upstream ion acoustic waves in potassium and caesium plasmas plotted as functions of frequency (after Wong, Motley, and D'Angelo[74])

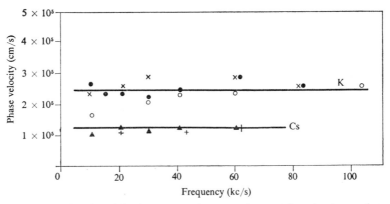

Fig. 8–7 Phase velocities of the downstream ion acoustic waves in potassium and caesium plasmas plotted as functions of frequency (after Wong, Motley, and D'Angelo[74])

In the same experiment wave damping was also examined. The damping distance δ is the distance over which the wave amplitude is attenuated by a factor e. The damping constant was calculated and found to be

$$\frac{\delta}{\lambda} = \frac{1}{2\pi}\frac{k_r}{k_i} = 0\cdot39 \pm 0\cdot19\frac{V_0}{\sqrt{(\kappa T^+/m^+)}}$$

that is, it depends on whether the wave propagates upstream ($+$) or downstream ($-$). The measured ion-wave damping in caesium is shown in Fig. 8–8. The attenuation is severe; over one wavelength the amplitude falls

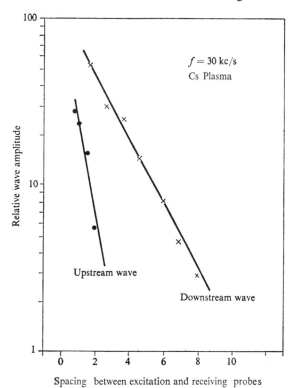

Fig. 8-8 Measured damping in upstream and downstream ion acoustic waves in a caesium plasma (after Wong, Motley, and D'Angelo[74])

exponentially by a factor of 6 downstream and by a factor of 50 upstream. Figures 8–9 and 8–10 show the frequency dependence of the damping distance δ. Observe that δ^{-1} varies linearly with ω in both potassium and caesium; that is, the damping distance is a constant fraction of a wavelength. Further, δ was found to be independent of plasma density between $2 \times 10^{10} - 3 \times 10^{11}$ ions/cm³.

As already mentioned, the ions in this experiment are to a good approximation, collision-free. Ion–electron collisions are ineffective in damping ion waves because of the large value of m^+/m^-; for the conditions in this experiment $\delta/\lambda \sim 10^{11}$ for a frequency of 100 kc/s so that the damping is wholly

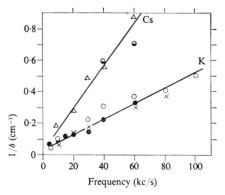

Fig. 8–9 Frequency dependence of the reciprocal of the damping distance δ for downstream ion acoustic waves in potassium and caesium plasmas (after Wong, Motley, and D'Angelo[74])

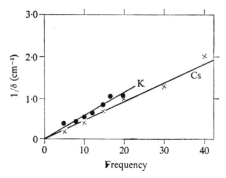

Fig. 8–10 Frequency dependence of the reciprocal of the damping distance δ for upstream ion acoustic waves in potassium and caesium plasmas (after Wong, Motley, and D'Angelo[74])

negligible. Collisions between ions and atoms (charge-exchange collisions) lead to values of δ/λ which, typically, are 10^3; again, damping is negligible. Ion–ion collisions (or in macroscopic terms, viscosity) are less readily ruled out but none the less Wong, Motley, and D'Angelo believe that the plasma may be represented by the collisionless theory. The ion waves are subject

Table 8–1   Ion-wave damping

	$\left(\dfrac{\delta}{\lambda}\right)_{K}^{\text{theory}}$	$\left(\dfrac{\delta}{\lambda}\right)_{K}^{\text{expt}}$	$\left(\dfrac{\delta}{\lambda}\right)_{Cs}^{\text{theory}}$	$\left(\dfrac{\delta}{\lambda}\right)_{Cs}^{\text{expt}}$
Downstream wave	0·64	0·65	0·51	0·55
Upstream wave	0·14	0·14	0·27	0·25

to the same sort of collisionless damping found by Landau for electron plasma oscillations; that is, these waves are damped through interaction with ions moving with velocities close to the wave phase velocity. Collisionless theory predicts an exponential decay of ion acoustic waves in plasmas having $T^+ = T^-$ and that the damping distance should be a constant fraction of a wavelength. The excellent agreement between theory and experiment is shown in Table 8–1.

8–6–2   *Electron plasma waves*

Many experiments going back over several decades provide evidence for the existence of longitudinal plasma waves. In this discussion, only one investigation by Malmberg and Wharton[75, 76] is considered since this examined not

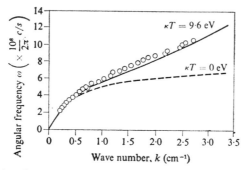

Fig. 8-11   Comparison between experimental results (circles) and the dispersion plot for electron plasma waves; the solid line corresponds to a calculation using the measured plasma temperature while the dashed line corresponds to a cold plasma. Note that as $k \to 0$, $\omega \to 0$ rather than ω_{pe} (cf. (8–39)); this is on account of the finite length of the apparatus which cuts off waves of long wavelengths (after Malmberg and Wharton[76])

only the dispersion relation but provided evidence of the Landau damping of these waves.

The plasma is created in a duoplasmatron arc and drifts from the source into a long uniform magnetic field of several hundred gauss. The magnetic field at the orifice of the duoplasmatron is typically a few kilogauss so that a powerful magnetic mirror exists at one end of the device. At the other end, a negatively charged plate reflects electrons back into the plasma; the ions on the other hand 'flow' through the device and provide a positive charge background. The plasma so contained is in a steady state.

The cylindrical plasma column is 230 cm long and the central density is typically 10^8 to 10^9 electrons/cm³. The background pressure is of the order of 10^{-5} torr (mostly hydrogen). Temperatures can range from 5 eV to 20 eV so that a characteristic Debye length is 0·1 cm, and the number of particles in a Debye sphere, $n\lambda_D^3 \sim 10^6$. The electron mean free path for electron–ion collisions is of the order of 10^5 cm, and that for electron–neutral collisions approximately 4.10^3 cm. Further, an electron velocity analyser was used to establish that the electron velocity distribution function was Maxwellian. Consequently, the plasma created in the device meets all the prerequisites for the collisionless theory to be valid.

The experiment was carried out using two probes one of which, the transmitter, was set at a series of fixed frequencies while the receiving probe, at each setting, was moved longitudinally. From the data, the real and imaginary parts of the wave number were obtained as functions of frequency. A dispersion plot is shown in Fig. 8–11. In comparing this plot with the theoretical dispersion relation, $k = k(\omega)$, Malmberg and Wharton chose a value of electron density which normalized the theoretical curve to the experimental data at low frequencies; in this region (high phase velocities) temperature corrections to the dispersion relation should be negligible. They followed this course rather than rely on the absolute values of electron density from Langmuir probe data though these were in fact consistent with those inferred from the normalization procedure. The curve so obtained is the solid line in Fig. 8–11 and shows excellent agreement with the experimental points. The small difference which does exist would be eliminated by a 5 per cent increase in the mean thermal velocity used over that found from the velocity analyser; this is within the accuracy of the calibration of the analyser. One should note that the theoretical curve is not simply a plot of (8–39) in this case; it will appear in Chapter 10 that the terms in (8–39) are just the leading terms in an asymptotic series and, under the conditions of this experiment, this series is only weakly convergent so that the first four or five terms have to be kept.

In Fig. 8–12, the measured damping is compared with that predicted by theory. The ordinate is the ratio of the imaginary to the real part of the longitudinal wave number and the abscissa that of $(\omega/kV_e)^2 \equiv (v_p/V_e)^2$. The ratio k_i/k_r and the phase velocity ω/k are found directly from the experiment; as the electron velocity distribution was shown to be Maxwellian and T^- measured, V_e is also known experimentally. It is clear that in a collisionless plasma, electron plasma waves exhibit severe exponential damping. The observed damping lengths range from 2 cm to 50 cm,

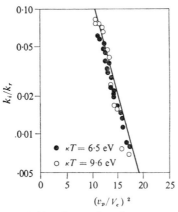

Fig. 8–12 Comparison between experimental results and theoretically predicted Landau damping of electron plasma waves (after Malmberg and Wharton[76])

while the electron mean free path characterizing the collision damping length is about 4.10^3 cm. The magnitude of this damping, and its dependence on phase velocity and on electron temperature, agree with the behaviour predicted for Landau damping.

8–7 General dispersion relation

The general dispersion relation gives six roots for ω^2 corresponding to each wave number, which fall naturally into a high-frequency group and a low-frequency one. The high-frequency group, for propagation along B_0, consists of right and left circularly polarized electromagnetic waves and the longitudinal electron plasma wave. The low-frequency group has been examined earlier in this chapter.

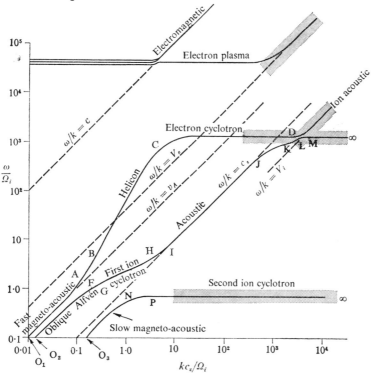

Fig. 8–13 Dispersion curves for waves in a hydrogen plasma with $\beta = 10^{-2}$ and having $v_A/c = 10^{-3}$, $v_A/c_s = 10$, $V_i/c_s = 0.33$, $\Omega_e < \omega_{pe}$ and $\theta = 45°$. Hatched areas represent regions in which waves are damped so that propagation does not take place; the damping is *not* described by the hydromagnetic equations (after Stringer[77])

We shall not draw the CMA diagram for warm plasma waves, as the wave-normal surfaces are now considerably more complicated than in the cold plasma limit.[64] Instead, an (ω, k) dispersion plot is shown in Fig. 8–13, the low-frequency part of which is due to Stringer[77]; the high-frequency curves come from the Appleton–Hartree dispersion relation (cf. Section 7–7–1) in which ion motion and electron pressure are ignored. This plot refers to an electron–proton plasma having $\beta = 10^{-2}$.† Other parameters have the values

† Recall that $\beta = \dfrac{8\pi P}{B_0{}^2} = \dfrac{2}{\gamma}\left(\dfrac{c_s}{v_A}\right)^2 \ll 1$ for many plasmas of interest in the laboratory and in astrophysics.

$$\frac{v_A}{c} = 10^{-3} \qquad \frac{V_i}{c_s} = 0\cdot 33 \qquad \theta = 45°$$

The value chosen for β ensures that the high- and low-frequency parts of the (ω, k) diagram are well separated. Tables 8–2 to 8–4 list the waves corresponding to each branch of the dispersion relation.

Table 8-2   Dispersion curves (Fig. 8–13): branch $O_3N\infty$

Section	Mode	Dispersion relation	Physical characteristics
O_3N	slow magneto-acoustic	$\omega^2 = k^2 c^2 \cos^2 \theta$ $kc_s \ll \Omega_i$	Similar to wave in IJ (Table 8–3); E almost longitudinal, coupling electron and ion fluids
$P\infty$	second ion cyclotron	$\omega^2 = \Omega_i^2 \cos^2 \theta$ $kc_s \gg \Omega_i$	longitudinal wave, $E \parallel k$; ions move in circular orbits $\perp k$, being inhibited from moving in Larmor orbits by the electron pressure gradient which prevents electrons from neutralizing the ion space charge.

Table 8-3   Dispersion curves (Fig. 8–13): branch $O_2J\infty$

Section	Mode	Dispersion relation	Physical characteristics
O_2F	oblique Alfvén	$\omega = kv_A \cos \theta$ $\omega \ll \Omega_i$	E elliptically polarized (in left-hand sense) in planes $\perp B_0$
GH	first ion cyclotron	$\omega^2 = \Omega_i^2 \left[1 + \dfrac{k^2 c_s^2}{\Omega_i^2} \sin^2 \theta \right.$ $\left. - \dfrac{\Omega_i^2}{k^2 v_A^2} \dfrac{(1 + \cos^2 \theta)}{\cos^2 \theta} \right]$ $\dfrac{c_s \Omega_i}{v_A \cos \theta} < kc_s \gtrsim \Omega_i$	E left-circularly polarized; existence of plasma pressure enables wave to propagate through the cold plasma resonance. Below Ω_i, magnetic energy drives the wave; above it, plasma pressure dominates
IJ	acoustic	$\omega^2 = k^2 c_s^2$ $\Omega_i < kc_s < \omega_{pi}$	longitudinal wave, $E \parallel k$; similar to acoustic wave in a collisional plasma except that here the ions and electrons are coupled by E rather than by collisions
KL	ion acoustic	$\omega^2 = k^2 c_s^2 \left[\dfrac{\gamma^+ \kappa T^+}{\gamma^+ \kappa T^+ + \gamma^- \kappa T^-} \right]$ $\omega_{pi}\left(1 + \dfrac{T^-}{T^+}\right) < kc_s < \Omega_e \cos \theta$	longitudinal wave, $E \parallel k$; electron pressure gradient increasingly inhibits electrons from neutralizing an ion space charge with $\lambda < \lambda_D$, so that ion and electron gases decouple
$M\infty$	electron cyclotron	$\omega^2 = \Omega_e^2 \cos^2 \theta \left[1 - \dfrac{\Omega_e^2 \sin^2 \theta}{\omega_{pe}^2} \right]$ $\Omega_e \cos \theta \ll kc_s$	similar to CD in Table 8–4

Table 8-4   Dispersion curves (Fig. 8–13): branch $O_1C\infty$

Section	Mode	Dispersion relation	Physical characteristics
O_1A	fast magneto-acoustic	$\omega^2 = k^2(v_A^2 + c_s^2 \sin^2 \theta)$ $\omega \ll \Omega_i$	E elliptically polarized (in right-hand sense) in planes $\perp \mathbf{B_0}$
BC	helicon or whistler	$\omega \simeq \dfrac{k^2 v_A^2}{\Omega_i} \cos \theta$ $\Omega_i \ll k v_A \cos \theta$	approximately right circularly polarized; as ω exceeds Ω_i the role of the ion gas decreases on account of its inertia
CD	electron cyclotron	$\omega^2 \simeq \Omega_e^2 \cos^2 \theta$ $\Omega_e < \omega_{pe}$	at approach to resonance at Ω_e, electron motion $\perp \mathbf{B_0}$ increases; preservation of charge neutrality demands an increase in the flow of electrons *along* $\mathbf{B_0}$ which is limited by inertia; thus growth of motion $\perp \mathbf{B_0}$ is limited, and in consequence, the approach to Ω_e

Figure 8–14 shows the effect on the dispersion relation due to choosing $\beta = 10^{-6}$ with

$$\frac{v_A}{c} = 10^{-1} \qquad \frac{V_i}{c_s} = 0{\cdot}33 \qquad \theta = 45°$$

In this case, the 'high'- and 'low'-frequency branches of the dispersion relation show some overlap.

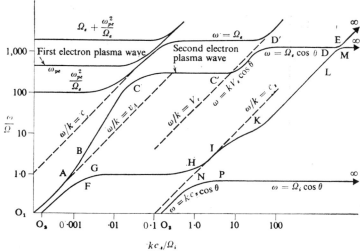

Fig. 8-14   Dispersion curves for low-frequency waves in a hydrogen plasma with $\beta = 10^{-6}$ and having $v_A/c = 10^{-1}$, $v_A/c_s = 10^3$, $\Omega_e > \omega_{pe}$ and $\theta = 45°$ (after Stringer[77])

(a) θ small

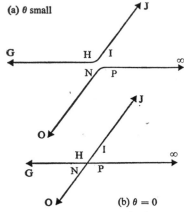

(b) $\theta = 0$

Fig. 8–15 First ion cyclotron/acoustic wave and slow magneto-acoustic/second ion cyclotron wave transition regions in passing to the limit $\theta = 0$ from small θ

In general, the dispersion curves do not change appreciably as θ is varied provided the values 0 and $\pi/2$ are avoided. As $\theta \to 0$, for example, the gap between the first ion cyclotron/acoustic wave transition (HI in Fig. 8–13) and the slow magneto-acoustic/second ion cyclotron wave transition (NP in Fig. 8–13) shrinks. In the limit, the points (H, P) and (N, I) become coincident, that is, the curves $O_2G\infty$ and O_3J now intersect. The transition from finite θ to $\theta = 0$ is shown in Fig. 8–15; the presence of a transverse magnetic field couples longitudinal and transverse wave components so that the transverse Alfvén wave passes into the longitudinal acoustic wave while on the lower frequency branch a longitudinal mode passes into a

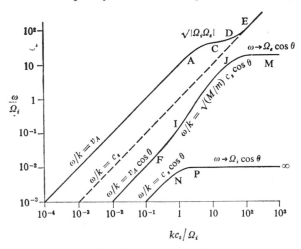

Fig. 8–16 Dispersion curves for low-frequency waves in a hydrogen plasma propagating almost orthogonally to the magnetic field for low β^- and having $v_A/c = 10^{-3}$, $v_A/c_s = 10$ and $\cos^2 \theta < m^-/m^+$

transverse mode. At $\theta = 0$, however, no such coupling occurs and the transverse shear Alfvén wave now becomes a transverse ion cyclotron wave, as in the cold plasma limit, while the other branch O_3XJ is now entirely longitudinal. A similar transition occurs between the electron cyclotron wave and the ion acoustic wave.

The situation as $\theta \to \pi/2$ is more complicated and will not be discussed; a typical dispersion plot for the three low-frequency branches is shown in Fig. 8–16. Observe that only the O_1CE branch survives in the limit $\theta = \pi/2$ and that the lower hybrid frequency (at which a resonance appeared for $\theta = \pi/2$ propagation in the cold plasma limit) now reappears as a pseudo-resonance.

Summary

See Tables 8–2 to 8–4.

Problems

8-1   Sketch the wave-normal surfaces for the magneto-acoustic and Alfvén waves described by (8–17) and (8–16b) respectively. Consider cases in which (i) $c_s < v_A$, (ii) $c_s > v_A$, and (iii) $c_s = v_A$.

8-2   Obtain the dispersion relation for waves in a warm plasma which will be valid close to the ion cyclotron frequency (i.e., (8–19)).

8-3   Obtain (8–21).

8-4   Show that the dispersion relation for ion cyclotron waves in the neighbourhood of $\omega = \Omega_i \cos \theta$ is approximately

$$\frac{k^2 c_s^2}{\omega^2} = \frac{2\Omega_i^2 c_s^2 + v_A^2(\Omega_i^2 - \omega^2)}{v_A^2(\Omega^2 \cos^2 \theta - \omega^2)}$$

8-5   Examine the behaviour of the oblique Alfvén wave as $\omega \to \Omega_i$, explaining how it is possible for this mode to persist above the ion cyclotron frequency in a warm plasma.

8-6   Check the implications of ignoring the last term in the dispersion relation (8–24).

8-7   Determine conditions under which the term

$$\frac{4(k^4 V_i^2 V_e^2 + k^2 V_i^2 \omega_{pe}^2 + k^2 V_e^2 \omega_{pi}^2)}{(\omega_{pe}^2 + \omega_{pi}^2 + k^2 V_e^2 + k^2 V_i^2)^2}$$

is *not* small compared with one.

8-8   Show that in ion acoustic waves, electron and ion velocities are of comparable magnitude.
   Hint: use (8–32) and (8–41).

8–9   Use (8–45) and (8–46) to plot the real and imaginary parts of the dispersion relation for $k^2 \ll k_D^2$.

8–10   Obtain the dispersion relation (8–50) for ion acoustic waves in a plasma which has a drift velocity V_0.

8–11   In Table 8–2 check the statement that ion cyclotron waves satisfying $\omega^2 = \Omega_i^2 \cos^2 \theta$ are electrostatic ($\mathbf{E} \parallel \mathbf{k}$).

8–12   Derive a general dispersion relation by retaining all the terms in the generalized Ohm's law (8–3). Show that for $|\,\Omega_e\,| < \omega_{pe}$ this dispersion relation will be valid for frequencies up to the electron cyclotron frequency and obtain the dispersion relation for electron cyclotron waves.

9

Plasma radiation

9–1 Introduction

A study of radiation processes in plasmas is important not only because of their intrinsic interest but also on account of their being a rich source of information about plasma conditions. Astrophysicists are almost wholly dependent on information contained in radiation from stars; plasma physicists also make use of radiation emitted from laboratory sources to determine, for example, plasma temperature and density. To those workers hopeful of achieving thermonuclear fusion by containing hot dense plasmas in various magnetic field configurations, the emission of radiation poses a problem since it represents a loss of energy from such systems.

Laboratory workers, of course, need not rely solely on measurements made using radiation *generated within* a plasma. In Chapter 7 we saw that a transverse wave of frequency ω could only propagate in an isotropic plasma provided $\omega > \omega_p$ so that a simple transmission experiment provides information about ω_p, i.e., about the electron density within the plasma. Another class of experiment which is valuable for diagnostic purposes uses the *scattering* of radiation by plasmas. The most important aspect of obtaining information about plasmas using radiation—whether by observing emitted radiation or by studying the absorption or scattering of radiation—is that one perturbs conditions in the plasma only very slightly, compared with such other diagnostic devices as the Langmuir and radio-frequency probes and the scattering of beams of particles (cf. Appendix 3).

A general treatment of radiation processes in plasmas is beyond the scope of the present discussion and we begin by imposing some sweeping limitations. In general, an ionized gas contains atomic and molecular ions which are not fully stripped of electrons (e.g., O^+, O_2^+, NO^+) so that any theory of radiation from such a system must take account of *line radiation* resulting from transitions between various atomic levels, and *band radiation* from the molecular ions. Such an account calls for a quantum mechanical treatment†
and to avoid this we shall restrict the processes considered in this chapter to those which occur in fully ionized plasmas (for instance, if the plasma is formed from helium then the *only* ions appearing are He^{++}). Our intention, then, is to discuss in a classical framework the radiation from free particles in a plasma which, in general, exists in a magnetic field. Finally, we will

† An exhaustive discussion of plasma line radiation appears in H. R. Griem, *Plasma Spectroscopy*. McGraw-Hill, New York, 1964.

deal mainly with thermal or equilibrium plasmas. This is not to say that the emission of radiation by non-thermal plasmas is unimportant or occurs infrequently; on the contrary, since instabilities seem to be almost endemic in plasmas, some understanding of non-thermal processes is vital if one is to interpret the radiation emitted. The reason for avoiding non-thermal emission is simply that it is not readily described mathematically.

9–2 Summary of electrodynamic results

The results of Maxwellian electrodynamics are conveniently summarized in the form[16]

$$\left.\begin{array}{c} \left[\nabla^2 - \dfrac{1}{c^2}\dfrac{\partial^2}{\partial t^2}\right]\mathbf{A} = -\dfrac{4\pi}{c}\mathbf{j} \\[3mm] \left[\nabla^2 - \dfrac{1}{c^2}\dfrac{\partial^2}{\partial t^2}\right]\phi = -4\pi q \end{array}\right\} \tag{9-1a}$$

together with the Lorentz condition on the potentials

$$\nabla . \mathbf{A} + \frac{1}{c}\frac{\partial \phi}{\partial t} = 0 \tag{9-1b}$$

The electric and magnetic fields are then given by

$$\mathbf{E} = -\nabla\phi - \frac{1}{c}\frac{\partial \mathbf{A}}{\partial t} \qquad \mathbf{B} = \nabla \times \mathbf{A} \tag{9-2}$$

To solve (9–1a) and (9–1b) requires a Green function satisfying

$$\left[\nabla^2 - \frac{1}{c^2}\frac{\partial^2}{\partial t^2}\right]G(\mathbf{r}, t; \mathbf{r}', t') = -4\pi\delta(\mathbf{r} - \mathbf{r}')\,\delta(t - t') \tag{9-3}$$

where the delta functions now replace the source terms in (9–1a). The Green function $G(\mathbf{r}, t; \mathbf{r}', t')$ will also be required to satisfy certain boundary conditions. Since the right-hand side of (9–3) has the representation

$$\delta(\mathbf{r} - \mathbf{r}')\,\delta(t - t') = \frac{1}{(2\pi)^4}\int d\mathbf{k}\int d\omega \exp\left[i\mathbf{k}.(\mathbf{r} - \mathbf{r}') - i\omega(t - t')\right]$$

it is convenient to write

$$G(\mathbf{r}, t; \mathbf{r}', t') = \frac{1}{(2\pi)^2}\int d\mathbf{k}\int d\omega\, G(\mathbf{k}, \omega) \exp\left[i\mathbf{k}.(\mathbf{r} - \mathbf{r}') - i\omega(t - t')\right] \tag{9-4}$$

One finds from the Fourier transform of (9–3)

$$G(\mathbf{k}, \omega) = \frac{1}{\pi}\frac{1}{k^2 - (\omega^2/c^2)} \tag{9-5}$$

Thus in attempting to determine $G(\mathbf{r}, t; \mathbf{r}', t')$ from (9–4), one is faced with singularities in the integrand at $\omega = \pm kc$; without some prescription for dealing with these singularities the solution has no meaning. To find such a prescription, consider (9–3) from a physical standpoint. G represents a wave created by a point source at $\mathbf{r}'$ which is pulsed at time t', and this wave propagates onwards from the source with a velocity c. It is therefore reasonable to require that (i) $G = 0$ over the whole domain for $t < t'$ and (ii) G should represent outward propagating waves for $t > t'$. With these demands, one can carry out the integrations in (9–4); one finds (cf. Problem 9–1)

$$G(\mathbf{r}, t; \mathbf{r}', t') = \frac{\delta\left(t' + \dfrac{|\mathbf{r} - \mathbf{r}'|}{c} - t\right)}{|\mathbf{r} - \mathbf{r}'|} \tag{9–6}$$

This is the *retarded Green function* which expresses the fact that the signal observed at some *field point* $\mathbf{r}$ at time t is due to a disturbance at a *source point* $\mathbf{r}'$ at an *earlier* (retarded) time $t' = t - \dfrac{|\mathbf{r} - \mathbf{r}'|}{c}$. Knowing the Green function, one may at once write the solutions to the electrodynamic equations (9–1a):

$$\mathbf{A}(\mathbf{r}, t) = \frac{1}{c} \int \int G(\mathbf{r}, t; \mathbf{r}', t') \mathbf{j}(\mathbf{r}', t') \, d\mathbf{r}' \, dt'$$

$$\left. = \frac{1}{c} \int \frac{[\mathbf{j}(\mathbf{r}', t')]_{t'=t-\frac{|\mathbf{r}-\mathbf{r}'|}{c}}}{|\mathbf{r} - \mathbf{r}'|} \, d\mathbf{r}' \right\} \tag{9–7}$$

and $$\phi(\mathbf{r}, t) = \int \frac{[q(\mathbf{r}', t')]_{t'=t-\frac{|\mathbf{r}-\mathbf{r}'|}{c}}}{|\mathbf{r} - \mathbf{r}'|} \, d\mathbf{r}'$$

These are known as the *retarded potentials*.

9–2–1 *Liénard–Wiechert potentials*

Consider now a rather special source, namely a single particle moving arbitrarily with velocity $\dot{\mathbf{r}}_0(t)$ at the point $\mathbf{r}_0(t)$. Then,

$$\mathbf{j}(\mathbf{r}, t) = e\dot{\mathbf{r}}_0(t) \, \delta(\mathbf{r} - \mathbf{r}_0(t)) \tag{9–8}$$

Using (9–8) in (9–7) gives†

$$\mathbf{A}(\mathbf{r}, t) = \frac{e}{c} \int \int \frac{\dot{\mathbf{r}}_0(t') \, \delta\left(t' + \dfrac{|\mathbf{r} - \mathbf{r}'|}{c} - t\right)}{|\mathbf{r} - \mathbf{r}'|} \, \delta(\mathbf{r}' - \mathbf{r}_0(t')) \, d\mathbf{r}' \, dt'$$

† Note that the dot implies differentiation with respect to the argument.

Now the integration over $\mathbf{r}'$ may be done at once to give

$$\mathbf{A}(\mathbf{r}, t) = \frac{e}{c} \int \frac{\dot{\mathbf{r}}_0(t')\, \delta\!\left(t' + \dfrac{R(t')}{c} - t\right) dt'}{R(t')} \tag{9-9}$$

where $R(t') = |\mathbf{r} - \mathbf{r}_0(t')|$. Integrating over t' is not quite so straightforward, because $R(t')$ appears in the argument of the delta function. However, by using the result† (cf. Problem 9–2)

$$\int_{-\infty}^{\infty} \delta\{g(x)\}\, \gamma(x)\, dx = \sum_i \frac{\gamma(x_i)}{\left|\dfrac{dg(x_i)}{dx}\right|}$$

where the x_i are the zeros of $g(x) = 0$, the integration over t' yields

$$\mathbf{A}(\mathbf{r}, t) = \frac{e}{c}\left[\frac{\dot{\mathbf{r}}_0(t')}{R\!\left(1 + \dfrac{1}{c}\dfrac{dR}{dt'}\right)}\right]_{t' = t - \frac{R(t')}{c}} \tag{9-10}$$

and

$$\frac{dR}{dt'} = \frac{(\mathbf{r}_0(t') - \mathbf{r}) \cdot \dot{\mathbf{r}}_0(t')}{|\mathbf{r} - \mathbf{r}_0(t')|} = -\hat{\mathbf{n}}(t') \cdot \mathbf{v}(t')$$

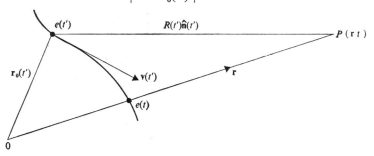

Fig. 9–1 Source-observer geometry for electron moving arbitrarily

where $\hat{\mathbf{n}}(t')$ is the unit vector from the source to the field point (Fig. 9–1) and $\mathbf{v}(t') \equiv \dot{\mathbf{r}}_0(t')$. Thus,

$$\mathbf{A}(\mathbf{r}, t) = \left[\frac{e\mathbf{v}}{cR - \mathbf{v} \cdot \mathbf{R}}\right]_{t' = t - \frac{R(t')}{c}} \tag{9-11a}$$

and likewise, since $q(\mathbf{r}, t) = e\, \delta(\mathbf{r} - \mathbf{r}_0(t))$,

$$\phi(\mathbf{r}, t) = \left[\frac{ec}{cR - \mathbf{v} \cdot \mathbf{R}}\right]_{t' = t - \frac{R(t')}{c}} \tag{9-11b}$$

These expressions are known as the *Liénard–Wiechert potentials*.

† Cf. D. S. Jones, *Generalized Functions*, p. 247. McGraw-Hill, New York, 1966.

9-2-2 *Fields from a point charge*

To obtain expressions for the fields **E** and **B**, one could use (9–11a) and (9–11b) in (9–2). However, it is more straightforward to return to (9–9) and the equivalent expression for $\phi(\mathbf{r}, t)$. Then,

$$
\mathbf{E}(\mathbf{r}, t) = -e\hat{\mathbf{n}}\frac{\partial}{\partial R}\int \frac{\delta\!\left(t' + \dfrac{R(t')}{c} - t\right)}{R(t')}\, dt'
$$

$$
- \frac{e}{c^2}\frac{\partial}{\partial t}\int \frac{\mathbf{v}(t')\,\delta\!\left(t' + \dfrac{R(t')}{c} - t\right)}{R(t')}\, dt'
$$

$$
= e\int \left[\frac{\hat{\mathbf{n}}}{R^2}\delta\!\left(t' + \frac{R(t')}{c} - t\right) + \frac{1}{cR}(\boldsymbol{\beta} - \hat{\mathbf{n}})\,\delta'\!\left(t' + \frac{R(t')}{c} - t\right)\right] dt'
$$

where $\boldsymbol{\beta}(t') = \mathbf{v}(t')/c$ and the prime on the delta function means differentiation with respect to the argument. On integrating the second term by parts (cf. Problem 9–3), it follows that

$$
\mathbf{E}(\mathbf{r}, t) = e\left[\frac{\hat{\mathbf{n}}}{gR^2} + \frac{1}{cg}\frac{d}{dt'}\!\left(\frac{\hat{\mathbf{n}} - \boldsymbol{\beta}}{gR}\right)\right]_{\text{ret}} \tag{9–12}
$$

where $g = 1 - \hat{\mathbf{n}}.\boldsymbol{\beta}$ and the label 'ret' means that the expression within the brackets is to be evaluated at the retarded time $t' = t - R(t')/c$.

It is instructive to rewrite (9–12) in a form due to Feynman.[78] From the result (cf. Problem 9–4)

$$
\frac{1}{c}\frac{d\hat{\mathbf{n}}}{dt'} = \frac{\hat{\mathbf{n}} \times (\hat{\mathbf{n}} \times \boldsymbol{\beta})}{R}
$$

it follows on rearranging that

$$
\frac{\boldsymbol{\beta}}{R} = \frac{(\hat{\mathbf{n}}.\boldsymbol{\beta})\hat{\mathbf{n}}}{R} - \frac{1}{c}\frac{d\hat{\mathbf{n}}}{dt'} \tag{9–13}
$$

Substituting (9–13) in (9–12) and using $dt/dt' = g$ gives

$$
\mathbf{E}(\mathbf{r}, t) = e\left[\frac{\hat{\mathbf{n}}}{R^2} + \frac{(\hat{\mathbf{n}}.\boldsymbol{\beta})\hat{\mathbf{n}}}{gR^2} + \frac{1}{c}\frac{d}{dt}\!\left(\frac{\hat{\mathbf{n}}}{gR}\right)\right.
$$

$$
\left. - \frac{1}{c}\frac{d}{dt}\!\left\{\frac{(\hat{\mathbf{n}}.\boldsymbol{\beta})\hat{\mathbf{n}}}{gR}\right\} + \frac{1}{c^2}\frac{d^2\hat{\mathbf{n}}}{dt^2}\right]_{\text{ret}} \tag{9–14}
$$

The second, third, and fourth terms of this expression for **E** combine (cf.

Problem 9–5) to give

$$\mathbf{E}(\mathbf{r}, t) = e\left[\frac{\hat{\mathbf{n}}}{R^2} + \frac{R}{c}\frac{d}{dt}\left(\frac{\hat{\mathbf{n}}}{R^2}\right) + \frac{1}{c^2}\frac{d^2\hat{\mathbf{n}}}{dt^2}\right]_{\text{ret}} \qquad (9\text{–}15)$$

The first term on the right-hand side represents the Coulomb field of the charge e at its retarded position. The second term is a correction to the retarded Coulomb field, being the product of the rate of change of this field and the retardation delay R/c. So the first two terms correspond to computing the retarded Coulomb field and extrapolating it forward in time by R/c, namely to the observer's time t; if this field changes slowly then the correction term compensates the effect of retardation. In other words, for slowly changing fields the first two terms approximate the *instantaneous* Coulomb field of the charge. The final term is the second time derivative of the unit vector from the retarded position of the charge to the observer; this term is vital to a study of radiation, since the first two terms are Coulombic ($\sim R^{-2}$). It is easy to show (cf. Problem 9–6) that the radiation terms ($\sim R^{-1}$) contained in $\dfrac{1}{c^2}\dfrac{d^2\hat{\mathbf{n}}}{dt^2}$ are just

$$\frac{(\hat{\mathbf{n}} - \boldsymbol{\beta})}{cg^3 R}(\hat{\mathbf{n}}\cdot\dot{\boldsymbol{\beta}}) - \frac{\dot{\boldsymbol{\beta}}}{cg^2 R} = \frac{\hat{\mathbf{n}} \times [(\hat{\mathbf{n}} - \boldsymbol{\beta}) \times \dot{\boldsymbol{\beta}}]}{cg^3 R} \qquad (9\text{–}16)$$

Thus the *radiation electric field* contained in (9–15) is

$$\mathbf{E}^{\text{rad}}(\mathbf{r}, t) = e\left[\frac{\hat{\mathbf{n}} \times \{(\hat{\mathbf{n}} - \boldsymbol{\beta}) \times \dot{\boldsymbol{\beta}}\}}{cg^3 R}\right]_{\text{ret}} \qquad (9\text{–}17)$$

9–3   Angular distribution from an accelerated charge

Once the radiation field is known, one can compute the Poynting vector $\mathbf{S}$ and so determine the instantaneous flux of energy (cf. Section 7–2)

$$\mathbf{S} = \frac{c}{4\pi} |\mathbf{E}^{\text{rad}}|^2 \,\hat{\mathbf{n}} \qquad (9\text{–}18)$$

with $\mathbf{E}^{\text{rad}}$ given by (9–17). Thus the power radiated per unit solid angle, $dP/d\Omega$, is

$$\frac{dP(t)}{d\Omega} = (\mathbf{S}\cdot\hat{\mathbf{n}})R^2 = \frac{c}{4\pi}|R\mathbf{E}^{\text{rad}}|^2 \qquad (9\text{–}19)$$

Note that $dP(t)$ is the energy per unit time measured by the observer at time t due to the emission by the charge at time t'. It is more useful to consider $dP(t')$, the energy per unit time at the charge. Then

$$\frac{dP(t')}{d\Omega} = (\mathbf{S}.\hat{\mathbf{n}})R^2 \frac{dt}{dt'} = g(\mathbf{S}.\hat{\mathbf{n}})R^2 \tag{9-20}$$

In the non-relativistic limit, $g \rightarrow 1$ and if θ is the angle between $\dot{\mathbf{v}}$ and $\hat{\mathbf{n}}$, (9–20) gives

$$\frac{dP}{d\Omega} = \frac{e^2\dot{v}^2}{4\pi c^3} \sin^2\theta \tag{9-21}$$

The relativistic case is more complicated; one has

$$\mathbf{S}.\hat{\mathbf{n}} = \frac{e^2}{4\pi c}\left[\frac{|\hat{\mathbf{n}} \times \{(\hat{\mathbf{n}} - \boldsymbol{\beta}) \times \dot{\boldsymbol{\beta}}\}|^2}{g^6 R^2}\right]_{\text{ret}} \tag{9-22}$$

Relativistic effects appear both in the numerator and, through g, in the denominator; obviously in the ultra-relativistic limit ($\beta \rightarrow 1$) the denominator will provide the dominant effect. Substituting (9–22) in (9–20) gives

$$\frac{dP(t')}{d\Omega} = \frac{e^2}{4\pi c} \frac{|\hat{\mathbf{n}} \times \{(\hat{\mathbf{n}} - \boldsymbol{\beta}) \times \dot{\boldsymbol{\beta}}\}|^2}{g^5} \tag{9-23}$$

As an example, consider the special case in which $\boldsymbol{\beta}$ and $\dot{\boldsymbol{\beta}}$ are collinear giving

$$\frac{dP(t')}{d\Omega} = \frac{e^2\dot{\beta}^2}{4\pi c} \frac{\sin^2\theta}{(1 - \beta\cos\theta)^5} \tag{9-24}$$

As $\beta \rightarrow 1$, the denominator begins to play a dominant role in determining the radiation pattern; the characteristic dipole distribution deforms with the lobes switching towards the forward direction and growing in size (Fig. 9–2).

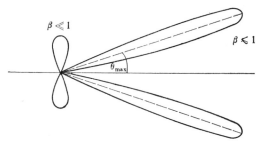

Fig. 9–2 Angular distribution of the power radiated by an accelerated charge whenever its velocity and acceleration are collinear

The angle at which the peak power is radiated is given by

$$\cos\theta_{\text{max}} = \frac{1}{3\beta}[(1 + 15\beta^2)^{1/2} - 1] \tag{9-25}$$

The case in which the charged particle executes instantaneously circular motion (so that $\dot{\beta} \perp \beta$) is the subject of Problem 9–9.

In the non-relativistic case, the power radiated in *all* directions is found simply by integrating (9–21), so that

$$P = \frac{2e^2\dot{v}^2}{3c^3} \tag{9-26}$$

which is the *Larmor formula*. The equivalent relativistic expression may be shown (cf. Problem 9–10) to be given by

$$P = \frac{2e^2}{3c} \frac{[(\dot{\beta})^2 - (\beta \times \dot{\beta})^2]}{(1 - \beta^2)^5} \tag{9-27}$$

From (9–27), the total power radiated in all directions for the case of β, $\dot{\beta}$ collinear is just

$$P(t') = \frac{2e^2}{3c} \frac{\dot{\beta}^2}{(1 - \beta^2)^3} \tag{9-28a}$$

(Alternatively, this follows on integrating (9–24).) For $\beta \perp \dot{\beta}$,

$$P(t') = \frac{2e^2}{3c} \frac{\dot{\beta}^2}{(1 - \beta^2)^2} \tag{9-28b}$$

9–4 Frequency spectrum of radiation from an accelerated charge

So far we have considered the angular distribution of radiated power. Next we ask how the radiated energy is distributed in *frequency*. It is appropriate to use time t (the observer's time) so that

$$\frac{dP(t)}{d\Omega} = \frac{e^2}{4\pi c} \left[\frac{|\hat{n} \times \{(\hat{n} - \beta) \times \dot{\beta}\}|^2}{g^6} \right]_{ret} \equiv |a(t)|^2 \tag{9-29}$$

Introducing the Fourier transform of $a(t)$,

$$a(\omega) = \frac{1}{\sqrt{(2\pi)}} \int_{-\infty}^{\infty} a(t)e^{i\omega t}\,dt$$

and using Parseval's theorem,[79] the total energy radiated per unit solid angle is

$$\frac{dW}{d\Omega} = \int_{-\infty}^{\infty} |a(t)|^2\,dt = \int_{-\infty}^{\infty} |a(\omega)|^2\,d\omega \tag{9-30}$$

Thus
$$\frac{dW}{d\Omega} = \int_0^{\infty} [\,|a(\omega)|^2 + |a(-\omega)|^2\,]\,d\omega$$

$$= \int_0^\infty \frac{d^2 W(\omega)}{d\Omega \, d\omega} \, d\omega \qquad (9\text{--}31)$$

where $d^2 W(\omega)/d\Omega \, d\omega$ is the energy radiated per unit solid angle per unit frequency interval. Calculating $\mathbf{a}(\omega)$, one finds

$$\mathbf{a}(\omega) = \sqrt{\left(\frac{e^2}{8\pi^2 c}\right)} \int_{-\infty}^\infty e^{i\omega t} \left[\frac{\hat{\mathbf{n}} \times \{(\hat{\mathbf{n}} - \boldsymbol{\beta}) \times \dot{\boldsymbol{\beta}}\}}{g^3}\right]_{\text{ret}} dt \qquad (9\text{--}32)$$

To simplify this expression, we change the variable of integration from t to t', so removing the calculation of the term in square brackets at the retarded time; then

$$\mathbf{a}(\omega) = \sqrt{\left(\frac{e^2}{8\pi^2 c}\right)} \int_{-\infty}^\infty \exp\left\{i\omega\left(t' + \frac{R(t')}{c}\right)\right\} \left[\frac{\hat{\mathbf{n}} \times \{(\hat{\mathbf{n}} - \boldsymbol{\beta}) \times \dot{\boldsymbol{\beta}}\}}{g^2}\right] dt' \qquad (9\text{--}33)$$

Since we wish to determine the spectrum in the radiation zone ($r \gg r_0$ in Fig. 9–1), $\hat{\mathbf{n}}$ is effectively time-independent and $R(t') \simeq r - \hat{\mathbf{n}} \cdot \mathbf{r}_0(t')$ so that

$$\frac{d^2 W(\omega)}{d\Omega \, d\omega} = \frac{e^2}{4\pi^2 c} \left| \int_{-\infty}^\infty \exp\left\{i\omega\left(t' - \frac{\hat{\mathbf{n}} \cdot \mathbf{r}_0(t')}{c}\right)\right\} \left[\frac{\hat{\mathbf{n}} \times \{(\hat{\mathbf{n}} - \boldsymbol{\beta}) \times \dot{\boldsymbol{\beta}}\}}{g^2}\right] dt' \right|^2 \qquad (9\text{--}34)$$

Thus the energy radiated per unit solid angle per unit frequency interval is determined as a function of ω and $\hat{\mathbf{n}}$ once $\mathbf{r}_0(t')$ is prescribed.

For purposes of calculation, we cast (9–34) in a slightly different form. Observing that (cf. Problem 9–11)

$$\frac{\hat{\mathbf{n}} \times \{(\hat{\mathbf{n}} - \boldsymbol{\beta}) \times \dot{\boldsymbol{\beta}}\}}{g^2} = \frac{d}{dt'}\left[\frac{\hat{\mathbf{n}} \times (\hat{\mathbf{n}} \times \boldsymbol{\beta})}{g}\right]$$

we integrate (9–34) by parts to find, in the radiation zone,

$$\frac{d^2 W(\omega)}{d\Omega \, d\omega} = \frac{e^2 \omega^2}{4\pi^2 c} \left| \int_{-\infty}^\infty \exp\left\{i\omega\left(t' - \frac{\hat{\mathbf{n}} \cdot \mathbf{r}_0(t')}{c}\right)\right\} [\hat{\mathbf{n}} \times (\hat{\mathbf{n}} \times \boldsymbol{\beta})] \, dt' \right|^2 \qquad (9\text{--}35)$$

9–5 Cyclotron radiation by an electron

We now examine some specific radiation problems. It should be clear from the last section (more obviously from (9–34) rather than the form (9–35) to be used in the calculations following) that a necessary criterion for the emission of radiation by a charged particle is the existence of particle acceleration. Consider a charged particle moving in a uniform static magnetic field as discussed in Chapter 2. The radiation from this source is known generally as *cyclotron radiation*; we consider only radiation from an electron since the intensity of ion cyclotron radiation is very much lower.

For low electron energies it will be shown that the cyclotron emission occurs as a 'line' at the electron Larmor frequency. For higher electron energies one finds radiation emitted at *harmonics* of the electron Larmor frequency in addition to the fundamental. The emission from highly relativistic electrons differs in that it is a smooth function of frequency, in contrast to the non-relativistic case, and is sometimes distinguished by the name *synchrotron radiation.*†

In discussing the motion of a charged particle in a static uniform magnetic field $\mathbf{B}_0$ in Chapter 2 we neglected radiation, so that the total energy of the particle was a constant of the motion. We shall now assume—and later justify—that the energy is effectively constant, that is, the energy change per complete orbit is negligible compared with $E = [(mc^2)^2 + (pc)^2]^{1/2}$, where m is the rest mass and $\mathbf{p}$ the electron momentum. Choosing the z-axis in the direction of $\mathbf{B}_0$, the motion is given by (cf. Problem 2–12)

$$\dot{x} = v_\perp \cos(\Omega t + \alpha) \qquad \dot{y} = -v_\perp \sin(\Omega t + \alpha) \qquad \dot{z} = v_\parallel \qquad (9\text{–}36)$$

$$\left.\begin{aligned} x &= (v_\perp/\Omega) \sin(\Omega t + \alpha) + x_0 \\ y &= (v_\perp/\Omega) \cos(\Omega t + \alpha) + y_0 \\ z &= v_\parallel t + z_0 \end{aligned}\right\} \qquad (9\text{–}37)$$

in which α, x_0, y_0, z_0 are determined by initial conditions. The electron

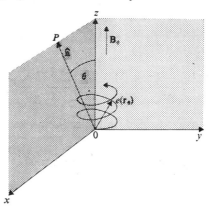

Fig. 9–3 Source–observer geometry for electron moving in a helix in a uniform magnetic field

moves on a helix with axis parallel to $\mathbf{B}_0$ (Fig. 9–3). Suppose $\alpha = 0$, $x_0 = 0 = y_0 = z_0$; then

$$\left.\begin{aligned} \mathbf{v} &= \hat{\imath} v_\perp \cos \Omega t - \hat{\jmath} v_\perp \sin \Omega t + \hat{k} v_\parallel \\ \mathbf{r}_0 &= \hat{\imath}(v_\perp/\Omega) \sin \Omega t + \hat{\jmath}(v_\perp/\Omega) \cos \Omega t + \hat{k} v_\parallel t \end{aligned}\right\} \qquad (9\text{–}38)$$

† In the Russian literature, the term *magnetic bremsstrahlung* is widely used in place of synchrotron radiation.

where
$$\Omega = \Omega_0(1 - \beta^2)^{1/2} \qquad \Omega_0 = -\frac{eB_0}{mc} \qquad (9\text{-}39)$$

(since $-e$ is the electronic charge). We wish to use (9–38) in (9–35) to calculate the energy radiated per unit solid angle per unit frequency interval. Suppose that the axes are so oriented that the observer is in the Oxz plane; then (Fig. 9–3)

$$\hat{\mathbf{n}} = \hat{\mathbf{i}} \sin \theta + \hat{\mathbf{k}} \cos \theta \qquad (9\text{-}40)$$

The exponential term in (9–35) becomes

$$\exp\left\{i\omega\left(t - \frac{\hat{\mathbf{n}}.\mathbf{r}_0(t)}{c}\right)\right\} = \exp\left\{i\omega\left[t - \frac{\beta_\perp}{\Omega}\sin\theta\sin\Omega t - \beta_\parallel t\cos\theta\right]\right\} \quad (9\text{-}41)$$

where the prime has now been dropped (since time is simply the variable of integration in (9–35)). From the theory of Bessel functions,[80] one has

$$\exp(ix\sin y) = \sum_{l=-\infty}^{\infty} J_l(x)\exp ily \qquad (9\text{-}42)$$

giving

$$\exp\left\{i\omega\left[t - \frac{\hat{\mathbf{n}}.\mathbf{r}_0(t)}{c}\right]\right\}$$
$$= \sum_{l=-\infty}^{\infty} J_l\left(\frac{\omega}{\Omega}\beta_\perp \sin\theta\right)\exp[i(\omega - l\Omega - \omega\beta_\parallel\cos\theta)t] \quad (9\text{-}43)$$

Straightforward vector algebra shows that

$$\hat{\mathbf{n}} \times (\hat{\mathbf{n}} \times \boldsymbol{\beta}) = \hat{\mathbf{i}}[-\beta_\perp\cos\Omega t\cos^2\theta + \beta_\parallel\sin\theta\cos\theta] + \hat{\mathbf{j}}[\beta_\perp\sin\Omega t]$$
$$+ \hat{\mathbf{k}}[\beta_\perp\cos\Omega t\sin\theta\cos\theta - \beta_\parallel\sin^2\theta]$$
$$\equiv X\hat{\mathbf{i}} + Y\hat{\mathbf{j}} + Z\hat{\mathbf{k}} \qquad (9\text{-}44)$$

and substitution of (9–43) and (9–44) in (9–35) gives

$$\frac{d^2W(\omega)}{d\Omega\,d\omega} = \frac{e^2\omega^2}{4\pi^2c}\left|\int_{-\infty}^{\infty}\sum_{l=-\infty}^{\infty} J_l\left(\frac{\omega}{\Omega}\beta_\perp\sin\theta\right)\exp[i(\omega - l\Omega - \omega\beta_\parallel\cos\theta)t]\right.$$
$$\left.\vphantom{\int}.[X\hat{\mathbf{i}} + Y\hat{\mathbf{j}} + Z\hat{\mathbf{k}}]\,dt\right|^2 \quad (9\text{-}45)\dagger$$

The integrals in (9–45) are of the form

$$\int_{-\infty}^{\infty}\begin{pmatrix}1\\\sin\Omega t\\\cos\Omega t\end{pmatrix}\exp[i(\omega - l\Omega - \omega\beta_\parallel\cos\theta)t]\,dt$$

† Observe that in all formulae in this chapter Ω is the relativistic Larmor frequency and $d\Omega$ the element of solid angle.

$$= 2\pi \begin{pmatrix} \delta(l\Omega - \omega[1 - \beta_\| \cos\theta]) \\ \frac{1}{2i}[\delta(\overline{l-1}\,\Omega - \omega[1 - \beta_\| \cos\theta]) - \delta(\overline{l+1}\,\Omega - \omega[1 - \beta_\| \cos\theta])] \\ \frac{1}{2}[\delta(\overline{l+1}\,\Omega - \omega[1 - \beta_\| \cos\theta]) + \delta(\overline{l-1}\,\Omega - \omega[1 - \beta_\| \cos\theta])] \end{pmatrix} \quad (9\text{-}46)$$

Writing $x = (\omega/\Omega)\beta_\perp \sin\theta$ and substituting (9-46) in (9-45), leads to summations of the form

$$\sum_{=-\infty}^{\infty} \delta(l\Omega - \omega[1 - \beta_\| \cos\theta]) \cdot \begin{pmatrix} J_l(x) \\ \frac{i}{2}[J_{l-1}(x) - J_{l+1}(x)] \\ \frac{1}{2}[J_{l-1}(x) + J_{l+1}(x)] \end{pmatrix}$$

$$= \sum_{l=-\infty}^{\infty} \delta(l\Omega - \omega[1 - \beta_\| \cos\theta]) \begin{pmatrix} J_l(x) \\ i J_l'(x) \\ \frac{l}{x} J_l(x) \end{pmatrix} \quad (9\text{-}47)$$

using simple recurrence formulae (cf. Appendix 2).
After some further reduction (cf. Problem 9-12), one finds

$$\frac{d^2 W}{d\Omega\, d\omega} = \frac{e^2\omega^2}{c}\left(\frac{T}{2\pi}\right) \sum_{l=1}^{\infty} \left| \begin{aligned} & \hat{\imath}(\beta_\| - \cos\theta)\cot\theta\, J_l(x) \\ & + \hat{\jmath} i \beta_\perp J_l'(x) \\ & + \hat{k}(\cos\theta - \beta_\|)J_l(x) \end{aligned} \right|^2 \delta(l\Omega - \omega[1 - \beta_\| \cos\theta])$$

$$= \frac{e^2\omega^2}{c}\left(\frac{T}{2\pi}\right) \sum_{l=1}^{\infty} \left[\left(\frac{\cos\theta - \beta_\|}{\sin\theta}\right)^2 J_l^2(x) + \beta_\perp^2 J_l'^2(x) \right]$$

$$\times\, \delta(l\Omega - \omega[1 - \beta_\| \cos\theta]) \quad (9\text{-}48)$$

where T is the radiation emission time.

The frequency spectrum of the emitted radiation is composed of *spectral lines* appearing at frequencies

$$\omega = \frac{l\Omega}{1 - \beta_\| \cos\theta} = \frac{l\Omega_0(1 - \beta_\perp^2 - \beta_\|^2)^{1/2}}{1 - \beta_\| \cos\theta} \quad (9\text{-}49)$$

Observe that the term $\beta_\| \cos\theta$ in the denominator displaces the spectral lines from exact integral multiples of Ω. This is the characteristic *Doppler shift*.

To calculate the power radiated into the element of solid angle $d\Omega$, integrate (9-48) over frequency and divide by T. The total power in a given harmonic l follows by integrating over all directions. The algebra is daunting if one attempts this directly but the situation is improved by making the calculation for a *guiding centre frame* (one moving along Oz with velocity $v_\parallel$); one may then use a Lorentz transformation to get the result in the observer's frame. The calculation with $v_\parallel = 0$ is the subject of Problem 9–13; one finds

$$P_l = \frac{2e^2 l\Omega_0^2(1 - \beta_\perp'^2)}{c\beta_\perp'}[\beta_\perp'^2 J_{2l}'(2l\beta_\perp') - l(1 - \beta_\perp'^2)\int_0^{\beta_\perp'} J_{2l}(2lt)\,dt] \quad (9\text{--}50)$$

where $c\beta_\perp'$ is the perpendicular velocity in the guiding centre frame. Applying a Lorentz transformation from the guiding centre frame to the observer's frame, the velocity transforms according to

$$v_\parallel = \frac{v_\parallel' + v_\parallel}{1 + \dfrac{v_\parallel' v_\parallel}{c^2}} \qquad v_\perp = \frac{v_\perp'(1 - \beta_\parallel^2)^{1/2}}{1 + \dfrac{v_\parallel' v_\parallel}{c^2}}$$

and $v_\parallel' = 0$ so that

$$v_\perp = v_\perp'(1 - \beta_\parallel^2)^{1/2} \quad (9\text{--}51)$$

From (9–50), using (9–51), one finds†

$$P_l = \frac{2e^2\Omega_0^2(1 - \beta^2)}{c\beta_\perp(1 - \beta_\parallel^2)^{3/2}}$$

$$\times \left[l\beta_\perp^2 J_{2l}'\left(\frac{2l\beta_\perp}{(1 - \beta_\parallel^2)^{1/2}}\right) - l^2(1 - \beta^2)\int_0^{\beta_\perp/(1-\beta_\parallel^2)^{1/2}} J_{2l}(2lt)\,dt \right] \quad (9\text{--}52)$$

By summing over all the P_l we find the total power radiated in all directions, P^{tot}. This result is also due to Schott and is the subject of Problem 9–14:

$$P^{\text{tot}} = \frac{2e^2\Omega_0^2}{3c}\left(\frac{\beta_\perp^2}{1 - \beta^2}\right) \quad (9\text{--}53)$$

This result may be used to check the assumption that the energy radiated is negligible compared with the total energy (Problem 9–15).

9–5–1 Cyclotron radiation by non-relativistic electrons

The expression found for the radiated power simplifies considerably in the limit of non-relativistic electron energy, $\beta \ll 1$. One can show in this case that

---

† The result in (9–52) was known long before cyclotron or synchrotron radiation had been observed. It is due to G. A. Schott, many of whose results were published as a monograph (G. A. Schott, *Electromagnetic Radiation.* Cambridge University Press, 1912). He obtained (9–52) by considering the radiation from a ring of n electrons (cf. Section 84 of his book). Perhaps inevitably, his results were forgotten 40 years later when synchrotron radiation had actually been detected from particle accelerators and they were rederived at that time by Schwinger, Feynman, and others.

$P_{l+1}/P_l \sim \beta^2$ for large l (Problem 9–16) so that one need consider only harmonics for which $l\beta \ll 1$. Then $J_l(x)$ may be replaced by its asymptotic form (cf. Appendix 2)

$$J_l(x) \simeq \frac{x^l}{2^l \Gamma(l+1)} \qquad x \ll 1 \qquad (9\text{--}54)$$

Substituting in (9–52) gives

$$P_l^{NR} = \frac{2e\Omega_0^2}{c} \frac{(l+1)l^{2l+1}}{(2l+1)!} \beta_\perp^{2l} \qquad (9\text{--}55)$$

Since

$$\frac{P_{l+1}^{NR}}{P_l^{NR}} \sim \beta_\perp^2 \sim \frac{v_\perp^2}{c^2} \qquad (9\text{--}56)$$

the emission spectrum is clearly composed of a sequence of discrete lines of rapidly diminishing intensity. The bulk of the radiation is emitted in the fundamental, known as the *cyclotron emission line*. The separation in frequency of the lines is now very close to Ω_0 (from (9–49)).

From (9–48), the angular distribution of the power radiated in the cyclotron emission line is

$$\frac{dP}{d\Omega} \simeq \frac{e^2 \Omega_0^2 \beta_\perp^2}{8\pi c}(1 + \cos^2 \theta) \qquad (9\text{--}57)$$

Thus an observer sited along $\mathbf{B}_0$ will detect *twice* as great an intensity as one at right angles to $\mathbf{B}_0$ (cf. Fig. 9–3). At $\theta = 0$, the radiation is circularly polarized (cf. Section 7–2–1) while at $\theta = \pi/2$ the polarization is linear with $\mathbf{E} \perp \mathbf{B}_0$; in general cyclotron emission is elliptically polarized.

Just as atomic line and molecular band spectroscopy help us understand basic atomic and molecular processes, the cyclotron emission line gives useful information about plasmas. Consider a plasma with electron density n and temperature T and suppose that n is so low that we can neglect any collective behaviour. Then (9–57), which applies to a single electron, may be generalized simply by multiplying by n; writing $v_\perp^2 = \kappa T/m$ one has

$$\frac{dP}{d\Omega} = \frac{e^2 \Omega_0^2}{8\pi c}\left(\frac{n\kappa T}{mc^2}\right)(1 + \cos^2 \theta) \qquad (9\text{--}58)$$

From this, we see that a measure of line intensity provides information about electron pressure.

Another important feature of line spectra is *line breadth*. In practice, the emission spectrum is not a set of discrete lines at Ω_0, $2\Omega_0$, and so on, but looks like Fig. 9–4. Valuable information about the plasma is contained in the line breadth. A number of mechanisms lead to line broadening; for example, any emission line has a certain *minimum width* on account of *radiation broadening*. This is because wave trains are not infinite, and hence a pure monochromatic line is not possible (Problem 9–19). However, in

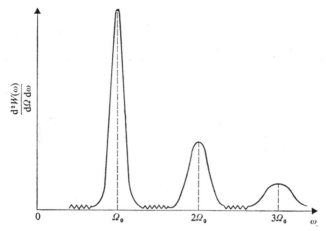

Fig. 9–4   Sketch of the cyclotron emission spectrum by a non-relativistic electron (not drawn to scale)

practice, the important mechanisms are *collision broadening* and *Doppler broadening*.

Collision broadening is due to the collisions of electrons with ions (and, in general, with atoms in a plasma which is not fully ionized). The effect for the cyclotron emission line is to replace the delta function determining the line profile, $\delta(\Omega_0 - \omega[1 - \beta_\parallel \cos \theta])$, by the Lorentzian line shape

$$\frac{\nu_c(v)}{[\Omega_0 - \omega(1 - \beta_\parallel \cos \theta)]^2 + \nu_c^2(v)} \tag{9–59}$$

(cf. Problem 9–20) where $\nu_c(v)$ is the collision frequency for electron–ion collisions. If collision broadening is the dominant effect, then the line width provides information about the electron–ion collision frequency.

Doppler broadening, on the other hand, derives from the electron thermal motion along the field lines; different values of $\beta_\parallel$ give different Doppler shifts. Adopting a Maxwellian distribution for the electrons one may show (cf. Problem 9–21) that the line profile is now given by the Gaussian shape characteristic of Doppler broadening,

$$\exp\left[-\left(\frac{mc^2}{2\kappa T}\right)\frac{(\omega - \Omega_0)^2}{\omega^2 \cos^2 \theta}\right] \tag{9–60}$$

From (9–60) it is apparent that the Doppler line width is $O\left(\dfrac{\omega}{c}\sqrt{\left(\dfrac{\kappa T}{m}\right)}\cos \theta\right)$.

Then, provided Doppler broadening dominates collision broadening, the line width gives information about *electron temperature*. Note that when $\theta = \pi/2$, Doppler broadening vanishes.

When both effects operate the line profile is more complicated and is known as a *Voigt profile*. As one might anticipate from (9–60), the Doppler profile is dominant at the centre of the line, while the Lorentz profile, from (9–59), dominates the wings.

All our discussion so far presupposes a tenuous plasma, so that self-absorption of the cyclotron emission has been neglected. The effect of this self-absorption is to broaden the emission line. Another important modifica-

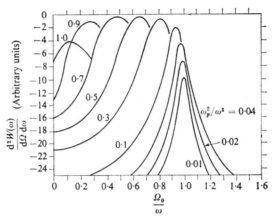

Fig. 9–5 Cyclotron emission for propagation orthogonal to $\mathbf{B}_0$; $(v_c/\omega)^2 = 10^{-3}$ (after Bekefi[88])

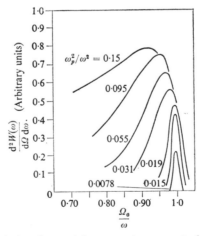

Fig. 9–6 Cyclotron emission detected from non-transparent plasmas of varying density for propagation orthogonal to $\mathbf{B}_0$ (after Hirshfield, J. L. and S. C. Brown, *Phys. Rev.*, **122**, 719, 1961)

tion when the plasma density increases, derives from collective effects. Obviously one can no longer discuss the emission in terms of a calculation based on particle orbit theory and one needs a kinetic theory treatment. The emission spectrum changes to the shapes shown in Fig. 9–5. Observe the shift of the resonance from Ω_0 to the upper hybrid frequency $(\omega_p^2 + \Omega_0^2)^{1/2}$ as the plasma density increases. This displacement from Ω_0 is shown clearly in the experimental emission profiles of Fig. 9–6.

9–5–2 *Emission spectrum from relativistic electrons*

As the electron energy increases, the emission spectrum changes radically from the single cyclotron line discussed in Section 9–5–1. One can get an expression for the power radiated in the limit of *ultra-relativistic* electrons (when $\beta \to 1$); for intermediate electron energies the problem has to be solved numerically. In the ultra-relativistic limit one may show for $\beta_{\parallel} = 0$ (cf. Problem 9–22) that

$$P_l^{UR} = \frac{e^2\Omega_0^2}{\pi\sqrt{3}c}l(1-\beta^2)^2 \int_{\frac{2l}{3}(1-\beta^2)^{3/2}}^{\infty} K_{5/3}(t)\,dt \qquad (9\text{–}61)$$

Note that $l \gg 1$ for (9–61) to hold. From (9–49) we see that the 'lines' are no longer separated by frequency bands of $O(\Omega_0)$ but are now only $O(\Omega_0(1-\beta^2)^{1/2})$ apart. One is no longer dealing with a discrete spectrum, as in the non-relativistic case, and so one may rewrite (9–61) to show the power radiated per unit frequency interval rather than per harmonic. One finds (Problem 9–23)

$$\frac{dP^{UR}(\omega)}{d\omega} = \frac{\sqrt{3}e^2\Omega_0}{2\pi c}\frac{\omega}{\omega_c} \int_{\omega/\omega_c}^{\infty} K_{5/3}(t)\,dt \qquad (9\text{–}62)$$

where $\omega_c = \frac{3}{2}\Omega_0\gamma^2$. This spectrum is plotted in Fig. 9–7.

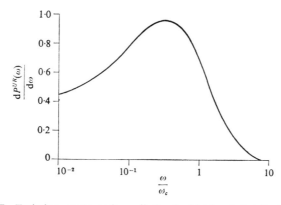

Fig. 9-7 Emission spectrum for radiation by highly relativistic electrons

At intermediate energies, Fig. 9–8 shows the type of emission spectrum obtained; it shows discrete structure at the low-frequency end, merging into a continuum. The emission lines are now subject to a *relativistic broadening* (due to the relativistic change of mass) which increases with increasing harmonic number.

The spectral distribution of synchrotron radiation from a hot plasma has been measured in the laboratory by Lichtenberg, Sesnic, and Trivelpiece.[81] The source of the radiation was a plasma in a magnetic mirror using a field

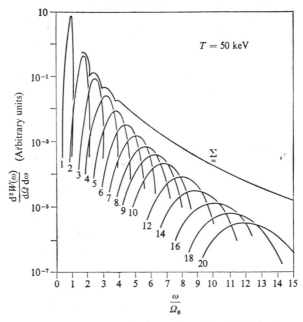

Fig. 9–8 Emission spectrum for radiation by moderately relativistic electrons before and after summation over the harmonics, neglecting self-absorption (after Bekefi[88])

of the order of 50 kG. The density of the plasma used and the dimensions of the mirror were such that the effect of self-absorption could be safely ignored, allowing the measured spectra to be compared with those computed from theory. The harmonic nature of the radiation showed clearly in the measured spectra. Figure 9–9 shows a comparison of the emitted power with theoretical spectra computed assuming a Maxwellian distribution at electron temperatures of 50 and 75 keV. A time-averaged temperature of 70 keV was found from X-ray pulse-height analysis. The theoretical points were normalized to the experimental second harmonic value and the results show a satisfactory measure of agreement although some structure not predicted theoretically appeared in the experimental spectrum.

Synchrotron radiation has been proposed as a mechanism to explain certain types of radio emission from the sun, the planet Jupiter, and from the Crab nebula, among other sources.[19, 82] Consider radio emission from the sun, in particular the emission known as type IV. Type IV noise is characterized by a very broad emission band devoid of structure (it has no

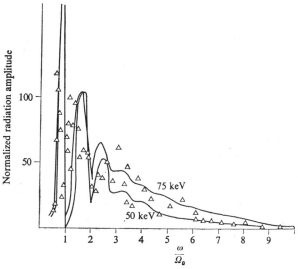

Fig. 9-9 Comparison between theoretical and experimental emission spectra for a plasma in which self-absorption was unimportant; the radiation was observed at right angles to $\mathbf{B_0}$
(after Lichtenberg, Sesnic, and Trivelpiece[81])

harmonic features like the type II bursts described in Section 9–7) and persisting for a few hours; the radiation is, in general, strongly circularly polarized. Typically intensities (per unit frequency interval) are of the order of 10^{-20} W/m² (c/s). In 1957 Boischot and Denisse suggested that type IV radiation might be interpreted in terms of synchrotron radiation. Using (9–28b) and the relation $\omega_c = \frac{3}{2}\Omega_0\gamma^2$, one may check the consequences of this suggestion. From (9–28b), one finds

$$P \sim 0\cdot6 . 10^{-26} B^2 \mathfrak{C}^2 \text{ erg/s}$$

and the emission is distributed over a band whose width is of the order of

$$\omega_c \sim 10^{-4} B\mathfrak{C}^2 \text{ rad/s}$$

(cf. Problem 9–24); in both expressions B is in gauss and the electron energy $\mathfrak{C}$ in electron volts. To obtain an expression to compare with the observed intensity, one must estimate $NP/\omega_c R^2 \Delta\Omega$, where R is the distance in metres from the sun, $\Delta\Omega$ the solid angle into which the radiation is emitted, and N the total number of emitting electrons. Then,

$$\frac{P}{\omega_c R^2 \Delta\Omega} \sim \frac{2\pi.0\cdot6.10^{-33}B^2\mathfrak{C}^2}{(10^{-4}B\mathfrak{C}^2)2.10^{22}\Delta\Omega} \quad \text{W/m}^2 \text{ (c/s)}$$

Taking $\Delta\Omega \sim 2\pi$, this gives

$$\frac{NP}{\omega_c R^2 \Delta\Omega} \sim 10^{-52} NB^2 \quad \text{W/m}^2 \text{ (c/s)}$$

Thus for magnetic fields of the order of 1 G, 10^{32} electrons will give the intensities observed in type IV emission. Assuming electron densities $\sim 10^{13}/\text{m}^3$, this gives a region of emitting electrons characterized by a scale length of $\sim 1,000$ km. Further, using a typical observed value for the bandwidth, 150 Mc/s, one may estimate the electron energy from the relation $\omega_c \sim 10^{-4} B\mathfrak{E}^2$, giving $\mathfrak{E} \sim 1$ MeV.

9–6 Bremsstrahlung from a plasma

We next examine a radiation process—bremsstrahlung—which is important in magnetic field-free plasmas (or indeed in a magnetized plasma at frequencies away from the cyclotron frequency and its harmonics). In this case, the radiation derives from the acceleration of an electron in the field of an ion.† The discussion in this section is restricted to non-relativistic energies. Consider first electron–ion bremsstrahlung from an elementary point of view. Suppose that the plasma density is low enough for us to consider the motion of an electron in the field of a single stationary ion, charge Ze. Then

$$|\dot{\mathbf{v}}| = \frac{Ze^2}{mr^2} \tag{9–63}$$

and substitution in Larmor's formula (9–26) gives for the power radiated per electron

$$P_e = \frac{2e^2}{3c^3}\left(\frac{Ze^2}{mr^2}\right)^2 \tag{9–64}$$

For a spatially uniform electron density n^-, the contribution to the bremsstrahlung from *all* electrons in encounters with this ion is

$$P = \frac{2Z^2e^6}{3m^2c^3}n^- \int_{r_{\min}}^{\infty} \frac{4\pi r^2 \, dr}{r^4} = \frac{8\pi}{3}\frac{n^- Z^2 e^6}{m^2 c^3 r_{\min}} \tag{9–65}$$

Observe that a cut-off at $r = r_{\min}$ is necessary to avoid a divergence. We may regard $r_{\min}$ as a measure of the distance at which encounters may no longer be discussed classically and a quantum mechanical treatment is required. If one takes the reciprocal of the de Broglie wave number as a measure of $r_{\min}$, i.e.,

$$r_{\min} \sim \frac{1}{k} = \hbar/p = \hbar/\sqrt{(m\kappa T^-)}$$

---

† In addition, electron–atom encounters produce bremsstrahlung; this effect is dominant in weakly ionized plasmas but will be ignored in this discussion.

where $\hbar$ is Planck's constant, then

$$P \simeq \frac{8\pi Z^2 e^6 n^-}{3mc^3\hbar} \sqrt{\frac{\kappa T^-}{m}} \tag{9-66}$$

If n^+ represents the ion density, then the total power radiated per unit volume of plasma is

$$P^{tot} \simeq \frac{8\pi Z^2 e^6 n^- n^+}{3mc^3\hbar} \sqrt{\frac{\kappa T^-}{m}} \tag{9-67}$$

It is of interest to compare (9–67) with the result of a quantum mechanical calculation[83]:

$$P^{tot}_{QM} = \sqrt{\frac{8}{3\pi}} \, \overline{g_{ff}} \, \frac{8\pi Z^2 e^6 n^- n^+}{3mc^3\hbar} \sqrt{\frac{\kappa T^-}{m}} \tag{9-67a}$$

in which $\overline{g_{ff}}$ is a number (the *Gaunt factor*) of order 1.†

So far, only electron–ion encounters have been considered. The reason for ignoring electron–electron collisions is that these do not alter the total electron momentum in the plasma and, consequently, no bremsstrahlung can be emitted (cf. Problem 9–25).

9–6–1   *Frequency spectrum of electron–ion bremsstrahlung*

If one considers just a single electron–ion system, the electron path in the encounter with the stationary ion will be hyperbolic. However, on account of the long range of the Coulomb force, an electron is much more likely to suffer an appreciable deflection as a result of *many weak interactions* rather than because of a *single close collision* (cf. Chapter 10). In practice, therefore, a significant contribution to the bremsstrahlung originates from distant encounters, so that to a first approximation the electron trajectory will be taken as a straight line. If L is a scale length characterizing the range of the interaction, then the duration of an encounter $\tau \sim L/v$, so that $\omega \sim v/L$. Since $L \gg b_{min}$, where b_{min} represents the value of the impact parameter below which the use of a straight line trajectory is a poor approximation, $\omega \ll v/b_{min}$ so that in this sense the frequency spectrum will be found at the *low-frequency* limit.

---

† The precise value of $\overline{g_{ff}}$ depends on which approximation is made in the quantum mechanical calculation; a Born approximation leads to a Gaunt factor of 1·10. However, the Gaunt factor is no longer of $O(1)$ at high energies or for low-energy plasmas at low frequencies.

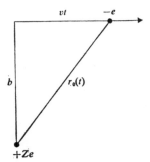

Fig. 9–10   Linear electron trajectory

For a linear trajectory (Fig. 9–10),

$$r_0^2(t) = (vt)^2 + b^2$$

where b is the impact parameter; the components of the acceleration normal and parallel to the trajectory are,

$$\left.\begin{aligned}
\dot{v}_\perp(t) &= \frac{Ze^2}{m} \frac{b}{[(vt)^2 + b^2]^{3/2}} \\
\dot{v}_\parallel(t) &= \frac{Ze^2}{m} \frac{vt}{[(vt)^2 + b^2]^{3/2}}
\end{aligned}\right\} \tag{9-68}$$

From (9–34), in the non-relativistic approximation, the energy radiated per unit frequency interval is

$$\begin{aligned}
\frac{dW(\omega)}{d\omega} &= \frac{2e^2}{3\pi c^3} \left| \int_{-\infty}^{\infty} \dot{\mathbf{v}}(t) e^{i\omega t}\, dt \right|^2 \\
&= \frac{2e^2}{3\pi c^3}\left(\frac{Ze^2}{m}\right)^2 \left| \int_{-\infty}^{\infty} \frac{b\hat{e}_\perp + vt\hat{e}_\parallel}{[(vt)^2 + b^2]^{3/2}} e^{i\omega t}\, dt \right|^2 \\
&= \frac{8e^2\omega^2}{3\pi c^3}\left(\frac{Ze^2}{mv^2}\right)^2 \left[K_1^2\left(\frac{\omega b}{v}\right) + K_0^2\left(\frac{\omega b}{v}\right) \right]
\end{aligned} \tag{9-69}$$

(cf. Problem 9–26) where the K's are the modified Bessel functions of the second kind (cf. Appendix 2). This frequency spectrum of the radiated energy is plotted in Fig. 9–11, as a function of $\omega b/v$. The bremsstrahlung emission drops off sharply for $\omega b/v \gtrsim 1$.

Thus far, only the bremsstrahlung emitted in a single encounter has been calculated. Now, as the electron moves through the plasma with velocity v, the number of encounters with ions per second having impact parameters between b and $b + db$ is just $2\pi b n^+ v\, db$, so that the total power radiated

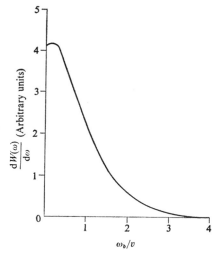

Fig. 9–11 Frequency spectrum of bremsstrahlung emission plotted as a function of $\omega b/v$

(per unit frequency interval) by a single electron in the low-frequency approximation is

$$2\pi \int_{b_{min}}^{\infty} \frac{dW(\omega)}{d\omega} n^+ vb \, db.$$

Next, one must sum over all the electrons in the plasma. If the electron velocity distribution function $f(v)$ is isotropic, the number of electrons per unit volume having velocity v is $4\pi v^2 f(v) \, dv$, so that the total power radiated per unit volume of the plasma in the frequency interval $d\omega$ is given by

$$dP^{tot}(\omega) = 8\pi^2 \int_0^{\infty} f(v) \left\{ \int_{b_{min}}^{\infty} \frac{dW(\omega)}{d\omega} n^+ vb \, db \right\} v^2 \, dv \, d\omega$$

$$= \frac{64\pi n^+ e^2 \omega^2}{3c^3} \left(\frac{Ze^2}{m}\right)^2 \int_0^{\infty} \int_{b_{min}}^{\infty} \frac{f(v)}{v} \left[K_1^2\left(\frac{\omega b}{v}\right) + K_0^2\left(\frac{\omega b}{v}\right) \right] b \, db \, dv \, d\omega$$

$$(9\text{--}70)$$

Writing $x = \omega b/v$ and using the identity

$$x[K_1^2(x) + K_0^2(x)] = -\frac{d}{dx}[xK_1(x)K_0(x)]$$

the integration with respect to b gives

$$dP^{tot}(\omega) = \frac{64\pi n^+ e^2}{3c^3} \left(\frac{Ze^2}{m}\right)^2 \int_0^{\infty} f(v) \, \omega b_{min} K_1\left(\frac{\omega b_{min}}{v}\right) K_0\left(\frac{\omega b_{min}}{v}\right) dv \, d\omega \quad (9\text{--}71)$$

Now, in the low-frequency approximation used, $\omega b_{min}/v \ll 1$, in which case one may replace K_1 and K_0 by their asymptotic forms,

$$\lim_{x \to 0} K_0(x) = -\ln x \qquad \lim_{x \to 0} K_1(x) = \frac{1}{x}$$

so that (9–71) becomes

$$dP^{tot}(\omega) = \frac{64\pi n^+ e^2}{3c^3}\left(\frac{Ze^2}{m}\right)^2 \int_0^\infty f(v)\, v \ln\left(\frac{v}{\omega b_{min}}\right) dv\, d\omega \qquad (9\text{–}72)$$

If one assumes that $f(v)$ is Maxwellian, (9–72) may be evaluated (cf. Problem 9–28) to give

$$\frac{dP^{tot}(\omega)}{d\omega} = \frac{16 n^+ n^- e^2}{3c^3}\left(\frac{Ze^2}{m}\right)^2 \left(\frac{m}{2\pi\kappa T^-}\right)^{1/2} \ln\left(\frac{2\kappa T^-}{m\omega^2 b_{min}^2}\right) \qquad (9\text{–}73)$$

Observe that different choices of b_{min} give only *slightly* different results since the variation of the bremsstrahlung emission with b_{min} is logarithmic.

9–6–2 *Plasma corrections to the bremsstrahlung*

Our calculation of the electron–ion bremsstrahlung frequency spectrum was based on a very simple model, which was entirely classical (apart from the fact that one might use the reciprocal of the de Broglie wave number to cut-off the b integration). However, there is effectively no difference between the classical result (9–73) and that from a quantum mechanical calculation for *microwave and radio frequencies*, which are of interest in the present context; at high frequencies the spectrum is quite different.

Besides this limitation in frequency, we ignored *plasma* effects in our calculation. Inquiring into the effect of the plasma on the bremsstrahlung raises two principal issues:

(i) the effect of the plasma on the generation of bremsstrahlung,
(ii) its effect on the emission of the bremsstrahlung from the plasma.

Dealing with the second question first, one recalls from Chapter 7 that the dielectric nature of a plasma modifies the propagation of radiation through it; in fact, for $\omega < \omega_p$, there can be no net bremsstrahlung emission. Consequently, the expressions in Section 9–6–1 are valid only for frequencies $\omega \gg \omega_p$; if $\omega \gtrsim \omega_p$, the power emission spectrum has to be multiplied by $\left(1 - \dfrac{\omega_p^2}{\omega^2}\right)^{1/2}$ (cf. Section 9–10). Hence, (9–73) is valid provided

$$\omega_p \ll \omega \ll (2\kappa T^-/mb_{min}^2)^{1/2}$$

Returning to (i) we recall that one of the basic characteristics of a plasma—

mentioned in Chapter 1 and discussed in Chapter 10—is that the field of an ion is shielded by plasma electrons beyond a distance λ_D, the Debye length. The effect of this shielding is to place an upper limit on the range chosen for the impact parameter b; thus in (9–70) $b_{min} \lesssim b \lesssim \lambda_D$.

Next, the result (9–73) is the outcome of a discussion assuming *binary* collisions between electrons and ions and a *linear* electron trajectory. In a real plasma (cf. Chapter 10) an electron moves in a field to which many ions contribute so that *multiple*, rather than binary, collisions are predominant. Notwithstanding, multiple encounters *can* be described as if the interaction between the electron and each ion were an independent binary collision. We see that this is plausible since the bremsstrahlung is proportional to $|\dot{\mathbf{v}}|^2$, that is, to $|\mathbf{E}|^2$, where $\mathbf{E}$ is the net electric field acting on the electron due to many ions. Thus if the ions are at positions $\mathbf{r}_j$

$$\mathbf{E} = - \sum_j \frac{Ze^2(\mathbf{r} - \mathbf{r}_j)}{|\mathbf{r} - \mathbf{r}_j|^3} = \sum \mathbf{E}_j$$

Now, assuming the positions of the ions are uncorrelated, the mean square resultant electric field is equal to the mean sum of the squares of the partial fields,

i.e.,

$$\overline{|E|^2} = \sum_j \overline{|E_j|^2}$$

where the averages are over the positions of the ions. In other words, the contributions to the bremsstrahlung from the deflections of the electron in the ion fields add *incoherently*. However, for frequencies in the neighbourhood of ω_p, we must take account of electron shielding and the contributions to the bremsstrahlung from the various ion fields are no longer uncorrelated.

It remains to obtain an expression for the power radiated as a result of *close* collisions. It may be shown (Problem 9–29) that the power radiated is comparable with (9–73) except in the limit of low frequencies.

9–7 Radiation from plasma oscillations

There is a further means of generating radiation in a plasma which, unlike those already discussed, involves collective electron motion. In an infinite homogeneous, isotropic plasma, the two basic modes of oscillation—the transverse electromagnetic wave and the longitudinal plasma oscillation— were uncoupled (cf. Chapter 7). It may be shown (cf. Problem 9–30) that in the presence of a *density or temperature gradient* in the plasma, this is no longer the case. The possibility then arises of converting some of the electrostatic energy of the plasma oscillation into radiation. The emission is in a narrow frequency band defined by $\omega_p \lesssim \omega \lesssim 1\cdot4\omega_p$ and is greatest when the plasma waves propagate in the direction of maximum density gradient.

In an experiment by Ben-Yosef and Kaufman,[84] radiation detected from the positive column of an arc discharge was compatible with that from plasma oscillations. These results (shown in Fig. 9–12, where frequency is plotted against arc-discharge current) have been interpreted on the hypothesis that radiation was detected over four discrete frequency bands and that it derived from two modes of plasma waves (on account of the geometry) with both radiation at the fundamental, that is, $\omega = \omega_p$ and the second harmonic $\omega = 2\omega_p$. For any given current, these frequencies are in the ratio $1/\sqrt{2} : 1 : \sqrt{2} : 2$; observe the good agreement with the experimental points although there is a discrepancy between the line B at $\omega = \omega_p$ and the dotted line representing the calculated electron plasma frequency.

In the arc-discharge plasma used in this experiment there are a number of density gradients, any or all of which might provide the coupling required to convert electrostatic into electromagnetic energy. There is a smooth, radial variation in electron density across the discharge and a much sharper one

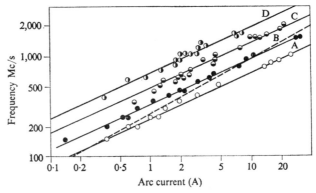

Fig. 9–12 Plot of frequency against arc discharge current in experiment detecting radiation compatible with that from plasma oscillations (after Ben-Yosef and Kaufman[84])

at the wall; in addition, there are sporadic, low-frequency fluctuations, along the axis of the discharge. Ben-Yosef and Kaufman associate the emission band at ω_p with the latter mechanism and that at $\omega_p/\sqrt{2}$ with the electron density discontinuity at the wall, but do not suggest the origin of the second harmonic bands, the intensities of which are comparable with those of the two fundamental bands.

In practice, this coupling is not very efficient so that for plasmas in thermal equilibrium this process is relatively unimportant. However, for non-equilibrium plasmas, radiation from plasma oscillations may be significant. In connection with non-equilibrium plasmas, this conversion mechanism has been proposed as the origin of type II (and type III) solar radio noise.[19] Type II bursts are comparatively rare events which show a strong correlation with the occurrence of solar flares. Typically they persist for about ten minutes and the main spectral features drift from high to low frequencies (hundreds to tens of megacycles per second) at rates of up to 1 Mc/s[2].

Intensities as great as 10^{-15} W/m² (c/s) have been reported, though two or three orders of magnitude less is more usual.

The spectral bands, shown in Fig. 9–13, can be as narrow as a few megacycles per second though at times are much greater than this. One of the

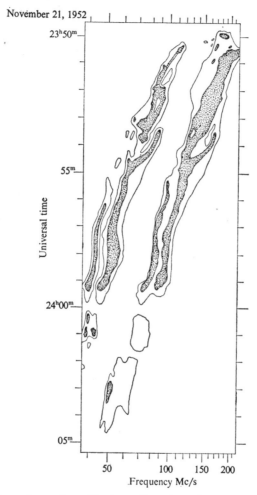

Fig. 9–13   Dynamic spectrum of type II solar radio noise emission showing features at the fundamental and second harmonic frequencies (after Kuiper[19])

features most characteristic of type II bursts is the appearance of harmonic structure. The intensities of both bands are often comparable and in many cases the structure of the fundamental is duplicated in the second harmonic. However, spectra showing *third* or higher harmonics have not been recorded although in some cases these could have been detected had their intensities been as low as one tenth that of the second harmonic. Many bursts reveal a *splitting* of the noise bands, in both fundamental and harmonic, small in comparison with the separation of the bands themselves.

The interpretation of solar type II bursts in terms of plasma oscillations was proposed first by Shklovsky. An order of magnitude estimate shows that the plasma hypothesis can explain the average energy radiated during a type II event ($\sim 10^{23}$ ergs), assuming an efficiency factor $\sim 10^{-5}$ for the conversion of electrostatic to electromagnetic energy. This model involves some disturbance—possibly a shock wave—propagating outwards through the corona and exciting, as it travels, large amplitude plasma oscillations; hence the drift from high to low frequencies observed in type II spectra. The observed frequency drift rates may be converted into radial velocities of the disturbance exciting the plasma oscillations, provided the radial electron density distribution is known. Using the Baumbach–Allen model corona,[19] a mean drift rate close to 500 km/s has been derived. One would not expect such a model to have much relevance in flare regions and other measurements indicate velocities up by a factor of 2 or 3.

The narrow bandwidth is, of course, consistent with the plasma mechanism. Moreover, the appearance of a second harmonic may be explained by the interaction of two longitudinal plasma waves.[85]

9–8 Scattering of radiation in plasmas

So far we have dealt only with the *generation* of radiation in a plasma, ignoring the *transport* of radiation *through* the plasma (other than to observe, in the case of bremsstrahlung, that there can be no net emission for frequencies $\omega \lesssim \omega_p$). Radiation propagating through a plasma may be scattered by the plasma, or it may suffer absorption. We shall treat these separately and begin by considering the scattering of radiation. As in the discussion of bremsstrahlung, we look first at scattering by free electrons before considering cooperative plasma effects.

9–8–1 *Thomson scattering*

The scattering of radiation by a free (i.e. unbound) electron is known as *incoherent* or *Thomson scattering*. A plane, monochromatic, electromagnetic wave incident on a free electron at rest will, in classical terms,† accelerate the electron and so give rise to radiation. This emission will be in all directions and will have the same frequency as the incident radiation *provided* the electron energy is non-relativistic. One can picture this process of absorption and re-emission of electromagnetic energy as scattering of the incident wave.

From (9–21), the instantaneous power radiated per unit solid angle in the non-relativistic limit is

$$\frac{\mathrm{d}P}{\mathrm{d}\Omega} = \frac{e^2 \dot{v}^2}{4\pi c^3} \sin^2 \theta$$

† In quantum mechanical terms, the incident photons are absorbed by the electron and re-emitted.

where θ is the angle between $\dot{\mathbf{v}}$ and $\hat{\mathbf{n}}$, the unit vector in the source–observer direction. Taking for the electric field of the incident wave

$$\mathbf{E}(\mathbf{r},\, t) = \mathbf{E}_0 \exp i(\mathbf{k}_0 . \mathbf{r} - \omega_0 t)$$

one has from the non-relativistic Lorentz equation

$$\dot{\mathbf{v}}(\mathbf{r},\, t) = \frac{e\mathbf{E}_0}{m} e^{-i\omega_0 t} \qquad (9\text{–}74)$$

The average power radiated per unit solid angle $\langle dP/d\Omega \rangle$ is found by time-averaging $\dot{v}^2$, and one finds from (9–21)

$$\left\langle \frac{dP}{d\Omega} \right\rangle = \frac{c}{8\pi} E_0^2 \left(\frac{e^2}{mc^2} \right)^2 \sin^2 \theta \qquad (9\text{–}75)$$

Defining a differential *scattering cross section*, $d\sigma/d\Omega$, as the energy radiated per unit time per unit solid angle divided by the incident energy flux, one has

$$\frac{d\sigma}{d\Omega} = \frac{(c/8\pi)E_0^2(e^2/mc^2)^2 \sin^2 \theta}{(c/8\pi)E_0^2}$$

$$= (e^2/mc^2)^2 \sin^2 \theta \qquad (9\text{–}76)$$

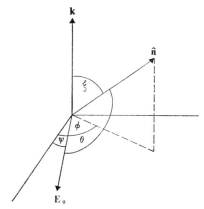

Fig. 9–14 Geometry for Thomson scattering

From Fig. 9–14,

$$\cos \theta = \sin \xi \cos (\phi - \psi)$$

and (9–76) becomes

$$\frac{d\sigma}{d\Omega} = (e^2/mc^2)^2 [1 - \sin^2 \xi \cos^2 (\phi - \psi)] \qquad (9\text{–}77)$$

If the incident radiation is *unpolarized*, one may average over ψ, giving

$$\frac{d\sigma}{d\Omega} = (e^2/mc^2)^2 \tfrac{1}{2}(1 + \cos^2 \xi) \qquad (9\text{-}78)$$

To find the *Thomson scattering cross-section*, σ_T, we integrate (9-78) over the solid angle: thus

$$\sigma_T = \frac{8\pi}{3}\left(\frac{e^2}{mc^2}\right)^2 \qquad (9\text{-}79)$$

For electrons, this has the value $0 \cdot 665 \times 10^{-24}$ cm^2; we are not, in general, interested in scattering by ions as this is at least some six orders of magnitude smaller. It should be emphasized that Thomson scattering is a classical result valid at frequencies such that $\hbar\omega \ll mc^2$.

Note that Thomson scattering results in a force driving the electron in the direction of the incident wave.[16] The electron 'absorbs' energy from the wave at an average rate $\langle S \rangle \sigma_T$. Thus, the electron gains momentum from the wave at a rate $\langle S \rangle \sigma_T/c$. On the other hand, the rate at which momentum is radiated by the electron is $\dfrac{1}{c} \displaystyle\int \left\langle \dfrac{dP}{d\Omega} \right\rangle \hat{n}\, d\Omega$, so that from (9-75) the total momentum of the scattered wave is zero. The final result, therefore, is equivalent to a force on the electron of magnitude

$$\frac{\langle S \rangle \sigma_T}{c} = \frac{2}{3}\left(\frac{e^2 E_0}{mc^2}\right)^2$$

in the direction of the incident wave. This phenomenon is known as *radiation pressure* and provides a small correction to the Lorentz equation (cf. Problem 9-31).

If the electron has a finite initial velocity, $\mathbf{v}_0$, one finds a Doppler shift of frequency in the scattered radiation (cf. Problem 9-32) given by

$$\omega = \omega_s - \omega_0 = (\mathbf{k}_s - \mathbf{k}_0)\cdot\mathbf{v}_0 \equiv \mathbf{k}\cdot\mathbf{v}_0 \qquad (9\text{-}80)$$

ω_s, $\mathbf{k}_s$ being the frequency, wave number of the scattered radiation. The frequency shift is determined solely by the velocity component in the direction of $\mathbf{k} = \mathbf{k}_s - \mathbf{k}_0$.

Thus far, only Thomson scattering by a single electron has been discussed; what of Thomson scattering by electrons in the plasma? Provided no correlations exist between the electrons, one may add the contributions from all electrons, giving a differential scattering cross-section per unit volume

$$\frac{d\sigma}{d\Omega} = n\left(\frac{e^2}{mc^2}\right)^2 \sin^2 \theta \qquad (9\text{-}81)$$

To obtain the frequency spectrum of Thomson scattered radiation from a plasma, we add the contributions from all electrons with the same velocity component in the direction **k**. For an equilibrium plasma, the electrons have a Maxwellian distribution and (cf. Problem 9–33) the differential scattering cross-section per unit volume per unit frequency interval is

$$\frac{d^2\sigma}{d\Omega\, d\omega} = \left(\frac{e^2}{mc^2}\right)^2 S(\mathbf{k}, \omega) \sin^2 \theta \tag{9–82}$$

where the profile $S(\mathbf{k}, \omega)$ is Gaussian. That is,

$$S(\mathbf{k}, \omega) = n\left(\frac{m}{2\pi k^2 \kappa T}\right)^{1/2} \exp\left[-\frac{m\omega^2}{2k^2\kappa T}\right] \tag{9–83}$$

The half-width of this profile is a function of electron temperature only; hence Thomson scattering from a laboratory plasma, if detectable, provides a measure of electron temperature. The half-width is just

$$\Delta\omega = 4\omega_0 \sin\frac{\xi}{2}\left(\frac{2\kappa T}{mc^2}\ln 2\right)^{1/2} \tag{9–84}$$

in which ξ is the scattering angle. The experiment has been carried out using a laser. The difficulties are formidable and stem principally from the smallness of the Thomson cross-section so that scattered radiation is liable to be swamped by unwanted stray light (such as scattering of the laser beam by the walls of the plasma container). In addition, one has to contend with

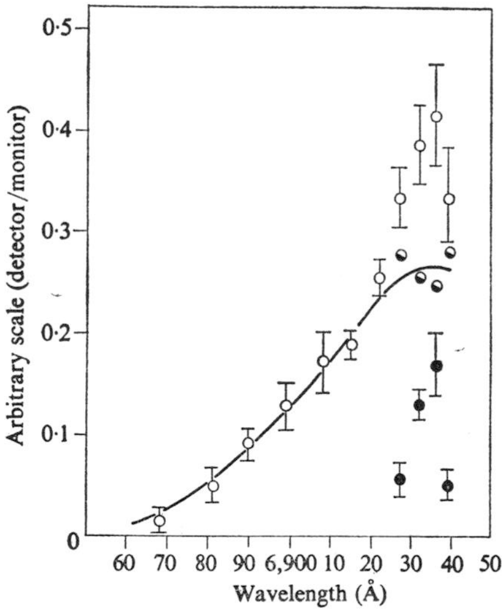

Fig. 9–15 Spectrum of Thomson scattered laser light; points ⧳ denote the stray light spectrum (after De Silva, Evans, and Forrest[86])

radiation from the plasma itself and—since a fully ionized plasma is less easily achieved by experimenters than it is conjured up by theoreticians—the possibility of Rayleigh and Raman scattering.

Despite the difficulties, Thomson scattering of laser light by electrons in a plasma has been observed in several experiments. The spectrum of scattered light obtained by De Silva, Evans, and Forrest[86] is shown in Fig. 9–15. Measurement of the half-width of this feature gives an electron temperature $T^- = 2 \cdot 2$ eV (cf. Problem 9–34).

9-8-2   Cooperative scattering

Next, consider scattering of radiation by a plasma when *collective* effects become important. In dealing with Thomson scattering the ions played no role in the scattering; however, in cooperative scattering the spectrum of scattered radiation is largely determined by the ion motion. To discuss cooperative scattering quantitatively is beyond our present scope since plasma kinetic theory is needed to determine the cross-section. However, the physics of the scattered spectrum may be outlined.

Writing the electron density

$$n(\mathbf{r}, t) = \sum_{i=1}^{N} \delta(\mathbf{r} - \mathbf{r}_i(t)) \qquad (9\text{–}85)$$

and using the non-relativistic limit of (9–17) for the scattered field

$$\mathbf{E}(\mathbf{r}, t) = \frac{e}{c^2 R}\left[\hat{\mathbf{n}} \times (\hat{\mathbf{n}} \times \dot{\mathbf{v}})\right]_{t' = t - \frac{R}{c}}$$

the total electric field of the scattered wave, using the Lorentz equation for $\dot{\mathbf{v}}$, is

$$\mathbf{E}(\mathbf{r}, t) = \frac{e^2 e^{ik_0 r}}{mc^2 r} \int n(\mathbf{r}', t) \exp\left[i(\mathbf{k}_0 - \mathbf{k}_s)\cdot\mathbf{r}' - i\omega_0 t\right] [\hat{\mathbf{n}} \times (\hat{\mathbf{n}} \times \mathbf{E}_0)]\, d\mathbf{r}' \qquad (9\text{–}86)$$

Proceeding as in Section 9–4, one may calculate (cf. Problem 9–35) the time-averaged power scattered per unit solid angle per unit frequency interval:

$$\left\langle \frac{d^2 P}{d\Omega\, d\omega_s} \right\rangle = \frac{c}{4\pi}\left(\frac{e^2}{mc^2}\right)^2 \lim_{T \to \infty} \frac{1}{\pi T} \left| \int\int n(\mathbf{r}', t) \right.$$
$$\left. \exp\left[i(\mathbf{k}_0 - \mathbf{k}_s)\cdot\mathbf{r}' - i(\omega_0 - \omega_s)t\right][\hat{\mathbf{n}} \times (\hat{\mathbf{n}} \times \mathbf{E}_0)]d\mathbf{r}'\, dt \right|^2 \qquad (9\text{–}87)$$

in which $\mathbf{k}_s$, ω_s are the wave number and frequency of the scattered wave, respectively, and $\omega_0 \gg \omega_p$. Fourier analysing the electron density

$$n(\mathbf{r}, t) = \frac{1}{(2\pi)^2} \int\int n(\mathbf{k}, \omega) \exp\left[i(\mathbf{k}\cdot\mathbf{r} - \omega t)\right] d\mathbf{k}\, d\omega \qquad (9\text{–}88)$$

and substituting in (9–87) gives

$$\left\langle \frac{d^2P}{d\Omega\,d\omega} \right\rangle = \frac{c}{4\pi}\left(\frac{e^2}{mc^2}\right)^2 \left| \hat{\mathbf{n}} \times (\hat{\mathbf{n}} \times \mathbf{E}_0) \right|^2$$

$$\times \lim_{T\to\infty} \frac{1}{\pi T} \left| n(\mathbf{k}_s - \mathbf{k}_0; \omega_s - \omega_0) \right|^2 \qquad (9\text{–}89)$$

Writing this in a form analogous to (9–82),

$$\frac{d^2\sigma}{d\Omega\,d\omega} = \left(\frac{e^2}{mc^2}\right)^2 S(\mathbf{k}_s - \mathbf{k}_0; \omega_s - \omega_0) \sin^2\theta \qquad (9\text{–}90)$$

with the *spectral power density* of the fluctuations defined by

$$S(\mathbf{k}, \omega) = \lim_{\substack{T\to\infty \\ V\to\infty}} \frac{2}{TV} \frac{|n(\mathbf{k}, \omega)|^2}{n} \qquad (9\text{–}91)$$

where V is the plasma volume. Thus the scattering cross-section in this case is determined now by the factor $S(\mathbf{k}, \omega)$ which embraces the effects of all the interacting particles in the system. To proceed beyond this formal expression requires methods of plasma kinetic theory. Qualitatively, however, since the scattering is determined by the spectrum of electron density fluctuations, $n(\mathbf{k}, \omega)$, one would expect to see in the spectrum of scattered radiation the effects of *collective* plasma behaviour on these fluctuations. Kinetic theory shows that the density fluctuations may be expressed as the sum of two contributions, one deriving from the fluctuations of electrons correlated with the motion of the ions (and sometimes called the 'ion component'), the other from the cooperative motion of electrons in electron plasma oscillations. In other words, correlations between electrons now give a degree of *coherence* in the various scattered waves, as opposed to Thomson scattering, which is totally incoherent. An important parameter in the theory is

$$\alpha = \frac{1}{k\lambda_D} \qquad (9\text{–}92)$$

where λ_D is the Debye length. Writing (9–92) in terms of the scattering angle $\xi = \cos^{-1}(\hat{\mathbf{k}}_0.\hat{\mathbf{k}}_s)$,

$$\alpha = \frac{1}{(k_0^2 - 2k_0 k_s \cos\,\xi + k_s^2)^{1/2}\,\lambda_D} \simeq \frac{\lambda_0}{4\pi\lambda_D \sin \xi/2} \qquad (9\text{–}93)$$

where λ_0 is the wavelength of the incident radiation. For $\alpha \ll 1$, plasma electrons scatter incoherently and the total intensity of scattered radiation is found by summing over the electrons. However, as $\alpha \to 1$ the scattered amplitudes start to interfere producing a degree of coherence. Note that $\alpha = 1$ implies

$$2\lambda_D \sin \xi/2 = \frac{\lambda}{2\pi}$$

which is just the *Bragg condition* for coherent scattering from planes in a crystal of spacing λ_D.† However, λ_D is not a crystal lattice periodicity but a plasma correlation length, so that the coherence in the case of scattering of radiation by a plasma is not nearly as pronounced as, say, in X-ray scattering by crystals. Cooperative scattering, then, arises whenever $\alpha \gtrsim 1$.

A schematic spectrum of scattered radiation in this case is shown in Fig. 9–16, showing the low-frequency (ion) and high-frequency (electron)

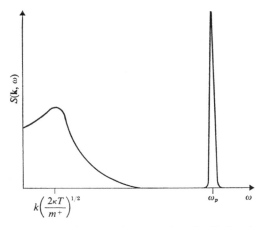

Fig. 9–16 Schematic spectrum of cooperative scattering of radiation showing both the ion feature at $\omega = k(2\kappa T/m^+)^{1/2}$ in a thermal plasma and that at $\omega \simeq \omega_p$ corresponding to scattering of radiation by electron plasma oscillations

contributions. The scattering in each case is done by the electrons—obviously so in the high-frequency feature—while for the low-frequency part of the spectrum the scattering derives from the electrons involved in the ion-acoustic waves (cf. Section 8–4). To picture this low-frequency feature in the scattered radiation spectrum, one may imagine the scattering to be due to fictitious particles having the thermal velocity of the ions but with a scattering cross-section that of the electrons. For a thermal plasma, the low-frequency feature shows the change from Thomson scattering with a Doppler broadened line (the line width being determined by the *electron* thermal speed, cf. (9–84)) to a line of width

$$\Delta\omega \sim k(2\kappa T/m^+)^{1/2} \tag{9–94}$$

† The factor 2π is absent in the Bragg condition in optics but arises here simply on account of the definition of λ_D.

The high-frequency feature appears just above the electron plasma frequency in the form of a sharp spike.

For plasmas in which the electron and ion temperatures are distinct and $T^+ < T^-$, the low-frequency component is significantly modified. The ion-acoustic mode is severely damped for $T^+ \gtrsim T^-$ (cf Section 8–5) and this is reflected in the absence of a well-defined peak in the spectrum at $\omega \simeq k(\kappa T^+/m^+)^{1/2}$. However, if $T^+ < T^-$ this damping is much reduced and a clear resonance appears in the spectrum.

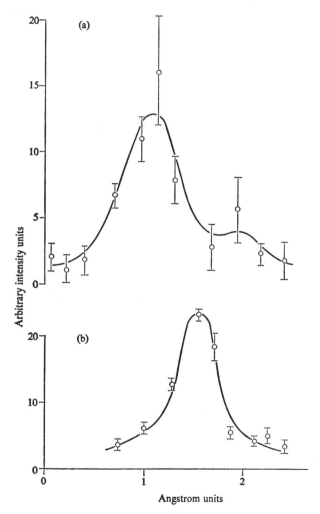

Fig. 9–17 Cooperative scattering of radiation at 5° to the forward direction from a theta-tron plasma; the ion feature shown here displays an asymmetry which has been interpreted in terms of a relative drift velocity in the thetatron plasma, leading to the onset of an ion wave instability. The stray-light spectrum (b) is shown to mark the position of the central wavelength from which scattered light is shifted (after Evans, Forrest, and Katzenstein[87])

Turning now to experimental work on the scattering of radiation by plasmas, one sees from (9–93) that by adjusting the angle of observation one can observe both Thomson and cooperative scattering from the same plasma. Substituting typical numerical values for λ_0, λ_D (cf. Problem 9–36) one finds that one must look close to the *forward direction* ($\xi = 0$) to observe cooperative scattering. If one can resolve low- and high-frequency features one can obtain valuable information on the ion temperature T^+, electron

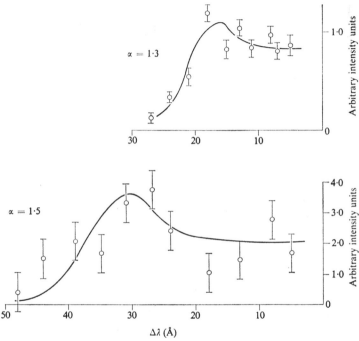

Fig. 9–18 Cooperative scattering of radiation at 5° to the forward direction from a thetatron plasma. The electron feature for values of $\alpha \equiv k_D/k = 1\cdot3$ and $1\cdot5$; the curves are best-fit theoretical scattered light profiles (after Evans, Forrest, and Katzenstein[87])

temperature T^-, density n^-, and (with a collisionless plasma, from the width of the feature at $\omega \simeq \omega_{pe}$) the Landau damping of electron plasma waves.

Light scattering by plasmas has been studied in many laboratories; an example is a series of experiments by Evans[86, 87] and his associates at Culham Laboratory. Using a pulsed arc in hydrogen with a 5 kG field maintaining the plasma for ~350 μs, scattering measurements were made for $\xi = 170°$ and $\xi = 10°$. The first case corresponds to $\alpha < 1$ and Thomson scattering, giving the electron temperature measurement (Fig. 9–15); for $\xi = 10°$, $\alpha \sim 1\cdot7$, so that cooperative scattering ought to be observable. Unfortunately, the laser was not sufficiently powerful, nor the arc plasma dense enough, to allow resolution of the ion feature. De Silva, Evans, and Forrest were able, however, to show that they were *not* observing Thomson scattering at $\xi = 10°$; this would correspond to a line width of ~ 6 Å, whereas the experimental points were consistent with an instrumental profile of 2 Å. A line width of less than 2 Å is in accordance with scattering by ion-acoustic waves.

Evans and his associates repeated the experiment using both a more powerful laser and a thetatron plasma where electron densities of $\sim 10^{16}/cm^3$ are available (some two orders of magnitude greater than in an arc discharge). In this work, the ion feature was resolved and, moreover, showed interesting structure (Fig. 9–17) in that the wing on the short-wavelength side of the laser line was enhanced. This was interpreted as due to a relative drift velocity between the electrons and ions of the thetatron plasma, which leads to the onset of an ion-wave instability (cf. Section 8–6–1). Values found for the temperatures were $T^- = 220 \pm 80$ eV, $T^+ = 150 \pm 50$ eV.

In another experiment using the thetatron plasma Evans, Forrest, and Katzenstein have observed the feature at the electron plasma frequency; their results are shown in Fig. 9–18. Theoretical curves have been fitted to the experimental points, thus determining α. From this, and the value of $\Delta\lambda$ at the shoulder, one may find electron density n^- and electron temperature T^-. For the curves shown, $\alpha = 1·3$ gave $n^- = 6 \times 10^{15}/cm^3$, $T^- = 103^{+29}_{-17}$ eV while $\alpha = 1·5$ gave $n^- = 2 \times 10^{16}/cm^3$, $T^- = 278^{+40}_{-36}$ eV. In the second case, an independent Thomson scattering measurement gave electron temperature $T^- = 240 \pm 36$ eV, a value in reasonable agreement with that found from cooperative scattering.

9–9 Transport of radiation in a plasma

To conclude this chapter, we turn to the *transport* of radiation through the plasma. To use the diagnostic possibilities of radiation in plasmas, we must understand not only the production of radiation within a plasma but also the effect of the plasma on the radiation as it travels to the detector. Once again, interest is centred on radiation at microwave and radio frequencies for which the emission processes—bremsstrahlung and cyclotron radiation—are comparatively straightforward for a thermal plasma. However, the transport of radiation at these frequencies through the plasma is in some ways more complicated than at optical wavelengths, since the plasma may now be highly dispersive and (for cyclotron radiation) anisotropic. In discussing the transport of radiation, scattering will be ignored.

The equation of radiative transfer is simply an expression of energy conservation in terms of geometric optics.[88] If $\mathbf{F}_\omega$ denotes the spectral density of the radiation flux then, by energy conservation in a steady state,

$$\nabla.\mathbf{F}_\omega = 0 \qquad (9\text{-}95)$$

Recall that the principal assumption in geometric optics is that the properties of the medium vary *slowly* with position—that is, the scale length of the variations is very much greater than the wavelength of the radiation in the medium. In this approximation, one may envisage the radiation being transported along *rays* which, though they may have effectively the same direction, do not interact with one another. Referring to Fig. 9–19 in which an element

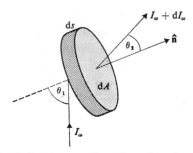

Fig. 9–19 Passage of ray through a plasma slab

of plasma of cross-section dA, thickness ds, is shown, $\hat{\mathbf{n}}$ being the unit normal out from the surface, one may write down the net radiation flux across this element, namely

$$\mathbf{dF}_\omega \cdot \hat{\mathbf{n}} = dF_\omega \cos\theta = I_\omega(\mathbf{s}) \cos\theta \, d\Omega \qquad (9\text{–}96)$$

$I_\omega(\mathbf{s})$ being the intensity† of the radiation. If dP_ω is the power in the frequency range $d\omega$ crossing dA within an element of solid angle $d\Omega$, then $I_\omega(\mathbf{s})$ is defined by

$$dP_\omega(\mathbf{s}) = I_\omega(\mathbf{s}) \cos\theta \, d\Omega \, d\omega \, dA$$

I_ω has units of watts per square centimetre per steradian per cycle per second. In general, the intensity is a function both of direction and position in the medium. If it is a function of position only, the radiation is isotropic, while if it is the same at all points, then the radiation is both isotropic and homogeneous.

Suppose the plasma through which the radiation is passing is loss-free and isotropic but slightly inhomogeneous so that a ray, on passing through the element of plasma shown in Fig. 9–19, suffers bending. Then, by energy conservation,

$$(I_\omega + dI_\omega) \cos\theta_2 \, d\Omega_2 \, d\omega \, dA - I_\omega \cos\theta_1 \, d\Omega_1 \, d\omega \, dA = 0 \qquad (9\text{–}97)$$

supposing no reflection of energy at the interface takes place. Now Snell's law requires that $n \sin\theta = \text{const.}$ (where n is the refractive index) along the

---

† In books on radiative transfer, $I_\omega(\mathbf{s})$ is sometimes called the *specific intensity* of radiation.

ray. Then, since

$$\frac{d\Omega_2}{d\Omega_1} = \frac{\sin\theta_2 \, d\theta_2}{\sin\theta_1 \, d\theta_1} = \frac{\sin\theta_2}{\sin\theta_1} \frac{n_1 \cos\theta_1}{n_2 \cos\theta_2}$$

$$= \left(\frac{n_1}{n_2}\right)^2 \frac{\cos\theta_1}{\cos\theta_2} \qquad (9\text{--}98)$$

it follows from (9–97) that

$$I_{\omega 2}\left(\frac{n_1}{n_2}\right)^2 \cos\theta_1 \, d\Omega_1 = I_{\omega 1} \cos\theta_1 \, d\Omega_1$$

i.e.,
$$\frac{I_\omega}{n^2} = \text{const.} \qquad (9\text{--}99)$$

Thus I_ω/n^2 is constant along a ray path in a slowly varying inhomogeneous, isotropic, transparent medium.† The result for optical frequencies simplifies to $I_\omega = \text{const.}$ along a ray path.

Now relax the requirement that the plasma be loss-free. In this case, (9–95) is replaced by

$$\mathbf{\nabla}.\mathbf{F}_\omega = \langle \mathbf{j}.\mathbf{E}\rangle_\omega \qquad (9\text{--}100)$$

where the source term now represents the spectral density of the product $\mathbf{j}(t).\mathbf{E}(t)$ and includes the effects of emission and absorption of radiation in the plasma. To obtain a useful expression from the formal equation, (9–100), look again at the geometrical optics representation and extend the previous result by introducing absorption and emission terms on the right-hand side in place of $\langle \mathbf{j}.\mathbf{E}\rangle_\omega$ in (9–100). Let α_ω be the coefficient of absorption per unit path length in the plasma, so that a radiative flux $I_\omega \, dA \, d\Omega$ suffers a loss $\alpha_\omega I_\omega \, dA \, d\Omega \, ds$ in travelling a distance ds. Similarly, introduce an emission coefficient η_ω, defined so that $\eta_\omega \, dA \, d\Omega \, ds$ is the emission from the elemental volume into solid angle $d\Omega$ in the direction of the ray. Then,

$$\frac{dI_\omega}{ds} = \frac{\partial I_\omega}{\partial s} - \alpha_\omega I_\omega + \eta_\omega \qquad (9\text{--}101)$$

Now, $\partial I_\omega/\partial s$ is the rate of change of I_ω due to the change in refractive index along a ray path; from (9–99)

$$\frac{\partial I_\omega}{\partial s} = \frac{2I_\omega}{n} \frac{dn}{ds}$$

† The case of an anisotropic plasma is less easy to analyse, since Snell's law is no longer obeyed in general but only for waves propagating in certain directions relative to that of the magnetic field.[88]

so that
$$\frac{dI_\omega}{ds} - \frac{2I_\omega}{n}\frac{dn}{ds} = \eta_\omega - \alpha_\omega I_\omega$$

i.e.,
$$n^2\frac{d}{ds}\left(\frac{I_\omega}{n^2}\right) = \eta_\omega - \alpha_\omega I_\omega \tag{9-102}$$

This is the *equation of transfer*. The neglect of scattering in deriving (9–102) is serious only at low plasma densities and high electron temperatures. If $\alpha_s I_\omega \, dA \, d\Omega \, ds$ represents a loss in the radiative flux due to scattering, for $\kappa T \sim 10$ eV and $\omega \sim 10^{10}$ c/s one finds (cf. Problem 9–37) $\alpha_s/\alpha_\omega \sim 10^9/n$.

To solve the transfer equation, one defines a *source function*

$$S_\omega = \frac{1}{n^2}\frac{\eta_\omega}{\alpha_\omega} \tag{9-103}$$

and an *optical depth* τ by

$$\tau = \int_0^\tau d\tau = -\int_s^s \alpha_\omega \, ds \tag{9-104}$$

in which the minus sign denotes that the optical depth is measured back into the plasma along the ray path. Then (9–102) reads

$$\frac{d}{d\tau}\left(\frac{I_\omega}{n^2}\right) = \frac{I_\omega}{n^2} - S_\omega \tag{9-105}$$

The integrating factor for (9–105) is $e^{-\tau}$ so that

$$\frac{d}{d\tau}\left(\frac{I_\omega}{n^2}e^{-\tau}\right) = e^{-\tau}S_\omega$$

Integrating over the total ray path in the plasma,

$$\frac{I_\omega(A)}{n^2(A)}e^{-\tau(A)} = \frac{I_\omega(B)}{n^2(B)}e^{-\tau(B)} + \int_{\tau(A)}^{\tau(B)} S_\omega(\tau)e^{-\tau} \, d\tau \tag{9-106}$$

in which $\tau(A)$ is the optical depth measured from the origin of τ to A. If A, B are points on a plasma–vacuum boundary then $\tau(A) = 0$, $\tau(B) = \tau_0$, the total optical depth of the plasma, and $n(A) = 1 = n(B)$. Thus, neglecting reflection at the boundaries, the emergent intensity is

$$I_\omega^{em} = I_\omega^{inc} e^{-\tau_0} + \int_0^{\tau_0} S_\omega(\tau)e^{-\tau} \, d\tau \tag{9-107}$$

The first term on the right-hand side represents the effect of the plasma on the incident radiation (i.e., absorption, since scattering has been ignored) while the second is the contribution from radiation generated (by bremsstrahlung, since only isotropic plasmas are considered) within the plasma, again allowing for absorption in transit from its origin to the point A. When $I_\omega^{inc} = 0$,

$$I_\omega^{em} = \int_0^{\tau_0} S_\omega(\tau) e^{-\tau} \, d\tau \qquad (9\text{--}108)$$

from which one sees immediately that for $\tau_0 \ll 1$, the radiation emitted within the plasma suffers negligible absorption in transit. In other words, the medium is transparent to radiation and is said to be *optically thin*. In the opposite limit, $\tau_0 \gg 1$, the plasma is opaque to radiation and is said to be *optically thick*.

9–10 Black-body radiation from a plasma

Black-body radiation is simply electromagnetic radiation in equilibrium. The processes which lead to a state of thermodynamic equilibrium being established are the absorption and emission of photons by the plasma electrons. The photons, being bosons, have a distribution function

$$\bar{n}_k = (\exp\left[\hbar\omega_k/\kappa T\right] - 1)^{-1} \qquad (9\text{--}109)$$

that is, the Planck distribution.

The spectral energy distribution of black-body radiation is given by *Planck's radiation law*,

$$dW(\omega) = \frac{Vh}{\pi^2 c^3} \frac{\omega^3 \, d\omega}{\exp\left[\hbar\omega/\kappa T\right] - 1} \qquad (9\text{--}110)$$

which, in the limit of low frequencies, $\hbar\omega \ll \kappa T$, reduces to the classical result, namely *Rayleigh–Jeans' law*

$$dW(\omega) = V\kappa T \frac{\omega^2 \, d\omega}{\pi^2 c^3} \qquad (9\text{--}111)$$

Since $V\omega^2 \, d\omega/\pi^2 c^3$ is simply the number of modes lying between ω and $\omega + d\omega$,† the right-hand side of (9–111) is just the product of the number of modes and the energy, κT, associated with each vibrating degree of freedom.

Suppose now that in each elemental volume of the plasma the emission and absorption of radiation are the same as if the whole system were in complete thermodynamic equilibrium. This is the concept of *local thermodynamic equilibrium* (LTE) and prevails when the particle distribution functions may be characterized by one temperature (which is, in general, space- and time-dependent). The radiation, however, will *not* normally be that from a black body at this temperature; consequently, for LTE one requires that

† Note that there are two independent directions of polarization.

collision processes dominate radiative ones. In LTE, the absorption and emission coefficients are related by *Kirchhoff's law* which states

$$\frac{\eta_\omega}{\alpha_\omega} = n^2 B(\omega, T) \qquad (9\text{-}112)$$

with $B(\omega, T)$ given by the Planck function (9-110) in general or, in the limit of radio and microwave frequencies, by the Rayleigh–Jeans form (9-111).

Thus in LTE the expression (9-108) for the radiation intensity outside the plasma becomes

$$I_\omega^{em} = \int_0^{\tau_0} B(\omega, T) e^{-\tau} d\tau \qquad (9\text{-}113)$$

For constant α_ω and a uniform plasma temperature

$$I_\omega^{em} = B(\omega, T)[1 - \exp(-\alpha_\omega L)] \qquad (9\text{-}114)$$

where L is the total path length of the ray in the plasma. For optically thin plasmas (where self-absorption of the radiation emitted is unimportant so that $\alpha_\omega L \ll 1$), $I_\omega^{em} \simeq \alpha_\omega L B(\omega, T)$. An optically thick plasma, on the other hand, is a black-body emitter. Plasmas do not radiate like black bodies unless they are both very dense and very thick.

It is perhaps worth emphasizing here that all of this discussion refers to an isotropic plasma. One must be careful about applying Kirchhoff's law to a plasma containing a magnetic field for which cyclotron radiation has to be considered in addition to bremsstrahlung[†] and in which the propagation of the radiation is anisotropic.

Consider the situation which arises when a vacuum–plasma–vacuum sandwich is contained between two black bodies at temperature T. As in the discussion in Section 9-9, the complication of reflection at the vacuum–plasma interface may be ignored. For the plasma under LTE conditions, at some interior point, assuming a constant absorption coefficient throughout the plasma, one has from (9-106)

$$\frac{I_\omega^{\text{plasma}}}{n^2} e^{-\tau} = I_\omega^{\text{vac}} e^{-\tau_0} + B(\omega, T)[e^{-\tau} - e^{-\tau_0}]$$

i.e.,
$$I_\omega^{\text{plasma}} = I_\omega^{\text{vac}} n^2 \qquad (9\text{-}115)$$

For isotropic plasmas, the spectral energy density is related to intensity by

$$u_\omega = \int \frac{I_\omega \, d\Omega}{v_g} = \frac{4\pi I_\omega}{v_g} \qquad (9\text{-}116)$$

[†] Plasma oscillations also couple with the radiation field in the presence of a magnetic field.

where v_g is the group velocity (cf. Chapter 7). Then, from (9–115) and (9–116),

$$u_\omega^{\text{plasma}} = u_\omega^{\text{vac}} \frac{c^3}{v_g v_p^2} \tag{9–117}$$

v_p being the phase velocity. Moreover, the dispersion relation for radiation in isotropic plasmas is simply $\omega^2 = \omega_p^2 + k^2 c^2$, so that

$$v_p^2 = \frac{\omega^2 c^2}{\omega^2 - \omega_p^2} \qquad v_g = \frac{d\omega}{dk} = \frac{c(\omega^2 - \omega_p^2)^{1/2}}{\omega}$$

and so

$$u_\omega^{\text{plasma}} = u_\omega^{\text{vac}} \left(1 - \frac{\omega_p^2}{\omega^2}\right)^{1/2} \tag{9–118}$$

as stated in Section 9–6–2.

Summary

1. *Electrodynamics*

   (i) Liénard–Wiechert potentials:

$$\mathbf{A}(\mathbf{r}, t) = \left[\frac{e\mathbf{v}}{cR - \mathbf{v}.\mathbf{R}}\right]_{t'=t-\frac{R(t')}{c}} \qquad \phi(\mathbf{r}, t) = \left[\frac{ec}{cR - \mathbf{v}.\mathbf{R}}\right]_{t'=t-\frac{R(t')}{c}}$$

   (ii) $\quad \mathbf{E}(\mathbf{r}, t) = e\left[\dfrac{\hat{\mathbf{n}}}{R^2} + \dfrac{R}{c}\dfrac{d}{dt}\left(\dfrac{\hat{\mathbf{n}}}{R^2}\right) + \dfrac{1}{c^2}\dfrac{d^2\hat{\mathbf{n}}}{dt^2}\right]_{\text{ret}}$

$$\mathbf{E}^{\text{rad}}(\mathbf{r}, t) = e\left[\frac{\hat{\mathbf{n}} \times \{(\hat{\mathbf{n}} - \boldsymbol{\beta}) \times \dot{\boldsymbol{\beta}}\}}{cR(1 - \hat{\mathbf{n}}.\boldsymbol{\beta})^3}\right]_{\text{ret}}$$

   (iii) Radiated power:

$$\frac{dP(t')}{d\Omega} = \frac{e}{4\pi c} \frac{|\hat{\mathbf{n}} \times \{(\hat{\mathbf{n}} - \boldsymbol{\beta}) \times \dot{\boldsymbol{\beta}}\}|^2}{(1 - \hat{\mathbf{n}}.\boldsymbol{\beta})^5}$$

$$P = \frac{2e^2}{3c} \frac{[(\dot{\boldsymbol{\beta}})^2 - (\boldsymbol{\beta} \times \dot{\boldsymbol{\beta}})^2]}{(1 - \beta^2)^3}$$

$$\rightarrow \frac{2e^2 \mathbf{v}^2}{3c^3} \quad \text{for } \beta \ll 1 \text{ (Larmor formula).}$$

   (iv) Energy radiated per unit solid angle per unit frequency interval:

$$\frac{d^2 W(\omega)}{d\Omega\, d\omega} = \frac{e^2 \omega^2}{4\pi^2 c} \left| \int_{-\infty}^{\infty} \exp\left\{i\omega\left(t' - \frac{\hat{\mathbf{n}}.\mathbf{r}_0(t')}{c}\right)\right\} [\hat{\mathbf{n}} \times (\hat{\mathbf{n}} \times \boldsymbol{\beta})]\, dt' \right|^2$$

2. Cyclotron radiation

(i) Frequency spectrum of cyclotron emission composed of spectral lines at

$$\omega = \frac{l\Omega_0(1 - \beta_\perp^2 - \beta_\parallel^2)^{1/2}}{1 - \beta_\parallel \cos\theta}$$

(ii) For non-relativistic electrons $\omega = l\Omega_0$ and $P_{l+1}/P_l \sim v_\perp^2/c^2$ so that the emission spectrum is a sequence of discrete lines of rapidly diminishing intensity. Line broadening mechanisms are collision and Doppler broadening, giving Lorentzian and Gaussian line profiles respectively.

(iii) for ultra-relativistic electrons ($\beta \to 1$) the separation of the 'lines' is only $O(\Omega_0(1 - \beta^2)^{1/2})$, i.e., the discrete spectrum of the non-relativistic case becomes a continuum.

3. Bremsstrahlung

(i) $P^{\text{tot}} = \sqrt{\dfrac{8}{3\pi}}\, \bar{g}_{\!f}\, \dfrac{8\pi Z^2 e^6 n^- n^+}{3mc^3\hbar}\left(\dfrac{\kappa T^-}{m}\right)^{1/2}$

(ii) $\dfrac{dP^{\text{tot}}(\omega)}{d\omega} = \dfrac{16 n^+ n^- e^2}{3c^3}\left(\dfrac{Ze^2}{m}\right)^2\left(\dfrac{m}{2\pi\kappa T}\right)^{1/2} \ln\left(\dfrac{2\kappa T^-}{m\omega^2 b_{\min}^2}\right)$

(iii) Plasma corrections to the bremsstrahlung:

$$\frac{dP^{\text{tot}}(\omega)}{d\omega} = [(ii)]\cdot\left(1 - \frac{\omega_{\text{pe}}}{\omega^2}\right)^{1/2}$$

Debye shielding imposes an upper limit on the impact parameter b; i.e., $b_{\min} \lesssim b \lesssim \lambda_D$.

4. Scattering of radiation

Thomson scattering is scattering by the free electrons in the plasma and is incoherent. When collective effects are important one has cooperative scattering (partially coherent) consisting of a contribution deriving from the fluctuations of electrons correlated with the ion motion and another due to the scattering of radiation by electron plasma oscillations.

5. Radiative transport in a plasma

(i) Equation of transfer:

$$n^2\frac{d}{ds}\left(\frac{I_\omega}{n^2}\right) = \underset{\downarrow}{\eta_\omega} - \underset{\downarrow}{a_\omega I_\omega}$$
$$\qquad\qquad\qquad\text{emission}\quad\text{absorption (in general includes effect of scattering)}$$

(ii) Optical depth $= -\displaystyle\int_{s_0}^{s} a_\omega\, ds$

Plasmas are optically thin (thick) if the total optical depth $\tau_0 \ll 1$ ($\gg 1$).

Problems

9–1 Show that

$$G(\mathbf{r}, t; \mathbf{r}', t') \equiv \left(\frac{1}{2\pi}\right)^2 \int\int d\mathbf{k}\, d\omega \frac{e^{i\mathbf{k}.(\mathbf{r} - \mathbf{r}') - i\omega(t - t')}}{\pi\left(k^2 - \frac{\omega^2}{c^2}\right)}$$

$$= \frac{\delta\left(t' + \frac{|\mathbf{r} - \mathbf{r}'|}{c} - t\right)}{|\mathbf{r} - \mathbf{r}'|}$$

by requiring that $G = 0$ over the entire domain for $t < t'$ and that G represent outward propagating waves for $t > t'$.

Hint: the first condition on G provides a prescription for dealing with the singularities in the integrand. In the complex ω-plane for $t > t'$ the integral along the real axis $= \int_C$ where closure takes place in the lower half-plane. The requirement that $G = 0$ everywhere for $t < t'$ means that the poles at $\omega = \pm ck$ must be displaced *below* the real axis. The second condition on G means that one considers only the retarded Green function.

9–2 Show that

$$\int_{-\infty}^{\infty} \delta\{g(x)\}\, \gamma(x)\, dx = \sum_i \gamma(x_i) \Big/ \left|\frac{dg(x_i)}{dx}\right|$$

in which the x_i are zeros of $g(x) = 0$.

Hint: this follows most simply by a change of variable from x to g.

9–3 Establish (9–12).

Hint: change the integration variable to $\tau(t') = t' + R(t')/c$ and the result follows.

9–4 Show that $\dfrac{1}{c}\dfrac{d\hat{\mathbf{n}}}{dt'} = \dfrac{\hat{\mathbf{n}} \times (\hat{\mathbf{n}} \times \boldsymbol{\beta})}{R}$.

Hint:
$$R\frac{d\hat{\mathbf{n}}}{dt'} = \frac{d}{dt'}(R\hat{\mathbf{n}}) - \hat{\mathbf{n}}\frac{dR}{dt'}$$

$$= \frac{d}{dt'}(\mathbf{r} - \mathbf{r}_0) + \hat{\mathbf{n}}(\hat{\mathbf{n}}.\mathbf{v})$$

(cf. Fig. 9–1).

9–5 Show that the expression for the electric field

$$\mathbf{E}(\mathbf{r}, t) = e\left[\frac{\hat{\mathbf{n}}}{R^2} + \frac{R}{c}\frac{d}{dt}\left(\frac{\hat{\mathbf{n}}}{R^2}\right) + \frac{1}{c^2}\frac{d^2\hat{\mathbf{n}}}{dt^2}\right]_{ret}$$

follows from (9–14).

9–6 Show that the contribution from $\left[\dfrac{e}{c^2}\dfrac{d^2\hat{\mathbf{n}}}{dt^2}\right]_{ret}$ to the radiation field is

$$\mathbf{E}^{rad}(\mathbf{r}, t) = e\left[\frac{\hat{\mathbf{n}} \times \{(\hat{\mathbf{n}} - \boldsymbol{\beta}) \times \dot{\boldsymbol{\beta}}\}}{cg^3 R}\right]_{ret}$$

9–7 In the non-relativistic limit show that, at large distances from the source, the radiation field from a particle of charge e moving with velocity $\dot{\mathbf{r}}_0(t)$ is given by $\dfrac{e}{c^2 R}\ddot{\mathbf{r}}_{0\perp}\left(t - \dfrac{R}{c}\right)$ where $\ddot{\mathbf{r}}_{0\perp}$ is the acceleration transverse to the line of sight.

9–8 Establish the expression for $\cos\theta_{max}$ in (9–25).

9–9 Using (9–23), consider the situation in which a charged particle executes instantaneously circular motion; show that

$$\frac{dP(t')}{d\Omega} = \frac{e^2\dot{\beta}^2}{4\pi c}\frac{1}{(1 - \beta\cos\theta)^3}\left[1 - \frac{(1 - \beta^2)\sin^2\theta\cos^2\phi}{(1 - \beta\cos\theta)^2}\right]$$

Compare this angular distribution with that found in the case in which particle velocity and acceleration are collinear; in particular observe that here too the peak power is radiated in the forward direction.

Hint: choose a coordinate system with $\boldsymbol{\beta}$ instantaneously along Oz, $\dot{\boldsymbol{\beta}}$ along Ox; θ, ϕ are the usual polar angles.

9–10 Establish the relativistic generalization of Larmor's formula, i.e.,

$$P = \frac{2e^2}{3c}\frac{[(\dot{\boldsymbol{\beta}})^2 - (\boldsymbol{\beta}\times\dot{\boldsymbol{\beta}})^2]}{(1 - \beta^2)^3}$$

Hint: while one might attempt to establish this result directly by integrating (9–23), such an approach entails lengthy algebra. Alternatively observe that since the non-relativistic Larmor formula is

$$P = \frac{2e^2}{3c^3}\dot{v}^2 = \frac{2e^2}{3c^3m^2}\left(\frac{d\mathbf{p}}{dt}\cdot\frac{d\mathbf{p}}{dt}\right)$$

one may write the Lorentz invariant generalization as

$$P = \frac{2e^2}{3m^2c^3}\left[\frac{dp_\mu}{d\tau}\frac{dp_\mu}{d\tau}\right]$$

where p_μ is the momentum-energy four vector and τ, the proper time. On evaluating the scalar product of the four vectors the result follows directly.

9–11 Show that

$$\frac{\hat{\mathbf{n}}\times\{(\hat{\mathbf{n}} - \boldsymbol{\beta})\times\dot{\boldsymbol{\beta}}\}}{g^2} \equiv \frac{d}{dt'}\left[\frac{\hat{\mathbf{n}}\times(\hat{\mathbf{n}}\times\boldsymbol{\beta})}{g}\right]$$

assuming that $\hat{\mathbf{n}}$ is not significantly time-dependent.

9–12 Establish (9–48) from (9–45) and (9–47); in particular, explain carefully the appearance of T, the radiation emission time, in (9–48).

9–13 By integrating (9–48) over all directions and dividing by the radiation time, establish (9–50).

Hint: to accomplish the integration over θ, some results from the theory of Bessel functions are needed. The argument is due to Schott and is given in Section 84 of his book *Electromagnetic Radiation* (Cambridge University Press, 1912). The procedure involves using Neumann's addition theorem

$$J_l^2(x) = \frac{1}{\pi} \int_0^\pi J_0(2x \sin \alpha) \cos 2l\alpha \, d\alpha$$

so that

$$\frac{dP_l}{d\Omega} = \frac{e^2 l^2 \Omega^2}{2\pi^2 c} \int_0^\pi J_0(2l\beta_\perp \sin \theta \sin \alpha)[\beta_\perp^2 \cos 2\alpha - 1]\cos 2l\alpha \, d\alpha$$

Using the result

$$\int_0^{\pi/2} J_0(y \sin \theta) \sin \theta \, d\theta = \sqrt{\frac{\pi}{2y}} J_{1/2}(y)$$

$$= \frac{\sin y}{y}$$

(9–50) follows.

9–14 By summing P_l in (9–52) over l, establish (9–53).

Hint: using the result $\displaystyle\sum_{=1}^\infty \frac{J_{2l}(2ly)}{l^2} = \frac{y^2}{2}$ one has from the differential equation for the J_{2l}

$$\sum_{l=1}^\infty l J_{2l}'(2ly) = \frac{y}{2(1-y^2)^2} \qquad \sum_{l=1}^\infty l^2 J_{2l}(2ly) = \frac{y^2(1+y^2)}{2(1-y^2)^4}$$

$$\sum_{l=1}^\infty l^2 \int_0^{\beta_\perp/\sqrt{(1-\beta_\parallel^2)}} J_{2l}(2ly) \, dy = \frac{\beta_\perp^3(1-\beta_\parallel^2)^{3/2}}{6(1-\beta^2)^3}$$

9–15 Using (9–53) check the assumption made in the development of the theory of cyclotron radiation that the energy radiated is negligible compared with the total energy of the radiating electron.

9–16 In the limit of non-relativistic electron energy show that $P_{l+1}/P_l \sim \beta^2$ for large l.

9–17 Establish (9–57).

9–18 Suppose that rather than a single electron, N electrons move on a given Larmor orbit. What effect does this have on the cyclotron radiation emitted assuming (i) these electrons are randomly distributed on the orbit, (ii) there is a uniform distribution of electrons on the orbit.

9–19 Obtain an estimate of the radiation broadening of an emission line in cyclotron radiation.

9–20 Consider the effect of collisions between electrons and ions on the cyclotron emission line. Show that the very narrow feature of the collision-free case broadens into the Lorentzian line profile

$$\frac{\nu_e(v)}{[\Omega_0 - \omega(1 - \beta_\parallel \cos \theta)]^2 + \nu_e^2(v)}$$

in which $\nu_e(v)$ is the electron–ion collision frequency.

What effect do electron–electron collisions have on the line profile?

9–21 Assuming that the electron distribution function is Maxwellian show that the cyclotron emission line profile is Doppler broadened according to

$$\exp\left[-\left(\frac{mc^2}{2\kappa T}\right)\frac{(\omega - \Omega_0)^2}{\omega^2\cos^2\theta}\right]$$

where T is the electron temperature.

$$Note: f(v) = n^-\left(\frac{m}{2\pi\kappa T}\right)^{3/2}\exp\left(-\frac{mv^2}{2\kappa T}\right)$$

9–22 Establish (9–61) in the limit $\beta_\perp \equiv \beta \to 1$ and $l \gg 1$.
Hint: the following results from the theory of Bessel functions are needed:

$$J_{2l}(2l\beta) = \frac{1}{\pi\sqrt{3}\gamma}K_{1/3}(R)$$

$$J'_{2l}(2l\beta) = \frac{1}{\pi\sqrt{3}\gamma^2}K_{2/3}(R)$$

$$\int_0^{2l\beta} J_{2l}(x)\,dx = \frac{1}{\pi\sqrt{3}}\int_{2l/3\gamma^3}^\infty K_{1/3}(t)\,dt$$

together with

$$2K_{2/3}\left(\frac{2l}{3\gamma^3}\right) - \int_{2l/3\gamma^3}^\infty K_{1/3}(t)\,dt = \int_{2l/3\gamma^3}^\infty K_{5/3}(t)\,dt$$

9–23 Establish (9–62).

9–24 Using (9–28b) and the relation $\omega_c = \frac{3}{2}\Omega_0\gamma^2$ check the numerical expressions given in Section 9–5–2 for synchrotron emission by electrons in connection with type IV solar radio noise:

$$P \sim 0\cdot 6.10^{-26}B^2\mathfrak{E}^2\text{erg/s}$$

distributed over a bandwidth of the order of

$$\omega_c \sim 10^{-4}B\mathfrak{E}^2 \text{ rad/s}$$

9–25 Show that electron–electron collisions do not contribute to the bremsstrahlung in the dipole approximation.

9–26 Obtain (9–69).

9–27 Use tables of $K_0(x)$, $K_1(x)$ to plot $dW(\omega)/d\omega$ as a function of $\omega b/v$ (cf. Fig. 9–11).

9–28 Using

$$f(v) = n^-\left(\frac{m}{2\pi\kappa T}\right)^{3/2}\exp\left(-\frac{mv^2}{2\kappa T}\right)$$

in (9–72) show that

$$\frac{dP^{tot}(\omega)}{d\omega} = \frac{16n^+n^-e^2}{3c^3}\left(\frac{Ze^2}{m}\right)^2\left(\frac{m}{2\pi\kappa T}\right)^{1/2}\left[\ln\left(\frac{2\kappa T}{m\omega^2 b^2_{min}}\right) - \frac{\gamma}{2}\right]$$

where γ is Euler's number, $\gamma \simeq 0.577$. In the low-frequency approximation $\omega b_{min}/v \ll 1$ so that one need only retain the logarithmic term in the bracket.

9–29 By considering close collisions show that the power radiated as bremsstrahlung as a result of these is comparable with that given by (9–73) *except* in the low-frequency limit.

9–30 Starting from the hydromagnetic equations for a warm, isotropic plasma show that a coupling exists between longitudinal and transverse waves in situations in which a density gradient exists in the plasma.

Hint: consider a plasma with an electron density $n(\mathbf{r})$ and ignore the motion of the ions and inter-particle collisions. Show that

$$\nabla \times (\nabla \times \mathbf{E}) + \frac{1}{c^2}\left(\frac{\partial^2}{\partial t^2} + \omega_p^2\right)\mathbf{E} - \frac{\kappa T^-}{mc^2}\nabla(\nabla.\mathbf{E}) = 0$$

and set $\mathbf{E} = \mathbf{E}_L + \mathbf{E}_T$ where $\nabla \times \mathbf{E}_L = 0$, $\nabla.\mathbf{E}_T = 0$; one finds

$$\left(\nabla^2 + \frac{\omega^2}{c^2}\mu^2\right)\nabla \times \mathbf{E}_T = -\frac{\omega^2}{c^2}(\nabla\mu^2 \times \mathbf{E}_L)$$

$$\left(\frac{\kappa T^-}{mc^2}\nabla^2 + \frac{\omega^2}{c^2}\mu^2\right)\nabla.\mathbf{E}_L = -\frac{\omega^2}{c^2}(\nabla\mu^2.\mathbf{E}_T)$$

where $\mu^2 = 1 - (\omega_p^2/\omega^2)$.

9–31 The non-relativistic equation of motion of a charge in an electromagnetic wave is[16]

$$m\dot{\mathbf{v}} = e\mathbf{E} + \frac{e}{c}\mathbf{v} \times \mathbf{B} + \frac{2e^2}{3c^3}\ddot{\mathbf{v}}$$

where the last term is a small correction to the Lorentz equation due to radiation. Show that the time average of the damping force is $\frac{2}{3}(e^2E_0/mc^2)^2$ where E_0 is the wave amplitude.

9–32 Show that the Thomson scattered radiation from electrons with initial velocity $\mathbf{v}_0$ is Doppler shifted by an amount $\omega = (\mathbf{k}_s - \mathbf{k}_0).\mathbf{v}_0$.

9–33 Show that in an equilibrium plasma for which the electron distribution function is Maxwellian, the line profile of Thomson scattered radiation is determined by the form factor

$$S(\mathbf{k}, \omega) = n\left(\frac{m}{2\pi k^2 \kappa T}\right)^{1/2}\exp\left[-\frac{m\omega^2}{2k^2\kappa T}\right]$$

9–34 Use the experimentally determined half-width from Fig. 9–15 to compute the electron temperature in the plasma.

9–35 Establish (9–87).

9–36 From $\alpha = \lambda_0/4\pi\lambda_D \sin \xi/2$ show that cooperative scattering may be observed close to the forward ($\xi = 0$) direction, by substituting appropriate numerical values for λ_0, λ_D.

9–37 Show that the neglect of scattering of radiation in deriving the transfer equation (9–102) is not serious provided the electron density in the plasma is sufficiently great; in particular for $\kappa T \sim 10$ eV, $\omega \sim 10^{10}$ c/s show that

$$\alpha / \alpha_\omega \sim 10^9 / n.$$

10

Kinetic theory

10–1 Introduction

In this last chapter an introduction to plasma kinetic theory is presented. The kinetic equation in its most general form

$$\frac{\partial f}{\partial t} + \mathbf{v} \cdot \frac{\partial f}{\partial \mathbf{r}} + \frac{\mathbf{F}}{m} \cdot \frac{\partial f}{\partial \mathbf{v}} = \left(\frac{\partial f}{\partial t}\right)_c \qquad (10\text{-}1)$$

was obtained in Chapter 1 and used in Chapter 3 to derive the hydromagnetic and cold plasma equations. In Section 3–6 it was observed that the hydromagnetic equations are valid whenever collisions are dominant while the cold plasma equations hold in the opposite limit of collisions negligible. Moreover it may be shown (Section 10–4) that particle orbit theory is equivalent to the description given by the collisionless kinetic equation. However, in kinetic theory lies the basic description of plasma motion and while simpler theories may often be used there are many situations where only a kinetic description is adequate. In particular, since microscopic fluctuations are averaged out in deriving fluid equations and orbit theory as presented in Chapter 2 ignores collective effects,† only kinetic theory includes both of these important aspects of plasma dynamics.

It is impossible to do justice to the usefulness of kinetic theory in one chapter of an introductory book. Since the theory is more comprehensive than the derivative orbit and fluid descriptions it is in general more complicated, frequently requiring the use of numerical computation. For this reason simple applications have been chosen; more appropriate and detailed calculations of plasma transport coefficients, for example, are available in the literature. Also for simplification, wherever possible two assumptions are made which have the effect of reducing notation and thereby clarifying procedure. First, a fully ionized plasma is considered and secondly, the motion of the ions is ignored. In essence this reduces the discussion to a one-component (electron) plasma. However, the ions cannot be ignored completely; rather it is assumed that they form a uniform background of positive charge. The electrons interact with the ions but this is treated as a 'field' interaction rather than as inter-particle. The extensions (which are cumbersome but not difficult in principle) of important formulae to multi-component plasmas are given in the Summary.

† Cf. the discussion of (2–23) in Section 2–4.

10–2 Equations for the distribution functions

One could present a discussion of kinetic theory on the basis of (10–1), giving meaning to the equation by specifying the form (or forms) of the collision term $(\partial f/\partial t)_c$. However, the arguments used to obtain (10–1) are deceptively simple; a number of mathematical and physical considerations are conveniently side-stepped. Our first aim, therefore, is to obtain (10–1) from the *Liouville equation* which describes the evolution of the N-particle distribution function and which is equivalent to the N Newtonian equations

$$m_i \ddot{\mathbf{R}}_i(t) = \mathbf{F}_i(t) \qquad (i = 1, 2, \ldots, N) \tag{10–2}$$

The procedure is not unlike the derivation of the moment equations in that a chain of equations is obtained and the deduction of a kinetic equation from this chain is clearest for plasmas close to thermal equilibrium.

The formal solutions of (10–2) are

$$\dot{\mathbf{R}}_i(t) = \dot{\mathbf{R}}_i(0) + \frac{1}{m_i} \int_0^t \mathbf{F}_i(t')dt'$$

$$\mathbf{R}_i(t) = \mathbf{R}_i(0) + \dot{\mathbf{R}}_i(0)t + \frac{1}{m_i} \int_0^t dt' \int_0^{t'} \mathbf{F}_i(t'')\, dt'' \tag{10–3}$$

$$(i = 1, 2, \ldots, N)$$

In principle, this completely determines the motion of the plasma. In practice, it is impossible to carry out the integrations in (10–3) since $\mathbf{F}_i$ is in general a function of the positions and velocities of *all* the plasma particles; the N vector equations (10–2) are coupled in a very complicated way. Moreover, even if the forces were sufficiently simple that one could do the integrations there would never be sufficient information to supply the required initial conditions $\mathbf{R}_i(0)$, $\dot{\mathbf{R}}_i(0)$ $(i = 1, 2, \ldots, N)$.

Turning now to a description of the plasma in terms of distribution functions, the *specific N-particle distribution function* $D(\mathbf{r}_1, \mathbf{v}_1, \mathbf{r}_2, \mathbf{v}_2, \ldots, \mathbf{r}_N, \mathbf{v}_N, t)$ is defined by the statement that $D \prod_{i=1}^{N} d\mathbf{r}_i\, d\mathbf{v}_i$ is the probability of finding particle i $(i = 1, 2, \ldots, N)$ within $d\mathbf{r}_i\, d\mathbf{v}_i$ of $(\mathbf{r}_i, \mathbf{v}_i)$ at time t. Since (10–3) formally prescribes the position and velocity of each particle, D is given by

$$D(\mathbf{r}_1, \mathbf{v}_1, \ldots, \mathbf{r}_N, \mathbf{v}_N, t) = \prod_{i=1}^{N} \delta(\mathbf{r}_i - \mathbf{R}_i(t))\, \delta(\mathbf{v}_i - \dot{\mathbf{R}}_i(t)) \tag{10–4}$$

Taking the partial derivative of D with respect to t and using (10–4) we get

$$\frac{\partial D}{\partial t} = \sum_{i=1}^{N} \left(\dot{\mathbf{R}}_i \cdot \frac{\partial D}{\partial \mathbf{R}_i} + \ddot{\mathbf{R}}_i \cdot \frac{\partial D}{\partial \dot{\mathbf{R}}_i} \right)$$

$$= -\sum_{i=1}^{N} \left(\dot{\mathbf{R}}_i \cdot \frac{\partial D}{\partial \mathbf{r}_i} + \ddot{\mathbf{R}}_i \cdot \frac{\partial D}{\partial \mathbf{v}_i} \right)$$

since $\partial\delta(x-y)/\partial y = -\partial\delta(x-y)/\partial x$. However, since D is zero unless $\dot{\mathbf{R}}_i = \mathbf{v}_i$, and $\ddot{\mathbf{R}}_i = \mathbf{F}_i/m_i$ from (10-2), this may be written

$$\frac{\partial D}{\partial t} + \sum_{i=1}^{N}\left(\mathbf{v}_i\cdot\frac{\partial D}{\partial\mathbf{r}_i} + \frac{\mathbf{F}_i}{m_i}\cdot\frac{\partial D}{\partial\mathbf{v}_i}\right) = 0 \qquad (10\text{-}5)$$

which is the *Liouville equation* (cf. Problem 10–1). So far the description of the plasma motion by (10–5) is equivalent to that given by (10–2). However, the lack of information concerning initial conditions leads to a *statistical* interpretation of D. Since the initial conditions are unknown, probability considerations may be applied to them: that is, instead of $D(t=0)$ being zero everywhere except at a single point in the $6N$ dimensional space $(\mathbf{r}_1, \mathbf{v}_1, \ldots, \mathbf{r}_N, \mathbf{v}_N)$, known as *phase space*, it will be given by a probability function over phase space. As time evolves, each initial point traces out a locus in phase space, determined by the dynamics of the system, thus prescribing $D(t)$; of course the normalization of D is still

$$\int D \prod_{i=1}^{N} d\mathbf{r}_i\, d\mathbf{v}_i = 1 \qquad (10\text{-}6)$$

The aim now is to integrate (10–5) over most of the coordinates $\mathbf{r}_i$ and velocities $\mathbf{v}_i$ to obtain equations for reduced distribution functions (containing less information) which one may hope to determine. Before embarking on this, one small refinement is required. If D is integrated over the positions and velocities of all particles except one, then the resulting distribution function gives the probability of finding that *particular* particle with a certain position and velocity. What one normally requires is the probability of finding *a* particle with a certain position and velocity. It is convenient, therefore, to define the *generic N-particle distribution function* $f_N(\mathbf{r}_1, \mathbf{v}_1, \mathbf{r}_2, \mathbf{v}_2, \ldots, \mathbf{r}_N, \mathbf{v}_N, t)$ by the statement that $f_N \prod_{i=1}^{N} d\mathbf{r}_i\, d\mathbf{v}_i$ is the probability of finding particle $i(i = 1, 2, \ldots, N)$ *or any particle of the same type as i* within $d\mathbf{r}_i\, d\mathbf{v}_i$ of $(\mathbf{r}_i, \mathbf{v}_i)$ at time t. Clearly f_N is obtained from D by summing over all permutations of like particles. It may be verified directly (as for D) that f_N obeys the Liouville equation (cf. Problem 10–2) but this can more easily be seen from the invariance of the Liouville operator

$$L_N \equiv \frac{\partial}{\partial t} + \sum_{i=1}^{N}\left(\mathbf{v}_i\cdot\frac{\partial}{\partial\mathbf{r}_i} + \frac{\mathbf{F}_i}{m_i}\cdot\frac{\partial}{\partial\mathbf{v}_i}\right) \qquad (10\text{-}7)$$

under permutations of like particles (cf. Problem 10–3). Thus, the equation we shall consider is

$$L_N f_N = 0 \qquad (10\text{-}8)$$

and adopting the one-component (electron) plasma described in Section 10–1 it follows on integrating f_N over phase space that

$$\int f_N \prod_{i=1}^{N} d\mathbf{r}_i \, d\mathbf{v}_i = N! \tag{10–9}$$

since there are $N!$ permutations of N electrons.

The reduced distribution functions are defined by

$$f_s(\mathbf{r}_1, \mathbf{v}_1, \mathbf{r}_2, \mathbf{v}_2, \ldots, \mathbf{r}_s, \mathbf{v}_s, t) = A \int f_N \prod_{i=s+1}^{N} d\mathbf{r}_i \, d\mathbf{v}_i \tag{10–10a}$$

where A is a normalization constant. Since there are $N!/(N-s)!$ permutations of s electrons from a total of N one requires

$$\int f_s \prod_{i=1}^{s} d\mathbf{r}_i \, d\mathbf{v}_i = N!/(N-s)!$$

so that (10–10a) becomes

$$f_s = \frac{1}{(N-s)!} \int f_N \prod_{i=s+1}^{N} d\mathbf{r}_i \, d\mathbf{v}_i \qquad (s = 1, 2, \ldots, N-1) \tag{10–10b}$$

It is then straightforward to integrate (10–8) over all but s spatial and velocity coordinates. Assuming f_N vanishes on the boundaries of phase space and the only velocity-dependent forces are Lorentzian, one obtains

$$\frac{\partial f_s}{\partial t} + \sum_{i=1}^{s} \mathbf{v}_i \cdot \frac{\partial f_s}{\partial \mathbf{r}_i} + \frac{1}{(N-s)! \, m} \sum_{i=1}^{N} \int \mathbf{F}_i \cdot \frac{\partial f_N}{\partial \mathbf{v}_i} \prod_{j=s+1}^{N} d\mathbf{r}_j \, d\mathbf{v}_j = 0 \tag{10–11}$$

where m is the electron mass (cf. the derivation of the general moment equation (3–14)). Separating $\mathbf{F}_i$ into its internal (inter-electron) and external (electron–ion together with any applied field) components, $\mathbf{F}_i^{\text{int}}$, $\mathbf{F}_i^{\text{ext}}$, it follows that

$$\frac{1}{(N-s)! \, m} \int \mathbf{F}_i^{\text{ext}} \cdot \frac{\partial f_N}{\partial \mathbf{v}_i} \prod_{j=s+1}^{N} d\mathbf{r}_j \, d\mathbf{v}_j = -\frac{1}{m} \mathbf{F}_i^{\text{ext}} \cdot \frac{\partial f_s}{\partial \mathbf{v}_i} \tag{10–12a}$$

$$(i = 1, 2, \ldots, s)$$

since $\mathbf{F}_i^{\text{ext}}$ may be taken outside the integral. In the non-relativistic approximation considered here, the only internal forces are electrostatic

$$\mathbf{F}_i^{\text{int}} = e^2 \sum_{\substack{j=1 \\ j \neq i}}^{N} \frac{\mathbf{r}_i - \mathbf{r}_j}{|\mathbf{r}_i - \mathbf{r}_j|^3} = -\sum_{\substack{j=1 \\ j \neq i}}^{N} \frac{\partial \phi_{ij}}{\partial \mathbf{r}_i}$$

where
$$\phi_{ij} = \frac{e^2}{|\mathbf{r}_i - \mathbf{r}_j|} \tag{10-13}$$

Hence (cf. Problem 10–4)

$$\frac{1}{(N-s)! \, m} \sum_{i=1}^{s} \int \mathbf{F}_i^{\text{int}} \cdot \frac{\partial f_N}{\partial \mathbf{v}_i} \prod_{j=s+1}^{N} d\mathbf{r}_j \, d\mathbf{v}_j$$

$$= -\frac{1}{m} \sum_{i=1}^{s} \sum_{\substack{j=1 \\ j \neq i}}^{s} \frac{\partial \phi_{ij}}{\partial \mathbf{r}_i} \cdot \frac{\partial f_s}{\partial \mathbf{v}_i} - \frac{1}{m} \sum_{i=1}^{s} \int \frac{\partial \phi_{is+1}}{\partial \mathbf{r}_i} \cdot \frac{\partial f_{s+1}}{\partial \mathbf{v}_i} d\mathbf{r}_{s+1} \, d\mathbf{v}_{s+1} \tag{10-12b}$$

Substituting (10–12a, b) in (10–11) gives the general equation for the reduced distribution functions. Using the definition (10–7) this may be written

$$L_s f_s = \frac{1}{m} \sum_{i=1}^{s} \int \frac{\partial \phi_{is+1}}{\partial \mathbf{r}_i} \cdot \frac{\partial f_{s+1}}{\partial \mathbf{v}_i} d\mathbf{r}_{s+1} \, d\mathbf{v}_{s+1} \tag{10-14}$$

$$(s = 1, 2, \ldots, N-1)$$

In (10–14) L_s is the Liouville operator for s electrons and the right-hand side represents the effect of the other electrons. One sees immediately the fundamental problem with this set of equations; the equation for f_s contains f_{s+1} so that the system closes only with the Liouville equation (10–8) and no simplification has yet been achieved. This is not surprising, since no approximations have been introduced thus far and the problem is therefore the one with which we started. The chain of equations represented by (10–14) is called the BBGKY hierarchy after Bogolyubov, Born and Green, Kirkwood, Yvon[89–93]. In principle we want some physical approximation enabling us to write f_{s+1} in terms of $f_1, f_2, \ldots, f_s$ for some small value of s, so obtaining a solvable set of equations—that is, a set of s equations, $s - 1$ of which may be used to eliminate $f_2, \ldots, f_s$ leaving a single (kinetic) equation for f_1. Even for $s = 2$ this is a formidable task and is discussed further in Section 10–13. Considering the case $s = 1$ in (10–14) one has

$$\frac{\partial f_1}{\partial t} + \mathbf{v}_1 \cdot \frac{\partial f_1}{\partial \mathbf{r}_1} + \frac{\mathbf{F}_1^{\text{ext}}}{m} \cdot \frac{\partial f_1}{\partial \mathbf{v}_1} = \frac{1}{m} \int \frac{\partial \phi_{12}}{\partial \mathbf{r}_1} \cdot \frac{\partial f_2}{\partial \mathbf{v}_1} d\mathbf{r}_2 \, d\mathbf{v}_2 \tag{10-15}$$

It is convenient to write

$$f_1(\mathbf{r}_1, \mathbf{v}_1, t) \equiv f(1) \tag{10-16a}$$

$$f_2(\mathbf{r}_1, \mathbf{v}_1, \mathbf{r}_2, \mathbf{v}_2, t) \equiv f(1) f(2) + P(1, 2) \tag{10-16b}$$

$$f_3(\mathbf{r}_1, \mathbf{v}_1, \mathbf{r}_2, \mathbf{v}_2, \mathbf{r}_3, \mathbf{v}_3 \, t) \equiv f(1) f(2) f(3) + f(1) P(2, 3) + f(2) P(3, 1)$$
$$+ f(3) P(1, 2) + T(1, 2, 3) \tag{10-16c}$$

defining f, P, and T; the expression for f_3 is included for future reference. The expansion represented by (10–16) is called the *cluster expansion*.[9] If $P = 0$ the two-particle distribution function is just the product of one-particle distribution functions; that is, particles 1 and 2 are uncorrelated. Thus P is that part of f_2 representing the *correlation* of the particles and is known as the *pair correlation function*. Substituting (10–16a, b) in (10–15) gives

$$\frac{\partial f(1)}{\partial t} + \mathbf{v}_1 \cdot \frac{\partial f(1)}{\partial \mathbf{r}_1} + \frac{\mathbf{F}_1^{\text{ext}}}{m} \cdot \frac{\partial f(1)}{\partial \mathbf{v}_1} - \frac{1}{m} \frac{\partial f(1)}{\partial \mathbf{v}_1} \cdot \int \frac{\partial \phi_{12}}{\partial \mathbf{r}_1} f(2) \, d\mathbf{r}_2 \, d\mathbf{v}_2$$

$$= \frac{1}{m} \int \frac{\partial \phi_{12}}{\partial \mathbf{r}_1} \cdot \frac{\partial P(1, 2)}{\partial \mathbf{v}_1} d\mathbf{r}_2 \, d\mathbf{v}_2 \quad (10\text{–}17)$$

The first three terms in (10–17) are familiar. The fourth term contains the average electric field experienced by one electron due to the other electrons and may be written

$$-\frac{1}{m} \frac{\partial f(1)}{\partial \mathbf{v}_1} \cdot \int \frac{\partial \phi_{12}}{\partial \mathbf{r}_1} f(2) \, d\mathbf{r}_2 \, d\mathbf{v}_2 = -\frac{e\mathbf{E}}{m} \cdot \frac{\partial f(1)}{\partial \mathbf{v}_1} \quad (10\text{–}18)$$

where the field $\mathbf{E}$ is given by

$$(-e)\mathbf{E}(\mathbf{r}_1, t) = -\int \frac{\partial \phi_{12}}{\partial \mathbf{r}_1} f(2) \, d\mathbf{r}_2 \, d\mathbf{v}_2 \quad (10\text{–}19)$$

and $(-e)$ is the electronic charge. Note, however, that $\mathbf{E}$ is computed assuming that the electrons are uncorrelated; the average is taken over all positions and velocities of particle 2 with no reference to any interaction between 1 and 2. This is in fact the electron contribution to the *self-consistent field* mentioned in Chapter 1 (see also the discussion in Section 3–6). Thus (10–17) may be written

$$\frac{\partial f(1)}{\partial t} + \mathbf{v}_1 \cdot \frac{\partial f(1)}{\partial \mathbf{r}_1} + \frac{\mathbf{F}}{m} \cdot \frac{\partial f(1)}{\partial \mathbf{v}_1} = \left(\frac{\partial f}{\partial t} \right)_c \quad (10\text{–}20)$$

where $\mathbf{F}$ includes all external and 'field' forces and

$$\left(\frac{\partial f}{\partial t} \right)_c = \frac{1}{m} \int \frac{\partial \phi_{12}}{\partial \mathbf{r}_1} \cdot \frac{\partial P(1, 2)}{\partial \mathbf{v}_1} d\mathbf{r}_2 \, d\mathbf{v}_2 \quad (10\text{–}21)$$

is the *collision term* or *collision integral*. This designation of (10–21) as the collision term is somewhat arbitrary since, if one regarded *all* inter-particle interactions as collisions, (10–18) would also be a collision term. The significance of this separation of particle interactions into a self-consistent field term and a collision term (brought about by the splitting of f_2 into its uncorrelated and correlated parts) will become clearer in the next section.

Two further points worth noting about (10–20) are:

(i) Even in the absence of collisions the equation is complicated by the non-linearity of the self-consistent field term (10–18).

(ii) The collision term (10–21) is not yet a function of f and some approximation must be introduced to make it so; then (10–20) will be the desired kinetic equation. This can be done either by 'solving' for P (Section 10–13), or by adopting some model for describing collisions which will give a collision integral in terms of f (Sections 10–7, 10–9).

10–3 Near-equilibrium plasmas

Before discussing various forms of the kinetic equation (that is, different models for the collision term $(\partial f/\partial t)_c$) we consider a plasma in, or near, thermal equilibrium. With no external forces acting, a plasma will settle into a state of thermal equilibrium for which (cf. Sections 10–8, 10–12) the distribution function is given by the Maxwell distribution

$$f_M(1) = n\left(\frac{m}{2\pi\kappa T}\right)^{3/2} \exp\left(-mv_1^2/2\kappa T\right) \tag{10–22}$$

Also, one may show (cf. Section 10–12) that the equilibrium value of the pair correlation function is given approximately by

$$P(1, 2) = -f_M(1) f_M(2)\frac{\phi_{12}}{\kappa T}\exp\left(-k_D\, r_{12}\right) \tag{10–23}$$

where $k_D = \lambda_D^{-1}$ is the reciprocal of the Debye length. Thus correlations occur over distances up to the order of the Debye length; for distances greater than this, Debye shielding makes the collision integrand vanishingly small. Thus, the collision term as defined by (10–21) refers to microscopic particle interactions within a Debye sphere as distinct from the macroscopic field represented by E.

Note further that the ratio of the correlated and uncorrelated parts of the two-particle distribution function is given by

$$\frac{P(1, 2)}{f(1) f(2)} \sim \frac{(e^2/r_{12})}{\kappa T}\exp\left(-k_D\, r_{12}\right) \tag{10–24}$$

This is small for $r_{12} \gg \lambda_D$; for $r_{12} \sim \lambda_D$

$$\frac{P}{ff} \sim \frac{e^2}{\kappa T\lambda_D} \sim \frac{1}{4\pi n\lambda_D^3} \tag{10–25}$$

Since the number of particles in a Debye sphere is usually very large one sees that the ratio is small within part of the Debye sphere as well as outside it. The distances for which the ratio is not small are clearly

$$r_{12} < \frac{e^2}{\kappa T} = \frac{1}{(4\pi n)^{1/3}(4\pi n \lambda_D^3)^{2/3}} \tag{10-26}$$

i.e., much less than the average inter-particle distance (provided the number of particles in a Debye sphere is large). Thus except for a small fraction of phase space the pair correlation function is small.† If these equilibrium, order of magnitude results can be extrapolated to more general non-equilibrium situations it may be permissible to ignore P in (10–16b). This approximation to the kinetic equation is the one we shall consider first (cf. Problem 10–5).

10–4 Vlasov equation

Setting $P = 0$ in (10–16b) leads to the collisionless kinetic equation

$$\frac{\partial f}{\partial t} + \mathbf{v}_1 \cdot \frac{\partial f}{\partial \mathbf{r}_1} + \frac{\mathbf{F}}{m} \cdot \frac{\partial f}{\partial \mathbf{v}_1} = 0 \tag{10-27}$$

as one can see directly from (10–20) and (10–21). Equation (10–27) is commonly referred to as the *Vlasov equation* or the *collisionless Boltzmann equation*. It is worth emphasizing that in the context of Section 10–2, (10–27) corresponds to the lowest-order approximation to the kinetic equation; we have taken $s = 1$ in (10–14) and written $f_2(1, 2) = f(1) f(2)$. Recall, too, that it is a non-linear equation for f in view of the dependence of the field $\mathbf{E}$ on f.

The equivalence of the Vlasov equation and particle orbit theory may be shown quite simply. The equation describing the latter is

$$m\ddot{\mathbf{r}} = \mathbf{F} \tag{10-28}$$

having the solution

$$\mathbf{r} = \mathbf{r}(\alpha_1, \alpha_2, \ldots, \alpha_6, t) \tag{10-29}$$

$$\mathbf{v} = \mathbf{v}(\alpha_1, \alpha_2, \ldots, \alpha_6, t)$$

where $\alpha_1, \alpha_2, \ldots, \alpha_6$ are the six constants of integration. Formally solving (10–29) for the α_i

$$\alpha_i = \alpha_i(\mathbf{r}, \mathbf{v}, t) \qquad (i = 1, 2, \ldots, 6) \tag{10-30}$$

Now any arbitrary function of the α_i

$$f = f(\alpha_1, \alpha_2, \ldots, \alpha_6)$$

† The small region, given by (10–26), in which this is not true is not without significance and its neglect leads to divergence in the collision integral (cf. Sections 10–7, 10–13).

is a solution of (10–27) as one can see by direct substitution. One gets

$$\sum_{i=1}^{6} \frac{\partial f}{\partial \alpha_i} \left(\frac{\partial \alpha_i}{\partial t} + \mathbf{v} \cdot \frac{\partial \alpha_i}{\partial \mathbf{r}} + \frac{\mathbf{F}}{m} \cdot \frac{\partial \alpha_i}{\partial \mathbf{v}} \right) = \sum_{i=1}^{6} \frac{\partial f}{\partial \alpha_i} \frac{d\alpha_i}{dt} = 0$$

since the α_i are constants of the motion. Thus the general solution of the Vlasov equation is an arbitrary function of the integrals of the motion of (10–28), the equation describing orbit theory. This was demonstrated by Jeans in connection with stellar dynamics and is sometimes referred to as *Jeans' theorem.*[94]

10–5 Landau damping

We now turn to Landau's solution[95] of the linearized Vlasov equation. With $\mathbf{F} = -e\mathbf{E}$ and $\mathbf{E}$ given by (10–19), any steady, homogeneous distribution function $f_0(\mathbf{v}_1)$ satisfies (10–27) identically. If a small perturbation $f'(\mathbf{r}_1, \mathbf{v}_1, t)$ is introduced

$$f(\mathbf{r}_1, \mathbf{v}_1, t) = f_0(\mathbf{v}_1) + f'(\mathbf{r}_1, \mathbf{v}_1, t)$$

and, since $\mathbf{E}$ is also a small quantity, the linearized Vlasov equation is

$$\frac{\partial f'}{\partial t} + \mathbf{v}_1 \cdot \frac{\partial f'}{\partial \mathbf{r}_1} - \frac{e\mathbf{E}}{m} \cdot \frac{\partial f_0}{\partial \mathbf{v}_1} = 0 \tag{10–31}$$

where
$$e\mathbf{E}(\mathbf{r}_1, t) = \int \frac{\partial \phi_{12}}{\partial \mathbf{r}_1} f'(\mathbf{r}_2, \mathbf{v}_2, t) \, d\mathbf{r}_2 \, d\mathbf{v}_2 \tag{10–32}$$

Solving (10–31) by means of Fourier and Laplace transforms, one writes

$$\left. \begin{aligned} f'(\mathbf{r}_1, \mathbf{v}_1, t) &= \frac{1}{(2\pi)^{3/2}} \int f'(\mathbf{k}, \mathbf{v}_1, t) e^{i\mathbf{k} \cdot \mathbf{r}_1} \, d\mathbf{k} \\[2mm] \mathbf{E}(\mathbf{r}_1, t) &= \frac{1}{(2\pi)^{3/2}} \int \mathbf{E}(\mathbf{k}, t) e^{i\mathbf{k} \cdot \mathbf{r}_1} \, d\mathbf{k} \end{aligned} \right\} \tag{10–33}$$

so that (10–31) gives for each Fourier component

$$\frac{\partial f'(\mathbf{k}, \mathbf{v}_1, t)}{\partial t} + i\mathbf{k} \cdot \mathbf{v}_1 f'(\mathbf{k}, \mathbf{v}_1, t) - \frac{e\mathbf{E}(\mathbf{k}, t)}{m} \cdot \frac{\partial f_0(\mathbf{v}_1)}{\partial \mathbf{v}_1} = 0 \tag{10–34}$$

Before taking the Laplace transform, (10–34) may be simplified by observing from (10–32) that

$$\mathbf{\nabla} \times \mathbf{E}(\mathbf{r}_1, t) = 0$$

and hence
$$\mathbf{k} \times \mathbf{E}(\mathbf{k}, t) = 0$$

Thus $\mathbf{E}(\mathbf{k}, t)$ is parallel to $\mathbf{k}$ so that if $\mathbf{v}_1 \cdot \hat{\mathbf{k}} = u$, (10–34) becomes

$$\frac{\partial f'(\mathbf{k}, \mathbf{v}_1, t)}{\partial t} + iku f'(\mathbf{k}, \mathbf{v}_1, t) - \frac{eE(\mathbf{k}, t)}{m} \frac{\partial f_0(\mathbf{v}_1)}{\partial u} = 0 \qquad (10–35)$$

Taking the Laplace transform of (10–35), that is, multiplying by e^{-pt} and integrating over t from 0 to ∞, one gets

$$(p + iku) f'(\mathbf{k}, \mathbf{v}_1, p) - \frac{eE(\mathbf{k}, p)}{m} \frac{\partial f_0(\mathbf{v}_1)}{\partial u} = f'(\mathbf{k}, \mathbf{v}_1, t = 0) \qquad (10–36)$$

where
$$\left.\begin{aligned}
f'(\mathbf{k}, \mathbf{v}_1, p) &= \int_0^\infty f'(\mathbf{k}, \mathbf{v}_1, t) e^{-pt} \, dt \\[2mm]
E(\mathbf{k}, p) &= \int_0^\infty E(\mathbf{k}, t) e^{-pt} \, dt
\end{aligned}\right\} \qquad (10–37)$$

From (10–36) f' is obtained as a function of E which we can now substitute in the Fourier–Laplace transform of (10–32) to obtain an equation for E alone. We take the divergence of (10–32) before transforming; using

$$\nabla^2(1/r) = -4\pi \, \delta(\mathbf{r}) \qquad (10–38)$$

we get

$$ikE(\mathbf{k}, p) = -4\pi e \int f'(\mathbf{k}, \mathbf{v}_2, p) \, d\mathbf{v}_2$$

Substituting for f' from (10–36)

$$ikE(\mathbf{k}, p) = -4\pi e \int \frac{f'(\mathbf{k}, \mathbf{v}_1, t = 0)}{p + iku} d\mathbf{v}_1 - \frac{4\pi e^2}{m} E(\mathbf{k}, p) \int \frac{\partial f_0/\partial u}{p + iku} d\mathbf{v}_1$$

i.e.,

$$E(\mathbf{k}, p) = \frac{4\pi ie}{kD(k, p)} \int \frac{f'(\mathbf{k}, \mathbf{v}_1, t = 0)}{p + iku} d\mathbf{v}_1 \qquad (10–39)$$

where
$$D(k, p) \equiv 1 - \frac{4\pi ie^2}{mk} \int \frac{\partial f_0/\partial u}{p + iku} d\mathbf{v}_1$$

is called the *plasma dielectric function*; note that this is independent of initial conditions. Carrying out the inverse Laplace and Fourier transforms formally solves the problem. Unfortunately, this is in general no simple matter. The time dependence of the kth Fourier component of the electric field is given by

$$E(\mathbf{k}, t) = \frac{1}{2\pi i} \int_{\sigma - i\infty}^{\sigma + i\infty} E(\mathbf{k}, p) e^{pt} \, dp \qquad (10–40)$$

where the integration is along a line parallel to the imaginary p-axis and to the right of all singularities of the integrand (assuming such a line exists). The integration may be accomplished by closing the contour by the dotted line

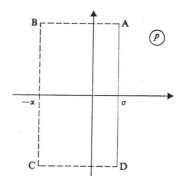

Fig. 10-1   Closure of contour for integration in Eqn (10-40)

ABCD as shown in Fig. 10–1. For reasonable choices of f_0 and $f'(t = 0)$ (the conditions that $\partial f_0/\partial u$ and $f'(t = 0)$ are analytic functions of u are sufficient) the only singularities of $E(\mathbf{k}, p)$ in the p plane are poles where the dielectric function vanishes; using the notation $u = v_x$, $F_0(u) = \int f_0(\mathbf{v}) \, dv_y \, dv_z$, that is where

$$\frac{4\pi i e^2}{mk} \int_{-\infty}^{\infty} \frac{dF_0/du}{p + iku} du = 1 \tag{10-41}$$

If the solutions of this equation are p_j and the residues of $E(\mathbf{k}, p)$ at the poles p_j are R_j,

i.e.,

$$R_j = \lim_{p \to p_j} (p - p_j) E(\mathbf{k}, p)$$

then from (10–40)

$$E(\mathbf{k}, t) = \sum_j R_j e^{p_j t} - \frac{1}{2\pi i} \int_{ABCD} E(\mathbf{k}, p) e^{pt} \, dp \tag{10-42}$$

where the summation is over all poles p_j inside the closed contour ABCDA. Provided $E(\mathbf{k}, p)e^{pt} \to 0$ sufficiently rapidly as $| p | \to \infty$, the contributions to the second term in (10–42) from the integration along AB and CD are negligible (cf. Problem 10–6) while that from BC is exponentially damped according to $e^{-\alpha t}$ and may be neglected for sufficiently large t or for any non-zero t by making α large. In general the poles p_j are complex, so writing

$$p_j(\mathbf{k}) = -i\omega_j(\mathbf{k}) - \gamma_j(\mathbf{k}) \tag{10-43}$$

where ω_j and γ_j are real, (10–42) becomes

$$E(\mathbf{k}, t) = \sum R_j e^{-i\omega_j t - \gamma_j} \qquad (10\text{–}44)$$

If any $\gamma_j < 0$ the field grows exponentially and the linear approximation breaks down, so we shall assume for the moment that all poles lie to the left of the imaginary p-axis (cf. Problem 10–7). Then all terms with $\gamma_j \neq 0$ in (10–44) are exponentially damped oscillations. Note that none are damped as strongly as $e^{-\alpha t}$; in general one is interested in the pole closest to the imaginary p-axis since this corresponds to the smallest damping decrement. However, complications arise when further poles exist very close to this pole.

We now investigate the limit of long wavelength waves, $k \to 0$. To lowest order in this limit (10–41) gives, on integration by parts and using $\int f_0 \, d\mathbf{v} = n$,

$$p = \pm i\omega_p \qquad (10\text{–}45)$$

that is, undamped plasma oscillations. To find the lowest-order k dependence we again integrate by parts and expand $(p + iku)^{-2}$ in powers of (iku/p), giving

$$-\frac{4\pi e^2}{mp^2} \int_{-\infty}^{\infty} du \, F_0(u) \left[1 - \frac{2iku}{p} - \frac{3k^2 u^2}{p^2} + \cdots \right] = 1 \qquad (10\text{–}46)$$

The imaginary term vanishes if $f_0(\mathbf{v})$ is isotropic; this is true for all the imaginary terms in the expansion (since they are all odd in u). The first correction to (10–45) arises from the term in k^2 giving for a Maxwellian distribution (cf. (8–45))

$$p = \pm i\omega_p (1 + \tfrac{3}{2}(k\lambda_D)^2) \qquad (10\text{–}47)$$

Since all the imaginary terms vanish in the expansion in powers of k no damping appears in such a solution. To find the damping decrement one must resort to the full expression (10–41). This presents a problem since the

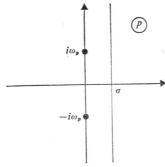

Fig. 10–2 $E(\mathbf{k}, p)$ is defined on the line $\mathrm{Re}\, p = \sigma$, to right of all singularities

integrand contains a pole at $u = ip/k$ which, with p given by (10–47), lies on the path of integration. However $E(\mathbf{k}, p)$ was originally defined on a line in the p plane to the right of all singularities, that is, for Re $p > 0$ (Fig. 10–2). Thus the integral in (10–41) is also defined for Re $p > 0$ and the integration

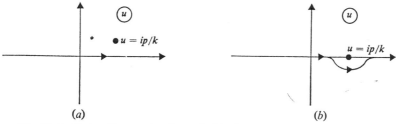

(a)   (b)

Fig. 10–3   Path of integration for pole (a) above Reu-axis, (b) on Reu-axis

along the real axis is below the pole in the u plane (Fig. 10–3a). Hence the path of integration must stay below the pole as shown in Fig. 10–3b. Passing half-way round the pole contributes $i\pi$ times the residue at ip/k while the rest of the integration (from $-\infty$ to $(ip/k) - \varepsilon$ and from $(ip/k) + \varepsilon$ to $+\infty$) may again be approximated by a power series as in (10–46). Thus

$$1 = -\frac{\omega_p^2}{p^2}\left[1 - \frac{3k^2}{p^2}\frac{\kappa T}{m}\right] + \sqrt{\frac{\pi}{2}}\left(\frac{m}{kT}\right)^{3/2}\frac{\omega_p^2\, p}{k^3}\exp\left(\frac{mp^2}{2k^2\kappa T}\right) \quad (10\text{–}48)$$

which on substituting (10–43) and using successive approximations gives for the *Landau damping decrement* (cf. Problem 10–8)

$$\gamma = \sqrt{\left(\frac{\pi}{8}\right)}\frac{\omega_p}{(k\lambda_D)^3}\exp\left[-\frac{1}{2(k\lambda_D)^2} - \frac{3}{2}\right] \quad (k\lambda_D \ll 1) \quad (10\text{–}49)$$

This result confirms what was said earlier concerning the vanishing of all imaginary terms in a power series expansion in small k; as $k \to 0$, $\gamma \to 0$ faster than any power of k. Numerical solution[96] of (10–41) shows that as $k\lambda_D \to 1$, $\gamma \to \omega_p$, that is, the damping time approaches the period of the oscillations. Thus λ_D, the Debye shielding distance, is the minimum wavelength at which longitudinal oscillations ($\mathbf{k} \parallel \mathbf{E}$) can occur.

A further observation to be made from (10–49) is the following. The damping decrement arose from the residue at the pole in (10–41). The sign of γ therefore depends critically on the slope of $F_0(u)$ at the pole. Since we considered a Maxwellian this was necessarily negative (for positive u as drawn in Fig. 10–3). This confirms the physical explanation of Landau damping given in Section 8–5. There it was argued that more particles moving slightly slower than the phase velocity of the wave than moving slightly faster (i.e., $dF_0/du < 0$ at $u = ip/k \simeq \omega_p/k$) leads to acceleration of particles at the expense of the wave energy, resulting in wave damping. Clearly, if dF_0/du had been positive at the pole (corresponding to more particles moving slightly faster than the phase velocity) γ would have had the opposite sign, leading to exponential growth of the wave (energy being transferred from particles to wave). Such an instability of electrostatic plasma oscillations is one of a class known as microinstabilities as opposed to the hydromagnetic instabilities discussed in Chapter 4.

10–6 Streaming instabilities

We now obtain a stability criterion for electrostatic plasma oscillations. The discussion follows closely a treatment due to Penrose.[97]

For convenience we write the plasma dispersion relation (10–41) in the form

$$Z(\zeta) = \alpha^2 k^2 \tag{10-50}$$

where
$$Z(\zeta) = \int_{-\infty}^{\infty} \frac{F_0'(u)}{u - \zeta} du \tag{10-51}$$

and $\zeta = ip/k$, $\alpha^2 = m/4\pi e^2$. For $k > 0$, given some distribution function $F_0(u)$, can $Z(\zeta)$ take a real positive value such that $\text{Im}\,\zeta > 0$? If so, then $F_0(u)$ is unstable with respect to electron plasma oscillations. Within the upper half plane the analyticity of $Z(\zeta)$ is ensured by requiring that

$$\int_{-\infty}^{\infty} |F_0'(u)|\, du < \infty$$

namely, that $F_0'(u)$ is an absolutely integrable function. The behaviour of $Z(\zeta)$ on the upper side of the cut along the real axis is

$$Z(\xi + i0) = \lim_{\eta \to 0} \int_{-\infty}^{\infty} \frac{F_0'(u)\, du}{u - (\xi + i\eta)} = P \int_{-\infty}^{\infty} \frac{F_0'(u)\, du}{u - \xi} + i\pi F_0'(\xi) \tag{10-52}$$

where P denotes the Cauchy principal value.

Now consider (10–51) as a transformation which maps the upper half of the ζ plane into some region of the Z plane. Let ζ move along the real axis from $-\infty$ to $+\infty$. The real axis of the ζ plane transforms into the boundary of this region in the Z plane, $Z(R)$, as shown in Fig. 10–4.† Since $Z(\pm\infty) = 0$ the curve starts and finishes at the origin. Suppose Z_0 is some point not on $Z(R)$; then, using the argument principle,[79] a variable point $Z(\xi + i0)$ tracing out $Z(R)$ completes a circuit (in an anticlockwise direction) of the point Z_0 as many times as $Z(\zeta)$ takes the value Z_0 in the upper half ζ plane. Thus the mapping of any point in the upper half of the ζ plane is either within $Z(R)$ or on it, so that the enclosed region in Fig. 10–4 represents the image of the

---

† Such curves, called Nyquist diagrams, were first used to discuss instabilities in electric circuits.

entire upper half ζ plane. Thus to satisfy (10–50) one requires that $Z(R)$ *enclose part of the positive real Z-axis.* Since $Z(R)$ is traced out anticlockwise

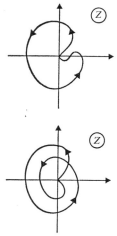

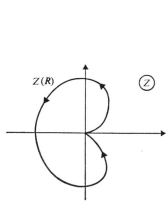

Fig. 10–4 Contour in Z plane corresponding to real axis in ζ plane

Fig. 10–5   Contours enclosing part of the real Z-axis

this means that the right-hand-most crossing of Re Z by the point $Z(\xi + i0)$ is in moving from Im $Z < 0$ into Im $Z > 0$ (cf. Fig. 10–5) so that, from (10–52), such a change in sign of Im $Z(\xi + i0)$ corresponds to a *minimum* in $F_0(u)$.

From (10–52) one has

$$\text{Re } Z(\xi + i0) = \lim_{\varepsilon \to 0} \left\{ \int_{-\infty}^{\xi - \varepsilon} + \int_{\xi + \varepsilon}^{\infty} \right\} \frac{d[F_0(u) - F_0(\xi)]}{u - \xi}$$

$$= \lim_{\varepsilon \to 0} \left\{ \int_{-\infty}^{\xi - \varepsilon} + \int_{\xi + \varepsilon}^{\infty} \right\} \frac{F_0(u) - F_0(\xi)}{(u - \xi)^2} du +$$

$$+ \lim_{\varepsilon \to 0} \left[\frac{F_0(\xi) - F_0(\xi - \varepsilon)}{\varepsilon} - \frac{F_0(\xi + \varepsilon) - F_0(\xi)}{\varepsilon} \right]$$

$$= P \int_{-\infty}^{\infty} \frac{F_0(u) - F_0(\xi)}{(u - \xi)^2} du$$

This gives the stability criterion: exponentially growing plasma modes will appear if, and only if, there is a minimum of $F_0(u)$ at a value $u = \xi$ such that

$$\int_{-\infty}^{\infty} \frac{F_0(u) - F_0(\xi)}{(u - \xi)^2} du > 0 \tag{10–53}$$

The principal value is unnecessary in (10–53) since $u = \xi$ is a minimum of $F_0(u)$.

As an example of a streaming instability, consider two electron beams travelling with equal speeds, V, but in opposite directions through a uniform background of positive charge. For a cold plasma, namely one with no thermal motion,

$$F_0(u) = \tfrac{1}{2}n[\delta(u - V) + \delta(u + V)]$$

This function has a minimum at $u = 0$ so that applying the Penrose criterion, (10–53), requires

$$\int_{-\infty}^{\infty} \frac{F_0(u)}{u^2} du > 0$$

for instability, and this is clearly the case. Having established the existence of an instability, its growth rate may be determined from (10–41). On integration by parts

$$-\frac{\omega_p^2}{n} \int \frac{F_0\, du}{(p + iku)^2} = 1$$

So, on substituting for F_0, one finds

$$-\frac{\omega_p^2}{2}\left[\frac{1}{(p + ikV)^2} + \frac{1}{(p - ikV)^2}\right] = 1$$

Again, in the limit $kV/\omega_p = 0$ one finds plasma oscillations, $p = \pm i\omega_p$. However, to the next order (cf. Problem 10–9) there exists a solution

$$p = kV$$

which is the growth rate of the instability.

A plausible physical explanation for this two-stream instability is as follows. If a perturbation of the initial homogeneous situation occurs such that one of the streams contains a region of increased electron density, then the electrons in the other stream approaching this region encounter a potential

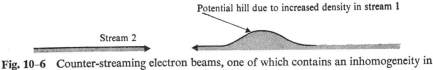

Potential hill due to increased density in stream 1

Stream 2

Fig. 10-6 Counter-streaming electron beams, one of which contains an inhomogeneity in density

hill (Fig. 10–6). They are slowed down while climbing the hill, reaching their minimum velocity at the peak. The tendency therefore is for electrons to spend a longer time passing through the region of increased electron density, thus augmenting the perturbation and leading to instability.

10–7 Boltzmann equation

We now return to the collisional kinetic equation (10–20) and consider how one might deal with the collision term $(\partial f/\partial t)_c$. As mentioned at the end of Section 10–2, one can either find some expression for P in terms of f or else forget about (10–21) and adopt some physical model for collisions, writing an expression for $(\partial f/\partial t)_c$ in terms of f. It is the latter approach we discuss first on the basis of two-body scattering theory. This heuristic formulation is due to Boltzmann who developed it as the basis of gas kinetic theory. The

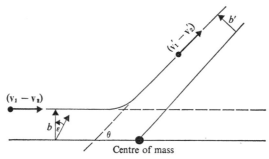

Fig. 10–7   Trajectory of particle 1 in centre of mass frame of reference

resulting kinetic equation has been investigated extensively by Hilbert, Enskog, and Chapman amongst others.[10, 98]

One assumes that a collision takes place within a volume $\delta \mathbf{r}$ and a time δt over which $f(\mathbf{r}_1, \mathbf{v}_1, t)$ is sensibly constant. Collisions will result in some particles being scattered out of the velocity volume element $d\mathbf{v}_1$ and some being scattered into $d\mathbf{v}_1$. The first type of event occurs when a particle (1) with velocity $\mathbf{v}_1$ collides with another (2) having a velocity $\mathbf{v}_2$, resulting in final velocities $\mathbf{v}'_1$, $\mathbf{v}'_2$. Figure 10–7 shows the trajectory of particle 1 in the centre of mass frame of reference. In this frame the collision is equivalent to the interaction of a particle of reduced mass $m_1 m_2/(m_1 + m_2)$ with a fixed scattering centre.[99] Then, by conservation of momentum and energy (cf. Problem 10–10),

$$| \mathbf{v}_1 - \mathbf{v}_2 | = | \mathbf{v}'_1 - \mathbf{v}'_2 | \qquad (10\text{–}54a)$$

Further, by conservation of angular momentum about the centre of mass

$$b = b' \qquad (10\text{–}54b)$$

The number of particles (1) passing through the element of area $b\, db\, d\varepsilon$ per second is $f(1) | \mathbf{v}_1 - \mathbf{v}_2 | b\, db\, d\varepsilon\, d\mathbf{v}_1$; ε is the angle the impact parameter makes with some fixed direction in the plane perpendicular to $(\mathbf{v}_1 - \mathbf{v}_2)$. The density of scattering particles (2) is $f(2)\, d\mathbf{v}_2$ so that the number of collisions per unit volume per second is

$$f(1) f(2) | \mathbf{v}_1 - \mathbf{v}_2 | b\, db\, d\varepsilon\, d\mathbf{v}_1\, d\mathbf{v}_2 \qquad (10\text{–}55)$$

Integrating this expression with respect to b, ε, and $\mathbf{v}_2$ then gives the total rate at which collisions remove particles from the velocity volume element $d\mathbf{v}_1$. Thus, if $(\partial f/\partial t)_c^{\text{out}}$ represents the rate of decrease of f due to collisions

$$\left(\frac{\partial f}{\partial t}\right)_c^{\text{out}} = \int f(1)f(2)\,|\,\mathbf{v}_1 - \mathbf{v}_2\,|\,b\,db\,d\varepsilon\,d\mathbf{v}_2 \tag{10-56}$$

To compute the rate of increase of f due to collisions $(\partial f/\partial t)_c^{\text{in}}$ we reverse the roles of the initial and final velocities—the particles start off with $\mathbf{v}_1'$, $\mathbf{v}_2'$ and finish with $\mathbf{v}_1$, $\mathbf{v}_2$. Then by analogy with (10–55) the number of collisions per unit volume per second is

$$f(1')f(2')\,|\,\mathbf{v}_1' - \mathbf{v}_2'\,|\,b'\,db'\,d\varepsilon\,d\mathbf{v}_1'\,d\mathbf{v}_2'$$
$$= f(1')f(2')\,|\,\mathbf{v}_1 - \mathbf{v}_2\,|\,b\,db\,d\varepsilon\,d\mathbf{v}_1'\,d\mathbf{v}_2' \tag{10-57}$$

using (10–54a, b). Further, one may regard the collision as a linear transformation of variables

$$\mathbf{v}_i' = \mathbf{h}_i(\mathbf{v}_1, \mathbf{v}_2) \qquad (i = 1, 2) \tag{10-58a}$$

for which the Jacobian (cf. Problem 10–11) is

$$\frac{\partial(\mathbf{v}_1', \mathbf{v}_2')}{\partial(\mathbf{v}_1, \mathbf{v}_2)} = 1 \tag{10-58b}$$

in which case (10–57) may be written $f(1')f(2')\,|\,\mathbf{v}_1 - \mathbf{v}_2\,|\,b\,db\,d\varepsilon\,d\mathbf{v}_1\,d\mathbf{v}_2$. Hence

$$\left(\frac{\partial f}{\partial t}\right)_c^{\text{in}} = \int f(1')f(2')\,|\,\mathbf{v}_1 - \mathbf{v}_2\,|\,b\,db\,d\varepsilon\,d\mathbf{v}_2 \tag{10-59}$$

where the variables of integration are the same as in (10–56); combining (10–56) and (10–59)

$$\left(\frac{\partial f}{\partial t}\right)_c = \int [f(1')f(2') - f(1)f(2)]\,|\,\mathbf{v}_1 - \mathbf{v}_2\,|\,b\,db\,d\varepsilon\,d\mathbf{v}_2$$

It is customary to write the collision integral using the differential scattering cross-section σ. To define this, note that particles passing through the element of area $b\,db\,d\varepsilon$ are scattered into the element of solid angle $d\Omega = \sin\theta\,d\theta\,d\varepsilon$; then[†]

$$\sigma \equiv -\frac{b\,db\,d\varepsilon}{d\Omega} = -\frac{b\,db}{\sin\theta\,d\theta}$$

[†] A minus sign is included in the definition of σ since increasing b decreases θ (cf. Fig. 10–7).

The differential scattering cross-section may be regarded as a function of any two of the three dependent variables b, $|\mathbf{v}_1 - \mathbf{v}_2|$, and θ; it is usually taken as a function of the latter two. The precise form of σ is determined by the nature of the interaction force. For Coulomb collisions between particles of charge and mass e_1, m_1 and e_2, m_2 one finds[99]

$$b = b_0 \cot \frac{\theta}{2} \tag{10–60a}$$

where

$$b_0 = \frac{e_1 e_2 (m_1 + m_2)}{m_1 m_2 |\mathbf{v}_1 - \mathbf{v}_2|^2} \tag{10–60b}$$

is the impact parameter for scattering through $\pi/2$. Hence

$$\sigma(|\mathbf{v}_1 - \mathbf{v}_2|, \theta) = \frac{b_0^2}{4 \sin^4 (\theta/2)} \tag{10–60c}$$

One has then for the *Boltzmann equation*

$$\frac{\partial f}{\partial t} + \mathbf{v}_1 \cdot \frac{\partial f}{\partial \mathbf{r}_1} + \frac{\mathbf{F}}{m} \cdot \frac{\partial f}{\partial \mathbf{v}_1}$$

$$= \int [f(1')f(2') - f(1)f(2)] \, |\mathbf{v}_1 - \mathbf{v}_2| \, \sigma(|\mathbf{v}_1 - \mathbf{v}_2|, \theta) \, d\mathbf{v}_2 \, d\Omega \tag{10–61}$$

Note that in deriving the Boltzmann collision integral we made these assumptions:

(i) The interaction length is much less than the length over which f varies appreciably.

(ii) The duration of a collision is much less than the time over which f varies appreciably.

(iii) All collisions are binary.

(iv) The colliding particles are uncorrelated (other than by the collision) since the probability of finding 1 and 2 within an interaction length was written as $\propto f(1) f(2)$.

Assumptions (i) and (ii) mean essentially that the field forces $\mathbf{F}$ may be ignored during a collision. Since Debye shielding limits the interaction length and time, this may be reasonable provided strong magnetic fields or rapidly fluctuating forces are not present. Assumptions (iii) and (iv) are almost always invalid for a plasma and lead to divergences in the collision integral. Since particles 1 and 2 are interacting with other particles at the same time that they interact with each other they are therefore correlated (via the other particles). While the effect of correlations is not easy to assess in general, we have seen that the pair correlation function $P \sim ff$ for small values of $|\mathbf{r}_1 - \mathbf{r}_2|$ while at large values of $|\mathbf{r}_1 - \mathbf{r}_2|$, correlations limit the interaction by Debye shielding. For these reasons, the neglect of correlations leads to divergences in the Boltzmann collision integral.

10–8 Properties of the Boltzmann equation

Multiplying (10–61) by an arbitrary function of velocity $\psi(v_1)$ and integrating over v_1, one obtains for the right-hand side

$$\int \psi(v_1)[f(1')f(2') - f(1)f(2)] \mid v_1 - v_2 \mid \sigma(\mid v_1 - v_2 \mid, \theta) \, d\Omega \, dv_1 \, dv_2$$

$$= \int \psi(v_2)[f(1')f(2') - f(1)f(2)] \mid v_1 - v_2 \mid \sigma(\mid v_1 - v_2 \mid, \theta) \, d\Omega \, dv_1 \, dv_2$$

$$(10\text{–}62)$$

on interchanging the variables of integration v_1 and v_2. In (10–62) v_1', v_2' are functions of v_1, v_2 given by (10–58a); the inverse transformations are (cf. Problem 10–11)

$$v_i = h_i(v_1', v_2') \qquad (i = 1, 2)$$

Thus by changing the variables of integration to v_1', v_2' in (10–62) we have, in view of (10–54a) and (10–58b),

$$\int \psi(h_1)[f(1')f(2') - f(h_1)f(h_2)] \mid v_1' - v_2' \mid \sigma \, d\Omega \, dv_1' \, dv_2'$$

$$= \int \psi(h_2)[f(1')f(2') - f(h_1)f(h_2)] \mid v_1' - v_2' \mid \sigma \, d\Omega \, dv_1' \, dv_2' \qquad (10\text{–}63)$$

Changing the labels of the variables of integration in (10–63) to v_1 and v_2 and using (10–58a), one finds

$$\int \psi(v_1')[f(1)f(2) - f(1')f(2')] \mid v_1 - v_2 \mid \sigma \, d\Omega \, dv_1 \, dv_2$$

$$= \int \psi(v_2')[f(1)f(2) - f(1')f(2')] \mid v_1 - v_2 \mid \sigma \, d\Omega \, dv_1 \, dv_2 \qquad (10\text{–}64)$$

Thus the expressions in (10–62) and (10–64) are all equal, so that

$$\int \psi(v_1)\left(\frac{\partial f}{\partial t}\right)_c dv_1 = \tfrac{1}{4}\int [\psi(1) + \psi(2) - \psi(1') - \psi(2')][f(1')f(2') -$$

$$- f(1)f(2)] \mid v_1 - v_2 \mid \sigma \, d\Omega \, dv_1 \, dv_2 \qquad (10\text{–}65)$$

Clearly, if ψ is any quantity which is conserved in a collision, (10–65) vanishes since

$$\psi(1) + \psi(2) = \psi(1') + \psi(2')$$

In particular, the collision moment vanishes for $\psi = m,\ mv,\ \frac{1}{2}mv^2$. These results were used in deriving the moment equations in Section 3–3.

Another important property of (10–61) was proved by Boltzmann and is known as the H-theorem. The function H is defined by

$$H = \langle \ln f(1) \rangle = \int f(1) \ln f(1)\, d\mathbf{v}_1$$

Assuming homogeneity (cf. Problem 10–12) and no forces $\mathbf{F}$

$$\frac{dH}{dt} = \int \frac{\partial f(1)}{\partial t}(\ln f(1) + 1)\, d\mathbf{v}_1 = \int (\ln f(1) + 1)(\partial f/\partial t)_c\, d\mathbf{v}_1$$

Now, using (10–65)

$$\frac{dH}{dt} = -\frac{1}{4}\int \ln\left[\frac{f(1')f(2')}{f(1)f(2)}\right] \cdot [f(1')f(2') - f(1)f(2)]\,|\,\mathbf{v}_1 - \mathbf{v}_2\,|\,\sigma\, d\Omega\, d\mathbf{v}_1\, d\mathbf{v}_2$$

and since $(a - b)\ln(a/b) \geqslant 0$ for $a, b > 0$

$$\frac{dH}{dt} \leqslant 0 \qquad\qquad (10\text{–}66)$$

the equality holding only if $f(1')f(2') = f(1)f(2)$ for all $\mathbf{v}_1,\ \mathbf{v}_2$. Thus H is a quantity which can only decrease with time.

We now find the distribution function which makes H a minimum† under the restrictions that density, momentum, and energy are conserved:

$$\left.\begin{aligned} \int f\, d\mathbf{v} &= \text{const.} \\ \int f\, m\mathbf{v}\, d\mathbf{v} &= \text{const.} \\ \int f\, \tfrac{1}{2}mv^2\, d\mathbf{v} &= \text{const.} \end{aligned}\right\} \qquad\qquad (10\text{–}67)$$

By the method of Lagrange multipliers we vary a function made up of H plus multiples of the five scalar equations in (10–67). Thus

$$\int d\mathbf{v}[\ln f + 1 + A + \mathbf{B}\cdot\mathbf{v} + Cv^2]\, \delta f = 0$$

where δf is arbitrary. Hence the quantity in the square bracket must be zero and

$$f = \text{const. } e^{-Cv^2 - \mathbf{B}\cdot\mathbf{v}} \qquad\qquad (10\text{–}68)$$

---

† We assume that the extremum corresponds to a minimum.

The constants in (10–68) are determined by those in (10–67). Identifying the latter as n, $nm\mathbf{u}$, and $(\frac{3}{2})n\kappa T + \frac{1}{2}nmu^2$,

$$f = n(m/2\pi\kappa T)^{3/2} \exp\left[-m(\mathbf{v} - \mathbf{u})^2/2\kappa T\right] \tag{10–69}$$

which reduces to (10–22) if there is no mean flow ($\mathbf{u} = 0$).

According to the H theorem, starting with an arbitrary f, H decreases until an f is reached such that $f(1')f(2') = f(1)f(2)$. This is satisfied by (10–69) (cf. Problem 10–13), which also makes H a minimum. Hence, in the absence of external forces the distribution function tends to a Maxwellian as $t \to \infty$.

The similarity between (10–66) and the law of increase of entropy is noticeable. For the Maxwellian distribution (10–69) (cf. Problem 10–14)

$$S = -\kappa H \tag{10–70}$$

and this equation is used as an extension of the definition of entropy for non-equilibrium situations.

Also connected with the law of increase of entropy is a very useful property of the *linearized* Boltzmann equation. Suppose some weak external force is applied to the plasma and a steady, homogeneous state is attained. If the force is sufficiently weak the distribution function will differ only slightly from its equilibrium value (10–22). Writing

$$f = f_M(1 + \phi(\mathbf{v}_1))$$

where ϕ is a small quantity, the linearized Boltzmann equation is, from (10–61)

$$\frac{\mathbf{F}}{m} \cdot \frac{\partial f_M(1)}{\partial \mathbf{v}_1} = \int [\phi(1') + \phi(2') - \phi(1) - \phi(2)] f_M(1) f_M(2) \,|\, \mathbf{v}_1 - \mathbf{v}_2 \,|\, \sigma \, d\Omega \, d\mathbf{v}_2 \tag{10–71a}$$

having used $f_M(1')f_M(2') = f_M(1)f_M(2)$. We may write this symbolically as

$$L = K\phi \tag{10–71b}$$

Then multiplying by $\psi(\mathbf{v}_1)$ and integrating over $\mathbf{v}_1$,

$$\int \psi(\mathbf{v}_1) L(\mathbf{v}_1) \, d\mathbf{v}_1 = \int \psi(\mathbf{v}_1) K\phi \, d\mathbf{v}_1 \tag{10–72a}$$

for which we shall use the notation

$$\langle \psi L \rangle = \langle \psi K\phi \rangle \tag{10–72b}$$

Now, using (10–65),

$$\langle \psi K\phi \rangle = \langle \phi K\psi \rangle \tag{10–73}$$

and
$$\langle \phi K \phi \rangle \leqslant 0 \qquad (10\text{--}74)$$

Properties (10–73) and (10–74) are sufficient for the formulation of a *variational principle*. This states that if ϕ is the solution of (10–71) and ψ is any function for which

$$\langle \psi L \rangle = \langle \psi K \psi \rangle \qquad (10\text{--}75)$$

then, of all functions ψ satisfying (10–75), ϕ is the one which makes $\langle \psi K \psi \rangle$ a minimum. The proof is straightforward. From (10–74)

$$\langle (\phi - \psi) K (\phi - \psi) \rangle \leqslant 0$$

so that using (10–73)

$$\langle \phi K \phi \rangle + \langle \psi K \psi \rangle \leqslant 2 \langle \psi K \phi \rangle$$

Finally, from (10–72b) and (10–75)

$$\langle \phi K \phi \rangle \leqslant \langle \psi K \psi \rangle$$

which proves the principle.

The procedure for solving (10–71) is thus to make an intelligent guess for the functional form of ϕ containing parameters. The parameters are then varied to find the minimum of $\langle \phi K \phi \rangle$. The principle has been employed widely in the calculation of transport coefficients; this may be illustrated for electrical conductivity.

If the force $\mathbf{F}$ in (10–71) is due to a constant electric field $\mathbf{E}$

$$L = \frac{e \mathbf{E} . \mathbf{v}_1}{\kappa T} f_M$$

and
$$\langle \phi K \phi \rangle = \langle \phi L \rangle = -\mathbf{E} . \mathbf{j} / \kappa T$$

Since $\mathbf{j} = \sigma \mathbf{E}$,

$$\sigma = -\frac{\kappa T}{E^2} \langle \phi K \phi \rangle$$

and one sees that it is σ which is being maximized. This is similar to the use of the variational principle in quantum mechanics to find energy levels; there ϕ would be the wave function and K the Hamiltonian. It is well known (cf. Problem 10–15) that the error in $\langle \phi K \phi \rangle$ is always less than the error in ϕ.

The relationship between the variational principle and the law of increase of entropy is easily demonstrated. Differentiating (10–70) with respect to time and using the definition of H

$$\frac{\partial S}{\partial t} = -\kappa \int \frac{\partial f}{\partial t} (1 + \ln f) \, d\mathbf{v}_1$$

In the initial stage, before a steady state has been reached, $\partial f/\partial t$ has two contributions, one being the field term and the other the collision term. Taking just the latter

$$\left(\frac{\partial S}{\partial t}\right)_c = -\kappa \int \left(\frac{\partial f}{\partial t}\right)_c (1 + \ln f)\, d\mathbf{v}_1$$

$$= -\kappa \int \{1 + \ln f_M + \ln(1 + \phi)\}\left(\frac{\partial f}{\partial t}\right)_c d\mathbf{v}_1$$

$$= -\kappa \int \ln(1 + \phi) \cdot \left(\frac{\partial f}{\partial t}\right)_c d\mathbf{v}_1$$

by (10–65). Keeping the lowest-order non-vanishing term in ϕ

$$\left(\frac{\partial S}{\partial t}\right)_c = -\kappa \int \phi K\phi\, d\mathbf{v}_1 = -\kappa\langle \phi K\phi\rangle \tag{10–76}$$

Combining (10–74) and (10–76) shows that the effect of collisions is to increase entropy.

On the other hand, entropy is reduced by the field which tends to order the system; by a calculation similar to that leading to (10–76)

$$\left(\frac{\partial S}{\partial t}\right)_{\text{field}} = \kappa\langle L\phi\rangle$$

and the steady state is reached when entropy production by collisions balances entropy decrease due to the field, that is, when $\langle L\phi\rangle = \langle \phi K\phi\rangle$.

Note further from (10–76) that minimizing $\langle \phi K\phi\rangle$, as one does in applying the variational principle, corresponds to maximizing entropy production by collisions.

10–9 Fokker–Planck equation

The main criticism of using the Boltzmann collision integral for a plasma arises from the assumptions that collisions are short-range and binary. Since there are typically a thousand particles within a Debye sphere, any given particle is interacting with many others at any instant and a significant deflection of the particle is more likely to be due to the cumulative effect of many weak interactions than a single close collision. The average time taken for a $\pi/2$ deflection due to a single collision of a particle travelling with speed v is

$$t_1 = \frac{1}{nv\sigma_t(\pi/2)}$$

where $\sigma_t(\pi/2)$ is the total cross-section for a $\pi/2$ deflection. Let us compare this with the cumulative effect of weaker collisions (deflections $< \pi/2$). For simplicity assume that the scattering particles are infinitely heavy so that the centre of mass frame is also the laboratory frame and $\mathbf{v}_2 = 0$ (Fig. 10–7); also since most collisions are weak take $\sin \theta \approx \theta$ (cf. Problem 10–16). Thus the deflection suffered in one collision is $|\Delta \mathbf{v}| = v\theta$ and from (10–60c), the cross-section for this is $4b_0^2/\theta^4$. Hence the mean square deflection per second is (the mean deflection being zero)

$$\frac{\mathrm{d}\,\overline{|\Delta \mathbf{v}|^2}}{\mathrm{d}t} = nv \int |\Delta \mathbf{v}|^2 \, \sigma(\theta) \, \mathrm{d}\Omega = 8\pi b_0^2 nv^3 \int_{\theta_{\min}}^{\pi/2} \frac{\mathrm{d}\theta}{\theta}$$

where the θ integration has been cut off at some minimum value to avoid divergence. This divergence arises directly from the neglect of shielding. If one uses the shielded Coulomb potential in the collision integral this divergence no longer appears and a small θ cut-off is unnecessary. Since the shielded potential vanishes exponentially for distances greater than λ_D, $\theta_{\min}$ will be taken as the deflection for scattering at impact parameter $b = \lambda_D$, i.e., from (10–60a) $\theta_{\min} = 2b_0/\lambda_D$. Thus the time for a $\pi/2$ deflection (when $\sqrt{|\Delta \mathbf{v}|^2} \sim v$) due to multiple scattering is

$$t_m \simeq \frac{1}{8\pi b_0^2 nv \ln \left(\dfrac{\pi \lambda_D}{4b_0} \right)} \tag{10–77}$$

Comparing this with t_1

$$\frac{t_1}{t_m} = \frac{8\pi b_0^2 nv \ln (\pi \lambda_D/4b_0)}{nv . \pi b_0^2} \simeq 8 \ln (\lambda_D/b_0)$$

Since $\ln (\lambda_D/b_0)$ has a value around ten for most plasmas this ratio shows that significant deflections are much more likely to be caused by many weak interactions than by one single close collision. The Fokker–Planck equation is a kinetic equation with the collision integral based on such a model.

One supposes that a function $\psi(\mathbf{v}, \Delta \mathbf{v})$ may be defined such that ψ is the probability that a particle with velocity $\mathbf{v}$ acquires an increment $\Delta \mathbf{v}$ in a time Δt. Note that it is assumed that ψ is time-independent—that is, independent of the particle's history; such a process is called a *Markoff process*. With the definition of ψ it follows that

$$f(\mathbf{r}, \mathbf{v}, t) = \int f(\mathbf{r}, \mathbf{v} - \Delta \mathbf{v}, t - \Delta t) \, \psi(\mathbf{v} - \Delta \mathbf{v}, \Delta \mathbf{v}) \, \mathrm{d}(\Delta \mathbf{v}) \tag{10–78}$$

Since the increments $\Delta \mathbf{v}$ are small, (10–78) may be expanded to give

$$f(\mathbf{r}, \mathbf{v}, t) = \int d(\Delta \mathbf{v}) \left\{ f(\mathbf{r}, \mathbf{v}, t - \Delta t) \, \psi(\mathbf{v}, \Delta \mathbf{v}) - \Delta \mathbf{v} \cdot \left[\frac{\partial f}{\partial \mathbf{v}} \psi + \frac{\partial \psi}{\partial \mathbf{v}} f \right] \right.$$

$$\left. + \tfrac{1}{2} \Delta \mathbf{v} \, \Delta \mathbf{v} : \left[\frac{\partial^2 f}{\partial \mathbf{v} \, \partial \mathbf{v}} \psi + 2 \frac{\partial f}{\partial \mathbf{v}} \frac{\partial \psi}{\partial \mathbf{v}} + \frac{\partial^2 \psi}{\partial \mathbf{v} \, \partial \mathbf{v}} f \right] + \cdots \right\} \quad (10\text{--}79)$$

Clearly, the total probability of all possible deflections must be one:

$$\int \psi \, d(\Delta \mathbf{v}) = 1$$

Then defining the rate of change of f due to collisions by

$$\left(\frac{\partial f}{\partial t} \right)_c \equiv \frac{f(\mathbf{r}, \mathbf{v}, t) - f(\mathbf{r}, \mathbf{v}, t - \Delta t)}{\Delta t}$$

one finds from (10–79)

$$\left(\frac{\partial f}{\partial t} \right)_c = -\frac{\partial}{\partial \mathbf{v}} \cdot (f \langle \Delta \mathbf{v} \rangle) + \frac{1}{2} \frac{\partial^2}{\partial \mathbf{v} \, \partial \mathbf{v}} : (f \langle \Delta \mathbf{v} \, \Delta \mathbf{v} \rangle) \quad (10\text{--}80)$$

where
$$\left\{ \begin{matrix} \langle \Delta \mathbf{v} \rangle \\ \langle \Delta \mathbf{v} \, \Delta \mathbf{v} \rangle \end{matrix} \right\} = \frac{1}{\Delta t} \int \psi(\mathbf{v}, \Delta \mathbf{v}) \left\{ \begin{matrix} \Delta \mathbf{v} \\ \Delta \mathbf{v} \, \Delta \mathbf{v} \end{matrix} \right\} d(\Delta \mathbf{v})$$

Substitution of (10–80) in (10–20) gives the *Fokker–Planck equation*. The first term on the right-hand side of (10–80) is called the *coefficient of dynamical friction* since it gives rise to a slowing-down effect. The second term is the *coefficient of diffusion* since it has the effect of spreading out an initial stream of uni-directional particles. This behaviour is discussed in the next section.

Until the function ψ is specified, the right-hand side of (10–80) is just as formal as the left-hand side. Various forms of the Fokker–Planck coefficients have been derived, including attempts to describe many-particle collisions by using rapidly fluctuating electric fields.[100] Rosenbluth, MacDonald, and Judd[101] assumed that multiple collisions could be treated as sequences of binary collisions and hence, by the same process used to obtain the Boltzmann collision integral (10–61), defined

$$\left\{ \begin{matrix} \langle \Delta \mathbf{v}_1 \rangle \\ \langle \Delta \mathbf{v}_1 \, \Delta \mathbf{v}_1 \rangle \end{matrix} \right\} = \int d\mathbf{v}_2 \, d\Omega \, f(\mathbf{v}_2) \, |\mathbf{v}_1 - \mathbf{v}_2| \, \sigma(|\mathbf{v}_1 - \mathbf{v}_2|, \theta) \left\{ \begin{matrix} \Delta \mathbf{v}_1 \\ \Delta \mathbf{v}_1 \, \Delta \mathbf{v}_1 \end{matrix} \right\} \quad (10\text{--}81)$$

The deflection $\Delta \mathbf{v}$ is then given by the kinematics of the two-body collision. Defining the relative velocity

$$\mathbf{g} = \mathbf{v}_1 - \mathbf{v}_2 \quad (10\text{--}82a)$$

and the centre of mass velocity

$$\mathbf{V} = \frac{m_1 \mathbf{v}_1 + m_2 \mathbf{v}_2}{m_1 + m_2} \quad (10\text{--}82b)$$

one may write

$$\mathbf{v}_1 = \mathbf{V} + \frac{m_2}{m_1 + m_2}\mathbf{g} \qquad \mathbf{v}_1' = \mathbf{V} + \frac{m_2}{m_1 + m_2}\mathbf{g}' \qquad (10\text{-}82c)$$

and hence

$$\Delta\mathbf{v}_1 = \mathbf{v}_1' - \mathbf{v}_1 = \frac{m_2}{m_1 + m_2}(\mathbf{g}' - \mathbf{g}) = \frac{m_2}{m_1 + m_2}\Delta\mathbf{g} \qquad (10\text{-}82d)$$

Let $(\hat{\mathbf{e}}_1, \hat{\mathbf{e}}_2, \hat{\mathbf{e}}_3)$ be the unit vectors of a right-handed Cartesian coordinate system with $\hat{\mathbf{e}}_1 = \hat{\mathbf{g}}$; the zero azimuthal angle $(\varepsilon = 0)$ occurs when $\mathbf{g}'$ lies in

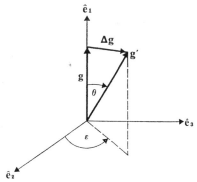

Fig. 10-8 Resolution of $\Delta\mathbf{g}$ in Cartesian coordinate system

the $(\hat{\mathbf{e}}_1, \hat{\mathbf{e}}_2)$ plane (cf. Fig. 10–8). One has

$$\Delta\mathbf{g} = 2g\sin\tfrac{1}{2}\theta[-\hat{\mathbf{e}}_1\sin\tfrac{1}{2}\theta + \hat{\mathbf{e}}_2\cos\tfrac{1}{2}\theta\cos\varepsilon + \hat{\mathbf{e}}_3\cos\tfrac{1}{2}\theta\sin\varepsilon] \qquad (10\text{-}82e)$$

It is then straightforward to substitute from (10–82) and (10–60) in (10–81) to obtain (cf. Problem 10–17)

$$\langle\Delta\mathbf{v}_1\rangle = -\Gamma_1\left(\frac{e_2}{e_1}\right)^2\frac{m_1 + m_2}{m_2}\int\frac{\hat{\mathbf{e}}_1}{g^2}f(\mathbf{v}_2)\,d\mathbf{v}_2$$

$$\langle\Delta\mathbf{v}_1\,\Delta\mathbf{v}_1\rangle = -\Gamma_1\left(\frac{e_2}{e_1}\right)^2\int\frac{(\hat{\mathbf{e}}_2\hat{\mathbf{e}}_2 + \hat{\mathbf{e}}_3\hat{\mathbf{e}}_3)}{g}f(\mathbf{v}_2)\,d\mathbf{v}_2$$

$$\Gamma_i = -\frac{4\pi e_i^4}{m_i^2}\ln\left[\sin\left(\frac{\theta_{\min}}{2}\right)\right] \qquad (i = 1, 2)$$

where only terms in $\ln[\sin(\theta_{\min}/2)]$ have been retained. These expressions may also be written, without reference to any particular frame,

$$\langle \Delta \mathbf{v}_1 \rangle = -\Gamma_1 \left(\frac{e_2}{e_1}\right)^2 \left(\frac{m_1 + m_2}{m_2}\right) \int \frac{\mathbf{g}}{g^3} f(\mathbf{v}_2) \, d\mathbf{v}_2$$

$$\langle \Delta \mathbf{v}_1 \, \Delta \mathbf{v}_1 \rangle = \Gamma_1 \left(\frac{e_2}{e_1}\right)^2 \int \frac{(g^2 \mathbf{I} - \mathbf{g}\mathbf{g})}{g^3} f(\mathbf{v}_2) \, d\mathbf{v}_2 \qquad \Bigg\} \quad (10\text{-}83a)$$

where $\mathbf{I}$ is the unit dyadic; alternatively in tensor notation

$$\langle (\Delta \mathbf{v}_1)_i \rangle = -\Gamma_1 \left(\frac{e_2}{e_1}\right)^2 \left(\frac{m_1 + m_2}{m_2}\right) \int \frac{g_i}{g^3} f(\mathbf{v}_2) \, d\mathbf{v}_2$$

$$\langle (\Delta \mathbf{v}_1)_i \, (\Delta \mathbf{v}_1)_j \rangle = \Gamma_1 \left(\frac{e_2}{e_1}\right)^2 \int \frac{g^2 \delta_{ij} - g_i g_j}{g^3} f(\mathbf{v}_2) \, d\mathbf{v}_2 \qquad \Bigg\} \quad (10\text{-}83b)$$

where δ_{ij} is the unit tensor. Finally, using

$$\frac{\mathbf{g}}{g^3} = -\frac{\partial}{\partial \mathbf{v}_1}\left(\frac{1}{g}\right) \qquad \frac{g^2 \mathbf{I} - \mathbf{g}\mathbf{g}}{g^3} = \frac{\partial^2 g}{\partial \mathbf{v}_1 \, \partial \mathbf{v}_1} \qquad (10\text{-}84a)$$

or

$$\frac{g_i}{g^3} = -\frac{\partial}{\partial v_{1i}}\left(\frac{1}{g}\right) \qquad \frac{g^2 \delta_{ij} - g_i g_j}{g^3} = \frac{\partial^2 g}{\partial v_{1i} \, \partial v_{1j}} \qquad (10\text{-}84b)$$

(10–80) becomes for $m_1 = m_2 = m$, $e_1 = e_2 = -e$,

$$\left(\frac{\partial f}{\partial t}\right)_c = \Gamma \left\{ -\frac{\partial}{\partial \mathbf{v}_1} \cdot \left[f(\mathbf{v}_1) \frac{\partial H(\mathbf{v}_1)}{\partial \mathbf{v}_1} \right] + \frac{1}{2} \frac{\partial^2}{\partial \mathbf{v}_1 \, \partial \mathbf{v}_1} : \left[f(\mathbf{v}_1) \frac{\partial^2 G(\mathbf{v}_1)}{\partial \mathbf{v}_1 \, \partial \mathbf{v}_1} \right] \right\} \qquad (10\text{-}85a)$$

where

$$H(\mathbf{v}_1) = 2 \int d\mathbf{v}_2 \frac{f(\mathbf{v}_2)}{|\mathbf{v}_1 - \mathbf{v}_2|}$$

and

$$G(\mathbf{v}_1) = \int d\mathbf{v}_2 f(\mathbf{v}_2) \, |\mathbf{v}_1 - \mathbf{v}_2| \qquad \Bigg\} \quad (10\text{-}86)$$

In tensor notation (10–85a) is

$$\left(\frac{\partial f}{\partial t}\right)_c = \Gamma \left\{ -\frac{\partial}{\partial v_{1i}} \left[f(\mathbf{v}_1) \frac{\partial H(\mathbf{v}_1)}{\partial v_{1i}} \right] + \frac{1}{2} \frac{\partial^2}{\partial v_{1i} \, \partial v_{1i}} \left[f(\mathbf{v}_1) \frac{\partial^2 G(\mathbf{v}_1)}{\partial v_{1j} \, \partial v_{1i}} \right] \right\} \qquad (10\text{-}85b)$$

Not surprisingly (10–85) can also be obtained from the Boltzmann collision integral (10–61) if it is assumed that collisions produce only small deflections $\Delta \mathbf{v}_1, \Delta \mathbf{v}_2$. Using (10–82) and the corresponding equations for $\mathbf{v}_2$ one expands $f(1') f(2')$, keeping only terms in Δg_i and $\Delta g_i \Delta g_j$ (cf. Problem 10–18). However, it should be emphasized that (10–85) is only one particular form of the Fokker–Planck coefficients obtained by adopting a binary collision model.

One may split up the range of collision impact parameters into three parts:

(i) $0 \leqslant b \leqslant$ several b_0.
(ii) Several $b_0 \leqslant b \leqslant d = n^{-1/3}$.
(iii) $d \leqslant b \leqslant \lambda_D$.

In (i) binary, large-angle scattering occurs; in (ii) scattering is again mainly binary but small-angle; finally, in (iii) scattering events are many-body and weak. The Boltzmann equation applies in regions (i) and (ii) while the Fokker–Planck equation may in general be applied in regions (ii) and (iii). It would appear that the assumption of binary collisions used in deriving (10–85) limits its validity to region (ii), as does the expansion of the Boltzmann collision integral in Problem 10–18. Nevertheless a derivation of the Fokker–Planck coefficients by Kaufman[102] in which scattering in region (iii) is caused by charge density fluctuations *also* yields (10–85).

10–10 Relaxation times

The Fokker–Planck equation, like the Boltzmann equation, is non-linear in f and usually one must resort to numerical integration in order to make use of it. However, in this section the equation is used to describe a simplified model in which a single *test particle* (either an electron or a proton) interacts with *field particles* which are in thermal equilibrium. Such a simple model permits the calculation of various time scales—for instance, the time for the test particle to be deflected through an angle $\pi/2$—and these provide a rough estimate for plasma *relaxation times* such as the time for the distribution function to become isotropic.

The distribution function of particle 1, the test particle, is

$$f(\mathbf{v}_1) = \delta(\mathbf{v}_1 - \mathbf{U}(t)) \tag{10–87}$$

where $\mathbf{U}(t)$ is its velocity at time t. Particle 2, the field particle or scatterer, may be either an electron or a proton and since the field particles are in thermal equilibrium

$$f(\mathbf{v}_2) = f_M(\mathbf{v}_2) = \frac{na^3}{\pi^{3/2}} \exp\left(-a^2 v_2^2\right) \tag{10–88}$$

where
$$a^2 = \frac{m_2}{2\kappa T}$$

Note that we are no longer considering the special case of a one-component plasma, so (10–86) must be modified slightly. This is trivial and it is easily seen from (10–83) that the factor 2 in the expression for $H(\mathbf{v}_1)$ must be replaced by $(m_1 + m_2)/m_2$ while $G(\mathbf{v}_1)$ is unaltered.

Substituting (10–88) in

$$H(\mathbf{v}_1) = \left(\frac{m_1 + m_2}{m_2}\right) \int d\mathbf{v}_2 \frac{f(\mathbf{v}_2)}{|\mathbf{v}_1 - \mathbf{v}_2|}$$

one finds (cf. Problem 10–19)

$$H(\mathbf{v}_1) = \left(\frac{m_1 + m_2}{m_2}\right) \frac{n}{v_1} \Phi(av_1) \qquad (10\text{–}89a)$$

where Φ is the error function

$$\Phi(x) = \frac{2}{\sqrt{\pi}} \int_0^x e^{-y^2} \, dy$$

Similarly

$$G(\mathbf{v}_1) = n\left[\left(v_1 + \frac{1}{2a^2 v_1}\right)\Phi(av_1) + \frac{e^{-a^2 v_1^2}}{a\sqrt{\pi}}\right] \qquad (10\text{–}89b)$$

The first time scale we consider is that characteristic of the slowing down of the test particle

$$\tau_s = -\frac{U}{\partial U/\partial t} \qquad (10\text{–}90)$$

Multiplying (10–85) by $\mathbf{v}_1$ and integrating over velocity space gives, using (10–87),

$$\frac{1}{\Gamma_1}\frac{\partial U}{\partial t} = \frac{\partial H(U)}{\partial U} = \frac{U}{U}\frac{\partial H}{\partial U} \qquad (10\text{–}91)$$

since H is an isotropic function of U. The term in G is zero on integrating twice by parts. This confirms the earlier statement (cf. Section 10–9) that the first of the Fokker–Planck coefficients represents the dynamical friction. From (10–89a) one has

$$\frac{\partial H}{\partial U} = -2\left(1 + \frac{m_1}{m_2}\right) na^2\, \Psi(aU) \qquad (10\text{–}92)$$

then, using (10–91) and (10–92), (10–90) gives for the *slowing-down time*

$$\tau_s = \frac{U}{[1 + (m_1/m_2)]A_D a^2\, \Psi(aU)} \qquad (10\text{–}93)$$

where $\qquad A_D = 2\Gamma_1 n = \dfrac{8\pi ne^4}{m_1^2}\ln \Lambda \qquad \Lambda = \dfrac{2}{\theta_{min}} = \dfrac{\lambda_D}{b_0}$

$$\Psi(x) = \frac{\Phi(x) - x\Phi'(x)}{2x^2} \qquad \Biggr\} \qquad (10\text{–}94)$$

Note that

$$\Psi(x) \to 2x/3\sqrt{\pi} \qquad \Phi(x) \to 2x/\sqrt{\pi} \qquad \text{as } x \to 0$$
$$\Psi(x) \to 1/2x^2 \qquad \Phi(x) \to 1 \qquad \text{as } x \to \infty$$

One may now make some general observations. First of all, for fixed U, lowering the density n or increasing the temperature T increases τ_s; in other words, collisions are less effective under low-density, high-temperature conditions. This also follows from (10–60); the collisional cross-section is greatest for $|\mathbf{U} - \mathbf{v}_2| \simeq 0$ but under low-density, high-temperature $[U \ll \sqrt{(\kappa T/m)}]$ conditions there are fewer scatterers with $\mathbf{v}_2 \simeq \mathbf{U}$. For very fast test particles, $aU \gg 1$, $\tau_s \propto U^3$ while in the opposite limit, $aU \ll 1$, τ_s is independent of U.

The *deflection time* τ_D is defined by†

$$\tau_D = \frac{U^2}{\partial U_\perp^2/\partial t} \tag{10–95}$$

and is found by taking the v_{1k}^2 moment of (10–85). One finds for each of the two components of $\mathbf{v}_1$ perpendicular to $\mathbf{U}$

$$\frac{1}{\Gamma_1} \frac{\partial U_k^2}{\partial t} = \frac{\partial^2}{\partial U_k^2} G(U) \tag{10–96}$$

In this case the first Fokker–Planck coefficient gives a zero contribution; only the second coefficient contributes to the lateral spread in velocity from the parallel direction, $\hat{\mathbf{U}}$. Hence this term is known as the coefficient of diffusion.

The sum of the two perpendicular moments is simplified by observing that $G(U)$ is isotropic, so that if $\hat{\mathbf{U}}$ defines the x direction

$$\left(\frac{\partial^2}{\partial U_y^2} + \frac{\partial^2}{\partial U_z^2}\right) G(U) = \left(\nabla_U^2 - \frac{\partial^2}{\partial U^2}\right) G = \frac{2}{U} \frac{\partial G}{\partial U}$$

Thus

$$\frac{1}{\Gamma_1} \frac{\partial U_\perp^2}{\partial t} = \frac{2}{U} \frac{\partial G}{\partial U} = \frac{2n}{U}(\Phi(aU) - \Psi(aU)) \tag{10–97}$$

and (10–95) becomes

$$\tau_D = \frac{U^3}{A_D[\Phi(aU) - \Psi(aU)]} \tag{10–98}$$

Note that τ_D, the *collision time*, was the quantity calculated very approximately in Section 10–9; in the limit $m_2 \to \infty$, $\tau_D \simeq t_m$. Clearly τ_D gives an

† We must define τ_D by taking the mean square deviation since the mean deviation of a test particle moving in a uniform plasma is zero (cf. Problem 10–20).

estimate for the relaxation to isotropy of an initially anisotropic distribution. For small U, $\tau_D \propto U^2$ while for large U, $\tau_D \propto U^3$.

Finally, consider the time scale for the exchange of energy. This too must be defined by a mean square deviation since the test particle is being slowed down in the forward (or parallel) direction, thereby losing energy, and accelerated in the sideways (or perpendicular) direction, thereby gaining energy. Thus, if W is the energy of the test particle the *energy exchange time* τ_E is defined by

$$\tau_E = \frac{W^2}{\partial(\Delta W)^2/\partial t} \tag{10-99}$$

and is obtained by taking the $(m_1^2/4)(v_1^2 - U^2)^2$ (i.e., $(\Delta W)^2$) moment of (10–85). One finds

$$\frac{1}{\Gamma_1}\frac{\partial(\Delta W)^2}{\partial t} = m_1^2 U^2 \frac{\partial^2 G}{\partial U^2} = 2m_1^2 n U\,\Psi(aU) \tag{10-100}$$

Substitution in (10–99) then gives

$$\tau_E \doteq \frac{U^3}{4A_D\,\Psi(aU)} \tag{10-101}$$

The energy exchange time gives an approximate measure of the time required for relaxation to a Maxwellian distribution. Putting $m_1 = m_2 = m^-$ gives the time for the electron distribution to become Maxwellian; for thermal velocities $U \sim a^{-1}$ this is

$$\tau_E^{--} \sim \frac{(m^-)^{1/2}(2\kappa T)^{3/2}}{32\pi n e^4 \ln \Lambda\,\Psi(1)} \tag{10-102}$$

Similarly, for protons

$$\tau_E^{++} \sim \frac{(m^+)^{1/2}(2\kappa T)^{3/2}}{32\pi n e^4 \ln \Lambda\,\Psi(1)} \tag{10-103}$$

Now suppose the electrons and protons have somewhat different temperatures. The time for equalization of temperatures may be estimated by putting $m_1 = m^-$, $m_2 = m^+$, $aU \sim (m^+/m^-)^{1/2}$ (or $m_1 = m^+$, $m_2 = m^-$, $aU \sim (m^-/m^+)^{1/2}$) and one finds

$$\tau_E^{-+} \sim \frac{m^+(m^-)^{-1/2}(2\kappa T)^{3/2}}{16\pi n e^4 \ln \Lambda} \sim \tau_E^{+-} \tag{10-104}$$

Since $\Psi(1) \simeq 0\cdot 2$, comparison of (10–102), (10–103), and (10–104) shows that

$$\tau_E^{--} : \tau_E^{++} : \tau_E^{-+} \sim 1 : \left(\frac{m^+}{m^-}\right)^{1/2} : \left(\frac{m^+}{m^-}\right) \qquad (10\text{--}105)$$

Thus if one starts with a non-equilibrium plasma having an average electron energy somewhat different from the average proton energy and allows it to relax to equilibrium, the electrons reach a Maxwellian distribution first, followed by the protons, and finally equipartition of energy takes place. These results have a simple explanation. The average electron velocity is of order $(m^+/m^-)^{1/2}$ times the average proton velocity; the electrons therefore have a higher collision frequency and relax more quickly by just such a factor. Equipartition of energy between electrons and protons relies on electron–proton collisions which are relatively inefficient in transferring energy; at most (m^-/m^+) of the kinetic energy involved in a collision may be transferred (cf. Problem 10–22).

10–11 Transport coefficients

Rigorous calculations of transport coefficients[103] usually involve numerical computation because of the complicated form of the collision integral. An expansion of the distribution function in powers of a small parameter is assumed and a variational principle (of which that given in Section 10–8 is the simplest form) is used to calculate the transport coefficients to the desired order of approximation. Here we present approximate calculations of plasma transport coefficients, in which only the first two terms in the expansion of the distribution function are kept, that is,

$$f = f_0 + f' \qquad (10\text{--}106)$$

Also, by a crude treatment of the collision term we avoid complicated integrations. We assume that a plasma in equilibrium is disturbed by various external perturbations causing small, steady state fluxes of matter, momentum, or energy. Then f_0 is given by (10–69), where in general n, $\mathbf{u}$, and T may be functions of position, and so

$$f_0 = n(\mathbf{r})\left(\frac{m}{2\pi\kappa T(\mathbf{r})}\right)^{3/2} \exp\left[-m(\mathbf{v} - \mathbf{u}(\mathbf{r}))^2/2\kappa T(\mathbf{r})\right] \qquad (10\text{--}107)$$

The collision term is not only linearized by using (10–106) but is simplified by assuming that it may be represented as the product of a collision frequency ν_c and a linear function of the distribution function; since the collision term must vanish for $f = f_0$, and collisions cause f to tend towards f_0, one takes

$$\left(\frac{\partial f}{\partial t}\right)_c = -\nu_c(f - f_0) = -\nu_c f' \qquad (10\text{--}108)$$

As we have seen in the last section the collision frequency, $\nu_c \sim \tau_D^{-1}$, is velocity dependent. However, this complication will be ignored† by assuming some mean value of ν_c, say its value for thermal particles.

The simplest transport coefficient to obtain is the *electrical conductivity* σ. For simplicity one may take

$$f_0 = f_M = n\left(\frac{m}{2\pi\kappa T}\right)^{3/2} \exp\left(-mv^2/2\kappa T\right)$$

and f' to be a function of velocity only. Then substitution in the kinetic equation (10–20) with $\mathbf{F} = e\mathbf{E}$, the (constant) electric force, and $(\partial f/\partial t)_c$ given by (10–108) leads to

$$\frac{e\mathbf{E}}{m} \cdot \frac{\partial f_0}{\partial \mathbf{v}} = -\nu_c f' \tag{10–109}$$

where the product of small quantities, $\mathbf{E}$ and f', has been neglected. Since

$$\sigma\mathbf{E} = \mathbf{j} = e\int \mathbf{v} f\, d\mathbf{v} = e\int \mathbf{v} f'\, d\mathbf{v} \tag{10–110}$$

it follows from (10–109), by multiplying by ev and integrating over velocity space, that

$$\nu_c\mathbf{j} = \frac{e^2}{\kappa T}\mathbf{E} \cdot \int \mathbf{v}\, \mathbf{v} f_M\, d\mathbf{v} \tag{10–111}$$

Taking $\mathbf{E}$ in the x direction, j_y and j_z are zero and one finds from (10–110) and (10–111)

$$\sigma = \frac{ne^2}{m\nu_c} \tag{10–112}$$

The electrical conductivity is the current per unit electric field. The *mobility* μ of a particle is defined as its velocity per unit field and hence

$$\mu = \frac{e}{m\nu_c} \tag{10–113}$$

Next consider the *diffusion coefficient* D. This is defined as the particle flux caused by unit density gradient. If temperature is constant one may take

$$f_0 = n(\mathbf{r})\left(\frac{m}{2\pi\kappa T}\right)^{3/2} \exp\left(-mv^2/2\kappa T\right)$$

† One could treat ν_c as velocity dependent. It would then appear inside the integrands of the integrations over velocity space which are performed in calculating the transport coefficients (cf. Problem 10-23).

so that with no external forces and steady state conditions the kinetic equation gives

$$\mathbf{v} \cdot \frac{\partial f_0}{\partial \mathbf{r}} = \frac{f_0}{n} \mathbf{v} \cdot \frac{\partial n}{\partial \mathbf{r}} = -\nu_c f' \qquad (10\text{–}114)$$

where the gradient in f' is assumed to be small compared with that in f_0. Since the particle flux $\mathbf{\Gamma} = \int \mathbf{v} f \, d\mathbf{v}$ one finds on multiplying (10–114) by $\mathbf{v}$ and integrating

$$\mathbf{\Gamma} = -\frac{1}{n\nu_c} \int \mathbf{v} f_0 \, \mathbf{v} \cdot \nabla n \, d\mathbf{v} = -\frac{\kappa T}{m\nu_c} \nabla n$$

and hence
$$D = \frac{\kappa T}{m\nu_c} \qquad (10\text{–}115)$$

From (10–113) and (10–115)

$$\frac{D}{\mu} = \frac{\kappa T}{e}$$

which is known as the *Einstein relation*.

The *thermal conductivity* K is usually defined for constant pressure so that, since

$$p = n\kappa T \qquad (10\text{–}116)$$

one takes both n and T inhomogeneous, i.e.,

$$f_0 = n(\mathbf{r}) \left(\frac{m}{2\pi\kappa T(\mathbf{r})} \right)^{3/2} \exp\left(-mv^2/2\kappa T(\mathbf{r}) \right)$$

Then the steady state force-free kinetic equation gives, using (10–116),

$$f_0 \mathbf{v} \cdot \frac{\partial T}{\partial \mathbf{r}} \left(\frac{mv^2}{2\kappa T^2} - \frac{5}{2T} \right) = -\nu_c f' \qquad (10\text{–}117)$$

The heat flux $\mathbf{q}$ is given by

$$\mathbf{q} = \int \tfrac{1}{2} mv^2 \mathbf{v} f \, d\mathbf{v} = \tfrac{1}{2} m \int v^2 \mathbf{v} f' \, d\mathbf{v}$$

so that multiplication of (10–117) by $\tfrac{1}{2} mv^2 \mathbf{v}$ and integration over velocity space gives

$$\nu_c \mathbf{q} = \frac{5m}{4T} \int v^2 \mathbf{v} \, \mathbf{v} \cdot \nabla T f_0 \, d\mathbf{v} - \frac{m^2}{4\kappa T^2} \int v^4 \mathbf{v} \, \mathbf{v} \cdot \nabla T f_0 \, d\mathbf{v}$$

$$= \frac{5n\kappa^2 T}{2m} \nabla T$$

Hence the thermal conductivity

$$K = \frac{5n\kappa^2 T}{2mv_c}$$

The *coefficient of viscosity* η is defined as the shear stress produced by unit velocity gradient. Taking the local mean velocity $\mathbf{u}$ in the x direction and its gradient in the z direction one may write

$$f_0 = n\left(\frac{m}{2\pi\kappa T}\right)^{3/2} \exp\left\{-\frac{m}{2\kappa T}[(v_x - u(z))^2 + v_y^2 + v_z^2]\right\}$$

and substitution into the kinetic equation gives

$$\frac{mv_z}{\kappa T}\frac{du}{dz}(v_x - u)f_0 = -v_c f' \tag{10–118}$$

From the definition of η

$$\eta\frac{du}{dz} = -\int mv_x v_z f\, d\mathbf{v} = -m\int v_x v_z f'\, d\mathbf{v} \tag{10–119}$$

Thus, multiplying (10–118) by $mv_x v_z$ and integrating gives, using (10–119),[†]

$$\eta = n\kappa T/v_c$$

From (10–98) observe that the proton–proton collision time is of order $(m^+/m^-)^{1/2}$ times the electron–electron collision time (cf. Problem 10–21); thus the viscosity of a plasma is principally that of the protons. However, in all the other transport coefficients the mass appears in the denominator so that the electrons dominate these transport processes.

So far we have considered the transport coefficients under the simplest conditions. To complete this section some important practical considerations are discussed briefly. For example, in the presence of a density gradient the diffusion of electrons and ions will in general occur at different rates; if the temperatures of the two species are approximately equal the electrons diffuse more rapidly than the ions and if the containing walls are insulated a space charge is set up due to the excess electrons near the wall. This has the effect of simultaneously decreasing electron mobility and increasing ion mobility. Writing the ion and electron fluxes as $\mathbf{\Gamma}^+$ and $\mathbf{\Gamma}^-$ one has

$$\begin{aligned}
\mathbf{\Gamma}^+ &= -D^+ \nabla n^+ + n^+\mu^+\mathbf{E}\\
\mathbf{\Gamma}^- &= -D^- \nabla n^- + n^-\mu^-\mathbf{E}
\end{aligned} \tag{10–120}$$

---

† Note that the kinematic viscosity used in Chapters 4 and 6 is obtained from η by dividing by the mass density.

where $\mathbf{E}$ is the electric field due to the space charge. When a steady state is reached $\Gamma^+ = \Gamma^- = \Gamma$ and eliminating $\mathbf{E}$ from (10–120) (assuming $n^+ \simeq n^- \simeq n$)

$$\Gamma = -D_a \nabla n$$

where the coefficient of *ambipolar diffusion*

$$D_a = \frac{D^- \mu^+ - D^+ \mu^-}{\mu^+ - \mu^-}$$

Using the Einstein relation and assuming $T^+ = T^-$ this becomes

$$D_a = \frac{2D^+ D^-}{D^+ + D^-} \simeq 2D^+$$

since $D^- \gg D^+$. Thus the resultant ambipolar diffusion takes place at the slower of the two rates.

The field set up by the space charge is, from (10–120),

$$\mathbf{E} = \frac{D^- - D^+}{\mu^- - \mu^+} \frac{\nabla n}{n}$$

$$= -\frac{\kappa T}{ne} \nabla n \tag{10–121}$$

Since
$$\nabla \cdot \mathbf{E} = 4\pi e (n^+ - n^-) \tag{10–122}$$

the condition of quasi-neutrality $n^+ \simeq n^-$ implies from (10–121) and (10–122)

$$\frac{n^+ - n^-}{n} \sim \frac{\kappa T}{4\pi n e^2 L^2} = \left(\frac{\lambda_D}{L}\right)^2 \ll 1$$

where L is the length scale over which the field and the density gradient exist. Thus quasi-neutrality does not apply within a region close to the wall and several Debye lengths thick; this defines the *sheath thickness*.

Another important practical consideration is diffusion across a magnetic field since this has a bearing on magnetic containment of plasmas. In Chapter 2 we saw that in the plane perpendicular to the magnetic field charged particles gyrate around the field lines. However, collisions disrupt this motion and allow particles to change orbits and thus to diffuse across the magnetic field $\mathbf{B}_0$. Instead of (10–114) one now has from the kinetic equation

$$\mathbf{v} \cdot \frac{\partial f}{\partial \mathbf{r}} + \frac{e}{mc} (\mathbf{v} \times \mathbf{B}) \cdot \frac{\partial f}{\partial \mathbf{v}} = -\nu_c f'$$

i.e.,
$$\mathbf{v} \cdot \frac{\partial f_0}{\partial \mathbf{r}} + \frac{e}{mc}(\mathbf{v} \times \mathbf{B}_0) \cdot \frac{\partial f'}{\partial \mathbf{v}} = -\nu_c f' \qquad (10\text{--}123)$$

ignoring spatial gradients in f' as before. Suppose $\mathbf{B}_0$ is in the z direction and the density gradient is in the z and x directions; the v_x, v_y, and v_z moments of (10–123) then give, using

$$f_0 = n(x, z) \left(\frac{m}{2\pi\kappa T_\perp}\right) \left(\frac{m}{2\pi\kappa T_\|}\right)^{1/2} \exp\left\{-\frac{m(v_x^2 + v_y^2)}{2\kappa T_\perp} - \frac{mv_z^2}{2\kappa T_\|}\right\}$$

since a plasma may have, in general, different temperatures along and across the magnetic field,

$$\left. \begin{array}{r} \dfrac{\kappa T_\perp}{m}\dfrac{\partial n}{\partial x} - \Omega\Gamma_y = -\nu_c\Gamma_x \\[2mm] \Omega\Gamma_x = -\nu_c\Gamma_y \\[2mm] \dfrac{\kappa T_\|}{m}\dfrac{\partial n}{\partial z} = -\nu_c\Gamma_z \end{array} \right\} \qquad (10\text{--}124)$$

where $\Omega = (eB_0/mc)$ is the Larmor frequency.

Defining $D_\perp$, $D_\|$ by

$$\Gamma_x = -D_\perp \frac{\partial n}{\partial x} \qquad \Gamma_z = -D_\| \frac{\partial n}{\partial z}$$

one finds from (10–124)

$$D_\perp = \frac{\nu_c \kappa T_\perp / m}{\nu_c^2 + \Omega^2} \qquad D_\| = \frac{\kappa T_\|}{m\nu_c}$$

that is, diffusion parallel to the field is unaffected by the field while that perpendicular is reduced by the factor $\nu_c^2/(\nu_c^2 + \Omega^2)$. Note that a flux exists in the y direction as well (perpendicular to the density gradient and the magnetic field) and in the case $\Omega \gg \nu_c$ this is the main flux perpendicular to the field being (Ω/ν_c) times the flux in the x direction.

There is an important limitation to the approximate arguments used in this section. Treating the collision frequency as a constant ignores the fact that $\nu_c \to 0$ as v becomes large. Thus even for very weak electric fields some very fast electrons never reach a steady balance between acceleration by the field and deceleration by collisions. These electrons which are accelerated by the field at a rate greater than they are slowed down by collisions are known as *runaway electrons* (cf. Problem 10–24). A beam of runaway electrons will however be disrupted by collective effects—for instance, the two-stream instability discussed in Section 10–6.

10–12   Equilibrium pair correlation function

In this section the results (10–22) and (10–23) are derived; the first of these has already been obtained in Section 10–8 but for completeness it is here derived from the equilibrium N-particle distribution function.

It is a well-known result of statistical mechanics that in thermodynamic equilibrium a solution of the Liouville equation (10–8) is the *Gibbs distribution function*†

$$f_N = C \exp(-H/\theta) \tag{10–125}$$

where C and θ are constants and H is the Hamiltonian of the plasma

$$H = \sum_{i=1}^{N} (\tfrac{1}{2}mv_i^2 + \sum_{\substack{j=1 \\ j<i}}^{N-1} \phi_{ij})$$

It is easily verified that (10–125) satisfies (10–8) and with the normalization of f_N given by (10–9) it follows that the constant C is given by

$$C = N! \Big/ \int \exp(-H/\theta) \prod_{i=1}^{N} d\mathbf{r}_i \, d\mathbf{v}_i$$

Knowing f_N, one may in principle calculate all the reduced equilibrium distribution functions though in practice only the calculation of f_1 is simple. From (10–10b) and (10–125)

$$f_1(1) = \frac{N \int \exp(-H/\theta) \prod_{i=2}^{N} d\mathbf{r}_i \, d\mathbf{v}_i}{\int \exp(-H/\theta) \prod_{i=1}^{N} d\mathbf{r}_i \, d\mathbf{v}_i} \tag{10–126}$$

The velocity integrations are separable and thus trivial. The substitutions

$$\mathbf{r}_i' = \mathbf{r}_i - \mathbf{r}_1 \qquad (i = 2, 3, \ldots, N) \tag{10–127}$$

remove $\mathbf{r}_1$ from both integrands in (10–126) with the result

$$f_1(1) = \frac{N \exp(-mv_1^2/2\theta)}{\int d\mathbf{r}_1 \, d\mathbf{v}_1 \exp(-mv_1^2/2\theta)}$$

$$= n\left(\frac{m}{2\pi\theta}\right)^{3/2} \exp(-mv_1^2/2\theta) \tag{10–128}$$

† This is known as the canonical ensemble but it is not the only solution; any function of the Hamiltonian of the system is a solution. For a fuller discussion see H. Margenau and G. M. Murphy, *The Mathematics of Physics and Chemistry*, p. 428. Van Nostrand, Princeton, N.J., 1948.

The constant θ may be identified by the definition of temperature T (cf. Section 3–3)

$$\tfrac{3}{2}n\kappa T = \int \tfrac{1}{2}mv_1^2 f_1 \, dv_1$$

giving

$$\theta = \kappa T$$

and the equilibrium distribution function (10–128) is thus the Maxwell distribution

$$f_1(1) = f_M(1) = n\left(\frac{m}{2\pi\kappa T}\right)^{3/2} \exp\left(-mv_1^2/2\kappa T\right)$$

Direct integration of (10–125) to obtain higher-order reduced distribution functions proves to be a formidable task and one has to resort to approximation techniques.[4] However, the equilibrium pair correlation function may be obtained indirectly by solving the truncated equation for f_2. From (10–14) this equation is

$$L_2 f_2 = \frac{1}{m}\sum_{i=1}^{2}\int \frac{\partial\phi_{i3}}{\partial r_i}\cdot\frac{\partial f_3}{\partial v_i}dr_3 \, dv_3 \qquad (10\text{–}129)$$

The truncation of (10–129) is achieved by assuming that in (10–16c)

$$T \ll Pf \ll \mathit{fff} \qquad (10\text{–}130)$$

which is an obvious extension of the argument $(P \ll ff)$ put forward in Section 10–3. Thus substituting (10–16a, b, c) in (10–129) and using (10–7), (10–17), and (10–130) one finds in the absence of external forces (cf. Problem 10–25)

$$\frac{\partial P(1,2)}{\partial t} + \left(v_1\cdot\frac{\partial}{\partial r_1}+v_2\cdot\frac{\partial}{\partial r_2}\right)P(1,2) - \frac{1}{m}\frac{\partial\phi_{12}}{\partial r_1}\cdot\left(f(2)\frac{\partial f(1)}{\partial v_1}-f(1)\frac{\partial f(2)}{\partial v_2}\right)$$

$$= \frac{1}{m}\int dr_3 \, dv_3 \left\{\frac{\partial\phi_{13}}{\partial r_1}\cdot\frac{\partial}{\partial v_1}[f(1)\,P(2,3)+f(3)\,P(1,2)] + \right.$$

$$\left. + \frac{\partial\phi_{23}}{\partial r_2}\cdot\frac{\partial}{\partial v_2}[f(2)\,P(1,3)+f(3)\,P(1,2)]\right\} \qquad (10\text{–}131)$$

This equation is to be solved for P in the equilibrium case, i.e., for $f = f_M$. Note first that P must be of the form (cf. Problem 10–26)

$$P(1,2) = f_M(v_1)\,f_M(v_2)\,p(r_{12}) \qquad (10\text{–}132)$$

Then (10–31) reduces to

$$(\mathbf{v}_1 - \mathbf{v}_2) \cdot \left[\frac{\partial p(r_{12})}{\partial \mathbf{r}_1} + \frac{1}{\kappa T} \frac{\partial \phi_{12}}{\partial \mathbf{r}_1} \right] = -\frac{n}{\kappa T} \int d\mathbf{r}_3 \left[\mathbf{v}_1 \cdot \frac{\partial \phi_{13}}{\partial \mathbf{r}_1} p(r_{23}) + \mathbf{v}_2 \cdot \frac{\partial \phi_{23}}{\partial \mathbf{r}_2} p(r_{13}) \right]$$

This equation is valid for arbitrary $\mathbf{v}_1$ and $\mathbf{v}_2$, so choosing $\mathbf{v}_2 = 0$ one has

$$\frac{\partial p(r_{12})}{\partial \mathbf{r}_1} + \frac{1}{\kappa T} \frac{\partial \phi_{12}}{\partial \mathbf{r}_1} = -\frac{n}{\kappa T} \int d\mathbf{r}_3 \, p(r_{23}) \frac{\partial \phi_{13}}{\partial \mathbf{r}_1} \tag{10–133}$$

The divergence of (10–133) with respect to $\mathbf{r}_1$ then gives, using (10–13) and (10–38),

$$\nabla_1^2 p(r_{12}) - \frac{4\pi e^2}{\kappa T} \delta(\mathbf{r}_{12}) = \frac{4\pi n e^2}{\kappa T} \int d\mathbf{r}_3 \, p(r_{23}) \, \delta(\mathbf{r}_{13})$$

or

$$(\nabla_1^2 - k_D^2) \, p(r_{12}) = \frac{4\pi e^2}{\kappa T} \delta(\mathbf{r}_{12}) \tag{10–134}$$

The solution of (10–134) is well known (cf. Problem 10–27)

$$p(r_{12}) = \frac{e^2}{\kappa T} \frac{e^{-k_D r_{12}}}{r_{12}}$$

and hence from (10–132) the result (10–23) for the equilibrium, pair correlation function is obtained.

Had one ignored the term on the right-hand side of (10–133), (10–134) would have read $\nabla_1^2 p = 4\pi e^2 \, \delta(\mathbf{r}_{12})/\kappa T$ with the solution $p = -e^2/r_{12} \kappa T$. Since this is equivalent to the neglect of the terms on the right-hand side of (10–131) one sees that in equilibrium the effect of these integral terms is to replace the Coulomb potential in the solution for P by the shielded potential. With hindsight this is not surprising; having discarded $T(1, 2, 3)$ the integral terms in the equation for $P(1, 2)$ are the only ones which retain any effect of particles other than 1 and 2. In fact these integrals 'sum up' such effects and describe the shielding which the other plasma particles provide between particles 1 and 2. This equilibrium result is used in the next section to give an approximate but relatively simple method for obtaining P in the general case.

10–13 Derivation of the Landau equation

In this section we return to the question of solving the truncated hierarchy of equations for the reduced distribution functions in order to obtain a kinetic equation. To be more specific, we are interested in solving (10–131) for P, in the general case, so that the result may be substituted in (10–17) to give an equation involving f only. From a mathematical point of view this is clearly more desirable than the *ad hoc* arguments used in Sections 10–7, 10–9, and 10–11 to obtain various expressions for the collision integral. Unfortunately this has been achieved only for two special cases, a homogeneous plasma[104, 105] and a near-equilibrium plasma,[106] and in both cases the resulting collision integral is considerably more complicated than either the Boltzmann or Fokker–Planck collision integrals.

Here we present an approximate solution of the problem based on an extension of the observation made in the last section concerning the integral terms in (10–131), to the general (non-equilibrium) case. It is assumed that the main effect of the integral terms in (10–131) is to replace the Coulomb potential (10–13) by the shielded potential

$$\phi_{ij} = \frac{e^2}{|\mathbf{r}_i - \mathbf{r}_j|} \exp\left(-k_D |\mathbf{r}_i - \mathbf{r}_j|\right) \qquad (10\text{–}135)$$

As pointed out in Section 10–12 these terms sum up the effect of the other particles on the correlation of particles 1 and 2, so one would expect shielding to take place and in equilibrium this is the only effect of the integral terms. The approximation being made, therefore, is to assume that the total effect of the other plasma particles is the same as in equilibrium.

Thus we put the right-hand side of (10–131) equal to zero and replace the Coulomb potential (10–13) by the shielded potential (10–135), giving

$$\frac{\partial P}{\partial t} + \mathbf{v}_1 \cdot \frac{\partial P}{\partial \mathbf{r}_1} + \mathbf{v}_2 \cdot \frac{\partial P}{\partial \mathbf{r}_2} = \frac{1}{m} \frac{\partial \phi_{12}}{\partial \mathbf{r}_1} \cdot \left[\frac{\partial f(1)}{\partial \mathbf{v}_1} f(2) + \frac{\partial f(2)}{\partial \mathbf{v}_2} f(1) \right]$$

This equation may be solved by Green function techniques; defining $G(\mathbf{r}_1, \mathbf{r}_1', \mathbf{r}_2, \mathbf{r}_2', t, t')$ as the solution of

$$\frac{\partial G}{\partial t} + \mathbf{v}_1 \cdot \frac{\partial G}{\partial \mathbf{r}_1} + \mathbf{v}_2 \cdot \frac{\partial G}{\partial \mathbf{r}_2} = \delta(t - t')\, \delta(\mathbf{r}_1 - \mathbf{r}_1')\, \delta(\mathbf{r}_2 - \mathbf{r}_2')$$

it is easily verified that G is given by

$$G = \theta(t - t')\, \delta(\mathbf{r}_1 - \mathbf{r}_1' - \mathbf{v}_1(t - t'))\, \delta(\mathbf{r}_2 - \mathbf{r}_2' - \mathbf{v}_2(t - t'))$$

where $\theta(t)$ is the step function

$$\theta(t) = \begin{cases} 1 & t > 0 \\ 0 & t < 0 \end{cases}$$

Then

$$P = \frac{1}{m} \int dt'\, d\mathbf{r}_1'\, d\mathbf{r}_2'\, G(\mathbf{r}_1, \mathbf{r}_1', \mathbf{r}_2, \mathbf{r}_2', t, t') \frac{\partial \phi(r_{12}')}{\partial \mathbf{r}_1'}.$$

$$\cdot \left[\frac{\partial f(\mathbf{r}_1', \mathbf{v}_1, t')}{\partial \mathbf{v}_1} f(\mathbf{r}_2', \mathbf{v}_2, t') - \frac{\partial f(\mathbf{r}_2', \mathbf{v}_2, t')}{\partial \mathbf{v}_2} f(\mathbf{r}_1', \mathbf{v}_1, t') \right]$$

$$= \frac{1}{m} \int_0^t dt' \frac{\partial \phi(|\mathbf{r}_{12} - (\mathbf{v}_1 - \mathbf{v}_2)(t - t')|)}{\partial \mathbf{r}_{12}} \cdot$$

$$\cdot \left[\frac{\partial f(\mathbf{r}_1 - \mathbf{v}_1(t - t'), \mathbf{v}_1, t')}{\partial \mathbf{v}_1} f(\mathbf{r}_2 - \mathbf{v}_2(t - t'), \mathbf{v}_2, t') \right.$$

$$\left. - \frac{\partial f(\mathbf{r}_2 - \mathbf{v}_2(t - t'), \mathbf{v}_2, t')}{\partial \mathbf{v}_2} f(\mathbf{r}_1 - \mathbf{v}_1(t - t'), \mathbf{v}_1, t') \right] \quad (10\text{--}136)$$

In carrying out the time integration in (10–136) it is assumed that f varies on a time scale much greater than ω_p^{-1} (cf. Section 10–10). Also, since $\phi(r) \simeq 0$ for $r \gg \lambda_D$, the integrand in (10–136) is very small except for (cf. Fig. 10–9)

$$t - t' \lesssim \lambda_D / |\mathbf{v}_1 - \mathbf{v}_2|$$

Since $\lambda_D / |\mathbf{v}_1 - \mathbf{v}_2| \sim \omega_p^{-1}$ [unless $\mathbf{v}_1 \simeq \mathbf{v}_2$; but $\mathbf{r}_1 \simeq \mathbf{r}_2$ (see below) so that

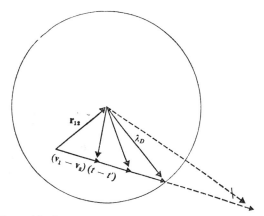

Fig. 10–9 Variation with time of the vector $\mathbf{r}_{12} - (\mathbf{v}_1 - \mathbf{v}_2)(t - t')$. Since this is the argument of the shielded potential its effective range lies only within the Debye sphere; $\mathbf{r}_{12}$ must lie within the sphere since it plays a similar role in Eqn (10–138)

in this case $f(2)\, \partial f(1)/\partial \mathbf{v}_1 \simeq f(1)\, \partial f(2)/\partial \mathbf{v}_2$] the integrand of (10–136) is non-vanishing only for $(t - t') \lesssim \omega_p^{-1}$. Hence (10–136) may be replaced by

$$P(1, 2) = \frac{1}{m} \left[f(2) \frac{\partial f(1)}{\partial \mathbf{v}_1} - f(1) \frac{\partial f(2)}{\partial \mathbf{v}_2} \right] \cdot \int_0^\infty \frac{\partial \phi(|\mathbf{r}_{12} - (\mathbf{v}_1 - \mathbf{v}_2)\tau|)}{\partial \mathbf{r}_{12}} d\tau$$

$$(10\text{--}137)$$

Substitution of (10–137) in (10–21) gives for the collision integral

$$\left(\frac{\partial f}{\partial t}\right)_c = \frac{1}{m^2} \int d\mathbf{r}_2 \, d\mathbf{v}_2 \frac{\partial \phi(r_{12})}{\partial \mathbf{r}_{12}} \cdot \frac{\partial}{\partial \mathbf{v}_1} \int_0^\infty d\tau \frac{\partial \phi(\,|\,\mathbf{r}_{12} - (\mathbf{v}_1 - \mathbf{v}_2)\tau\,|\,)}{\partial \mathbf{r}_{12}} \cdot$$

$$\cdot \left[f(2)\frac{\partial f(1)}{\partial \mathbf{v}_1} - f(1)\frac{\partial f(2)}{\partial \mathbf{v}_2} \right] \qquad (10\text{–}138)$$

Writing $\phi(r)$ in terms of its Fourier transform

$$\phi(r) = \frac{e^2}{2\pi^2} \int \frac{d\mathbf{k} \exp(i\mathbf{k}.\mathbf{r})}{k^2 + k_D^2}$$

(10–138) becomes

$$\left(\frac{\partial f}{\partial t}\right)_c = \frac{-e^4}{4\pi^4 m^2} \int d\mathbf{r}_2 \, d\mathbf{v}_2 \, d\mathbf{k} \, d\mathbf{l} \frac{\exp(i\mathbf{k}.\mathbf{r}_{12})}{(k^2 + k_D^2)} \mathbf{k} \cdot \frac{\partial}{\partial \mathbf{v}_1} \int_0^\infty d\tau$$

$$\frac{\exp\{i\mathbf{l}.[\mathbf{r}_{12} - (\mathbf{v}_1 - \mathbf{v}_2)\tau]\}}{(l^2 + k_D^2)} \mathbf{l} \cdot \left[\frac{\partial f(1)}{\partial \mathbf{v}_1} f(2) - \frac{\partial f(2)}{\partial \mathbf{v}_2} f(1) \right] \qquad (10\text{–}139)$$

It is convenient to integrate over $\mathbf{r}_2$ space first. In doing this we use a similar approximation to that for the time integration in (10–136) and assume f varies on a length scale much greater than λ_D; since the integrand in (10–139) contains $\phi(r_{12})$ it vanishes exponentially for $r_{12} \gg \lambda_D$ so that one may replace $\mathbf{r}_2$ by $\mathbf{r}_1$ in $f(2)$. The integration over $\mathbf{r}_2$ then involves $\exp[-i(\mathbf{k} + \mathbf{l}).\mathbf{r}_2]$ only and gives $(2\pi)^3 \, \delta(\mathbf{k} + \mathbf{l})$. The integration over $\mathbf{l}$ space is then trivial and (10–139) becomes

$$\left(\frac{\partial f}{\partial t}\right)_c = \frac{2e^4}{\pi m^2} \int \frac{d\mathbf{v}_2 \, d\mathbf{k}}{(k^2 + k_D^2)^2} \mathbf{k} \cdot \frac{\partial}{\partial \mathbf{v}_1} \int_0^\infty d\tau \exp[i\mathbf{k}.(\mathbf{v}_1 - \mathbf{v}_2)\tau]$$

$$\mathbf{k} \cdot \left(\frac{\partial f(\mathbf{v}_1)}{\partial \mathbf{v}_1} f(\mathbf{v}_2) - \frac{\partial f(\mathbf{v}_2)}{\partial \mathbf{v}_2} f(\mathbf{v}_1) \right) \qquad (10\text{–}140)$$

The integration over τ is given by the *Plemelj formula*[107]

$$\int_0^\infty d\tau \exp[i\mathbf{k}.(\mathbf{v}_1 - \mathbf{v}_2)\tau] = \pi\delta(\mathbf{k}.(\mathbf{v}_1 - \mathbf{v}_2)) - iP\frac{1}{\mathbf{k}.(\mathbf{v}_1 - \mathbf{v}_2)} \qquad (10\text{–}141)$$

where P stands for the principal part. However, the second term in (10–141) gives no contribution to (10–140), since it makes the integrand odd in $\mathbf{k}$ (its contribution must be zero since it is imaginary). This leaves an integration over $\mathbf{k}$ space of the form

$$\int \frac{\mathbf{k}.\mathbf{A} \, \delta(\mathbf{k}.\mathbf{g}) \, \mathbf{k}.\mathbf{B} \, d\mathbf{k}}{(k^2 + k_D^2)^2} \qquad (10\text{–}142)$$

which diverges at large **k**. This is a direct result of using (10–130), since this approximation breaks down for $r_{12} \lesssim e^2/\kappa T$ (cf. (10–26)). One cuts off the integration therefore at $k = \kappa T/e^2$ ($\sim b_0^{-1}$ for thermal particles) and finds for (10–142) (cf. Problem 10–28)

$$\pi \ln \Lambda \, A_i \left(\frac{\delta_{ij}}{g} - \frac{g_i g_j}{g^3} \right) B_j$$

where only terms in $\ln \Lambda$ have been retained (cf. Section 10–9).

Thus the kinetic equation may finally be written, using (10–17), (10–18), and (10–84b),

$$\frac{\partial f}{\partial t} + \mathbf{v} \cdot \frac{\partial f}{\partial \mathbf{r}} - \frac{e}{m} \mathbf{E} \cdot \frac{\partial f}{\partial \mathbf{v}} = \frac{2\pi e^4 \ln \Lambda}{m^2} \frac{\partial}{\partial v_i} \int d\mathbf{v}' \, \frac{\partial^2 \, | \, \mathbf{v} - \mathbf{v}' \, |}{\partial v_i \, \partial v_j}$$

$$\left(\frac{\partial f(\mathbf{v})}{\partial v_j} f(\mathbf{v}') - \frac{\partial f(\mathbf{v}')}{\partial v_j'} f(\mathbf{v}) \right) \quad (10\text{–}143)$$

where **E** is the self-consistent field (10–19). The form of the collision integral in (10–143) was first derived by Landau[108] in 1936. It is, however, precisely the same as the particular form (10–85b) of the Fokker–Planck equation derived in Section 10–9 (cf. Problem 10–29). Note that in this derivation a cut-off in the weak scattering limit (small k) was unnecessary because of the presence of the shielded potential.

Summary†

1. *Kinetic equation*
Liouville equation $\equiv$ BBGKY chain of equations
Cluster expansion with assumption $T \ll Pf \ll fff$ permits truncation.

(i) $P/ff = 0$: Vlasov equation

$$\frac{\partial f_i}{\partial t} + \mathbf{v}_1 \cdot \frac{\partial f_i}{\partial \mathbf{r}_1} + \frac{\mathbf{F}_i}{m_i} \cdot \frac{\partial f_i}{\partial \mathbf{v}_1} = 0$$

Self-consistent field

$$\mathbf{E}(\mathbf{r}_1, t) = \sum_j e_j \int \frac{(\mathbf{r}_1 - \mathbf{r}_2)}{|\mathbf{r}_1 - \mathbf{r}_2|^3} f_j(\mathbf{r}_2, \mathbf{v}_2, t) \, d\mathbf{r}_2 \, d\mathbf{v}_2$$

(ii) $T/Pf = 0$: Collisional kinetic equation

$$\frac{\partial f_i}{\partial t} + \mathbf{v}_1 \cdot \frac{\partial f_i}{\partial \mathbf{r}_1} + \frac{\mathbf{F}_i}{m_i} \cdot \frac{\partial f_i}{\partial \mathbf{v}_1} = \left(\frac{\partial f_i}{\partial t} \right)_c$$

† In the Summary, subscripts i and j are used only to denote particle species.

(a) Boltzmann collision integral

$$\left(\frac{\partial f_i}{\partial t}\right)_c = \sum_j \int [f_i(\mathbf{v}_1')f_j(\mathbf{v}_2') - f_i(\mathbf{v}_1)f_j(\mathbf{v}_2)] \,|\,\mathbf{v}_1 - \mathbf{v}_2\,|\,\sigma_{ij}(\,|\,\mathbf{v}_1 - \mathbf{v}_2\,|,\theta)\,d\mathbf{v}_2\,d\Omega$$

$$\sigma_{ij}(v,\theta) = \frac{e_i^2 e_j^2 (m_i + m_j)^2}{4m_i^2 m_j^2 v^4 \sin^4(\theta/2)}$$

(b) Fokker–Planck collision term

$$\left(\frac{\partial f_i}{\partial t}\right)_c = -\frac{\partial}{\partial \mathbf{v}_1} \cdot (f_i \langle \Delta \mathbf{v}_1 \rangle_i) + \frac{1}{2} \frac{\partial^2}{\partial \mathbf{v}_1\,\partial \mathbf{v}_1} : (f_i \langle \Delta \mathbf{v}_1 \Delta \mathbf{v}_1 \rangle_i)$$

coefficient of   coefficient of diffusion
dynamical friction

$$\left\{ \begin{array}{c} \langle \Delta \mathbf{v}_1 \rangle_i \\ \langle \Delta \mathbf{v}_1 \Delta \mathbf{v}_1 \rangle_i \end{array} \right\} = \frac{1}{\Delta t} \int \psi_i(\mathbf{v}_1, \Delta \mathbf{v}_1) \left\{ \begin{array}{c} \Delta \mathbf{v}_1 \\ \Delta \mathbf{v}_1 \Delta \mathbf{v}_1 \end{array} \right\} d(\Delta \mathbf{v}_1)$$

Rosenbluth, MacDonald, and Judd (and Kaufman) form of Fokker–Planck collision term

$$\left(\frac{\partial f_i}{\partial t}\right)_c = \Gamma_i \left\{ -\frac{\partial}{\partial \mathbf{v}_1} \cdot \left[f_i \frac{\partial H_i}{\partial \mathbf{v}_1} \right] + \frac{1}{2} \frac{\partial^2}{\partial \mathbf{v}_1\,\partial \mathbf{v}_1} : \left[f_i \frac{\partial^2 G_i}{\partial \mathbf{v}_1\,\partial \mathbf{v}_1} \right] \right\}$$

$$\Gamma_i = \frac{4\pi e_i^4}{m_i^2} \ln \Lambda$$

$$H_i = \sum_j \left(\frac{m_i + m_j}{m_j}\right) \left(\frac{e_j}{e_i}\right)^2 \int \frac{f_j(\mathbf{v}_2)}{|\mathbf{v}_1 - \mathbf{v}_2|} d\mathbf{v}_2$$

$$G_i = \sum_j \left(\frac{e_j}{e_i}\right)^2 \int f_j(\mathbf{v}_2) |\mathbf{v}_1 - \mathbf{v}_2| \, d\mathbf{v}_2$$

This is identical with the Landau collision integral (and with the expanded Boltzmann collision integral)

$$\left(\frac{\partial f_i}{\partial t}\right)_c = \frac{\Gamma_i}{2} \sum_j \left(\frac{e_j}{e_i}\right)^2 m_i \frac{\partial}{\partial \mathbf{v}_1} \cdot \int \frac{\partial^2\,|\,\mathbf{v}_1 - \mathbf{v}_2\,|}{\partial \mathbf{v}_1\,\partial \mathbf{v}_1} \cdot \left[\frac{f_j(\mathbf{v}_2)}{m_i} \frac{\partial f_i(\mathbf{v}_1)}{\partial \mathbf{v}_1} - \right.$$

$$\left. - \frac{f_i(\mathbf{v}_1)}{m_j} \frac{\partial f_j(\mathbf{v}_2)}{\partial \mathbf{v}_2} \right] d\mathbf{v}_2$$

2. *Relaxation times*

(i) Slowing-down time

$$\tau_s = \frac{U}{[1 + (m_1/m_2)]A_D a^2 \, \Psi(aU)}$$

(ii) Deflection time (collision time) $\sim$ time for establishing isotropy

$$\tau_D = \frac{U^3}{A_D[\Phi(aU) - \Psi(aU)]}$$

$$\tau_D^{--} : \tau_D^{++} : \tau_D^{-+} : \tau_D^{+-} \sim 1 : (m^+/m^-)^{1/2} : 1 : (m^+/m^-)$$

(iii) Energy exchange time $\sim$ time for establishing Maxwellian

$$\tau_E = \frac{U^3}{4A_D \, \Psi(aU)}$$

$$A_D = 2n\Gamma_1$$

$$\Psi(x) = \frac{\Phi(x) - x\Phi'(x)}{2x^2}$$

$$\Phi(x) = \frac{2}{\sqrt{\pi}} \int_0^x \exp(-y^2) \, dy$$

$$\tau_E^{--} : \tau_E^{++} : \tau_E^{-+} \ (\text{or } \tau_E^{+-}) \sim 1 : (m^+/m^-)^{1/2} : (m^+/m^-)$$

3. *Transport coefficients*

Electrical conductivity
$$\sigma = \frac{ne^2}{mv_c}$$

Mobility
$$\mu = \frac{e}{mv_c}$$

Diffusion
$$D = \frac{\kappa T}{mv_c}$$

Thermal conductivity
$$K = \frac{5n\kappa^2 T}{2mv_c}$$

Viscosity
$$\eta = \frac{n\kappa T}{v_c}$$

Ambipolar diffusion coefficient
$$D_a = \frac{2D^+ D^-}{D^+ + D^-} \simeq 2D^+$$

Diffusion in magnetic field

$$D_\perp = \frac{v_c(\kappa T_\perp/m)}{v_c^2 + \Omega^2} \qquad D_\parallel = \frac{\kappa T_\parallel}{mv_c}$$

4. *Thermodynamic equilibrium results*

$$f = f_M = n\left(\frac{m}{2\pi\kappa T}\right)^{3/2} \exp\left(-mv^2/2\kappa T\right)$$

$$P(1, 2) = -\frac{e^2}{\kappa T}f_M(1)\, f_M(2)\, \frac{\exp\left(-k_D r_{12}\right)}{r_{12}}$$

Problems

10–1　The Liouville equation for a classical system is usually written

$$\frac{\partial D}{\partial t} + [D, H] = 0 \tag{10–5a}$$

where H is the Hamiltonian of the system and the Poisson bracket

$$[D, H] \equiv \sum_{i=1}^{N} \left(\frac{\partial D}{\partial q_i}\cdot\frac{\partial H}{\partial p_i} - \frac{\partial D}{\partial p_i}\cdot\frac{\partial H}{\partial q_i}\right)$$

the q_i and p_i being the generalized coordinates and momenta. Show that (10–5a) reduces to (10–5) for

$$H = \sum_i \frac{p_i^2}{2m_i} + V(q_1, q_2, \ldots, q_n, t)$$

where $p_i = mv_i$, $q_i = r_i$, and $F_i = -\partial V/\partial r_i$.

10–2　Show by direct differentiation that the generic distribution function f_N obeys the Liouville equation (10–8).

Hint: note that the interchange of like particles j and k ($j \neq i \neq k$) leaves F_i unaltered while the interchange of like particles i and j changes F_i to F_j and vice versa.

10–3　Use the hint in the previous question to show that the Liouville operator (10–7) is invariant under permutations of like particles.

10–4　Derive (10–12b). Note that all the terms in the summation over j from $s + 1$ to N are identical.

10–5　In the case of no external forces compare the order of magnitude of the terms in the kinetic equation (10–17). Taking ω_p^{-1} and λ_D as characteristic time and length scales show that the collision term is of order P/ff compared with all the other terms.

10–6　If $E(p) \to |p|^{-1}$ as $|p| \to \infty$, show that the contributions to the integral in (10–42) along AB and CD are negligible.

10–7 From (10–41) show that $E(p)$ has no poles to the right of the imaginary p-axis for a Maxwellian $f_0(\mathbf{v})$.

10–8 Using (10–43), derive (10–49) from (10–48). Note that in all terms on the right-hand side of (10–48) except the first (i.e., the largest term) one may use the approximation (10–47).

10–9 By solving the dispersion relationship (cf. Section 10–6)

$$1 + \frac{\omega_p^2}{2}\left[\frac{1}{(p + ikV)^2} + \frac{1}{(p - ikV)^2}\right] = 0$$

as a quadratic in p^2 show that for $kV \ll \omega_p$ a streaming instability exists with a growth rate given by kV.

10–10 The initial and final velocities of colliding particles m_1 and m_2 may be written

$$\mathbf{v}_1 = \mathbf{V} + \frac{m_2}{m_1 + m_2}\mathbf{g} \qquad \mathbf{v}_2 = \mathbf{V} - \frac{m_1}{m_1 + m_2}\mathbf{g}$$

$$\mathbf{v}_1' = \mathbf{V}' + \frac{m_2}{m_1 + m_2}\mathbf{g}' \qquad \mathbf{v}_2' = \mathbf{V}' - \frac{m_1}{m_1 + m_2}\mathbf{g}'$$

where $\mathbf{V}$, $\mathbf{V}'$, $\mathbf{g}$, $\mathbf{g}'$ are the initial and final centre of mass velocities and relative velocities respectively. Use the conservation of momentum and energy to show that

$$\mathbf{V} = \mathbf{V}' \qquad |\mathbf{g}| = |\mathbf{g}'|$$

10–11 Prove that the transformation of velocities from $(\mathbf{v}_1, \mathbf{v}_2)$ to $(\mathbf{v}_1', \mathbf{v}_2')$ caused by the collision of particles m_1 and m_2 is its own inverse. Deduce that the Jacobian of the transformation equals one.
  Hint: let

$$m_1(\mathbf{v}_1' - \mathbf{v}_1) = A\hat{\mathbf{e}}$$

where $\hat{\mathbf{e}}$ is a unit vector. Then conservation of momentum and energy leads to

$$A = -\frac{2m_1m_2}{m_1 + m_2}(\mathbf{v}_1 - \mathbf{v}_2)\cdot\hat{\mathbf{e}} = -\frac{2m_1m_2}{m_1 + m_2}\mathbf{g}\cdot\hat{\mathbf{e}}$$

Thus

$$\left.\begin{aligned}
\mathbf{v}_1' &= \mathbf{v}_1 - \frac{2m_2}{m_1 + m_2}(\hat{\mathbf{e}}\cdot\mathbf{g})\hat{\mathbf{e}} \\
\mathbf{v}_2' &= \mathbf{v}_2 + \frac{2m_1}{m_1 + m_2}(\hat{\mathbf{e}}\cdot\mathbf{g})\hat{\mathbf{e}}
\end{aligned}\right\} \qquad (10\text{–}144)$$

To find the inverse transformation observe that from (10–144)

$$\mathbf{g}' \equiv \mathbf{v}_1' - \mathbf{v}_2' = \mathbf{v}_1 - \mathbf{v}_2 - 2(\hat{\mathbf{e}}\cdot\mathbf{g})\hat{\mathbf{e}}$$

so that

$$\hat{\mathbf{e}}\cdot\mathbf{g}' = -\hat{\mathbf{e}}\cdot\mathbf{g}$$

Thus from (10–144)

$$
\begin{aligned}
\mathbf{v}_1 &= \mathbf{v}_1' - \frac{2m_2}{m_1 + m_2}(\hat{\mathbf{e}} \cdot \mathbf{g}')\hat{\mathbf{e}} \\[2mm]
\mathbf{v}_2 &= \mathbf{v}_2' + \frac{2m_1}{m_1 + m_2}(\hat{\mathbf{e}} \cdot \mathbf{g}')\hat{\mathbf{e}}
\end{aligned}
\right\} \tag{10–145}
$$

i.e., the transformation from $(\mathbf{v}_1, \mathbf{v}_2)$ to $(\mathbf{v}_1', \mathbf{v}_2')$ is given by the same function as the inverse transformation from $(\mathbf{v}_1', \mathbf{v}_2')$ to $(\mathbf{v}_1, \mathbf{v}_2)$.

The Jacobian of the transformation obtained by two successive transformations having Jacobians J_1 and J_2 is given by $J_1 J_2$. Applying this to the transformations (10–144), (10–145) it follows that $J^2 = 1$ and hence $J = 1$ (the root $J = -1$ corresponds to an improper transformation).

10–12 Prove the H-theorem in the more general case of inhomogeneous f and $\mathbf{F} \neq 0$, defining H by[10]

$$
H = \int f \ln f \, d\mathbf{v} \, d\mathbf{r}
$$

Make the usual assumptions about the vanishing of f on the boundaries of $(\mathbf{r}, \mathbf{v})$ space. Deduce that in the general case (10–69) is replaced by (10–107).

10–13 Use conservation of momentum and energy in a collision to verify that (10–69) satisfies $f(1')f(2') = f(1)f(2)$.

10–14 If in (5–44) S represents the entropy per unit mass show that (10–70) defines the entropy per unit volume when f is given by (10–69).

10–15 Show that if ϕ is chosen such as to minimize $\langle \phi K \phi \rangle$ then the error in $\langle \phi K \phi \rangle$ is of the order of the square of the error in ϕ.

10–16 Repeat the calculation to obtain (10–77) without assuming $\sin \theta \approx \theta$ and show that the result is approximately the same.

10–17 Use (10–60) and (10–82) in (10–81) to obtain (10–83). Note that in the calculation of $\langle \Delta v_1 \, \Delta v_1 \rangle$ terms involving cross-products like $\mathbf{e}_1 \mathbf{e}_2$, $\mathbf{e}_2 \mathbf{e}_3$, and so on, vanish on integration over ε while the term in $\mathbf{e}_1 \mathbf{e}_1$ proves to be small compared with the $\mathbf{e}_2 \mathbf{e}_2$, $\mathbf{e}_3 \mathbf{e}_3$ terms on integration over θ.

10–18 Derive (10–85) from (10–61) by assuming that collisions produce only small deflections $\Delta \mathbf{v}_1$, $\Delta \mathbf{v}_2$. Use (10–82d) and the corresponding equation for $\mathbf{v}_2'$ (cf. Problem 10–10) and expand $f(\mathbf{v}_1')f(\mathbf{v}_2')$ keeping only terms in $\Delta \mathbf{g}$ and $\Delta \mathbf{g} \Delta \mathbf{g}$.

10–19 Derive (10–89a, b).

10–20 Deduce from (10–91) that the mean deviation of a test particle in a uniform plasma is zero.

10–21 From (10–98) show that the electron–electron collision time

$$
\tau_D^{--} \sim \frac{n \lambda_D^3}{\omega_p \ln \Lambda}
$$

and verify the relative magnitudes of the other collision times given in section 2(ii) of the Summary.

10–22 From (10–144) show that the fraction of kinetic energy transferred in a proton–electron collision is of the order of (m^-/m^+) times the total energy.

10–23 Repeat the calculation of electrical conductivity σ (cf. Section 10–11) taking the collision frequency as velocity dependent, $\nu_c = \tau_D^{-1}(v)$. Show that the result is

$$\sigma = \frac{e^2}{3\kappa T} \int v^2 f_M \, \tau_D(v) \, d\mathbf{v}$$

10–24 By taking the $\mathbf{v}_1$ moment of the Fokker–Planck equation,

$$\frac{\partial f(1)}{\partial t} - \frac{e\mathbf{E}}{m^-} \cdot \frac{\partial f(1)}{\partial \mathbf{v}_1} = \left(\frac{\partial f}{\partial t}\right)_c^{--} + \left(\frac{\partial f}{\partial t}\right)_c^{-+}$$

with $f(1)$, $f(2)$ given by (10–87) and (10–88) (compare the derivation of (10–90)) find the relationship between the velocity $\mathbf{U}$ and field strength $\mathbf{E}$ for electron runaway (i.e., such that $\partial U/\partial t > 0$).

10–25 Derive (10–131); note that two terms in ϕP on the left-hand side of the equation have been dropped as small compared with the remaining terms.

10–26 Verify (10–132).
  Hint: carry out a formal integration as for the derivation of (10–128), using the substitutions (10–127).

10–27 Solve (10–134) by Fourier transform.

10–28 Retaining only terms in $\ln \Lambda$ show that (10–142) with a cut-off at large $k = \kappa T/e^2$ is given by

$$\pi \ln \Lambda \, A_i \left(\frac{\delta_{ij}}{g} - \frac{g_i g_j}{g^3}\right) B_j$$

10–29 Show that (10–143) may be put in the form (10–85b).
  Hint: integrate by parts with respect to v_j' and use (10–84).

Appendix 1

Tensors and dyadics

Scalars and vectors are the simplest examples of more general quantities known as *tensors*. In fact, a scalar ψ is a zero-order tensor having just one component or element while a vector $\mathbf{A}$ is a first-order tensor having, in a three-dimensional coordinate system, $3^1 = 3$ components (A_1, A_2, A_3). Similarly, a second-order tensor like the pressure tensor has $3^2 = 9$ components

$$(P_{ij}) = \begin{pmatrix} P_{11} & P_{12} & P_{13} \\ P_{21} & P_{22} & P_{23} \\ P_{31} & P_{32} & P_{33} \end{pmatrix}$$

The pressure tensor is symmetric, $P_{ij} = P_{ji}$ $(i, j = 1, 2, 3)$, but this is *not* true of second-order tensors in general. Another second-order tensor is the Kronecker delta

$$(\delta_{ij}) = \begin{pmatrix} 1 & 0 & 0 \\ 0 & 1 & 0 \\ 0 & 0 & 1 \end{pmatrix}$$

which is also known as the *unit tensor* by analogy with matrix theory. The heat tensor Eqn (3–22) is an example of a third-order tensor and has $3^3 = 27$ components.

It is usually most convenient to represent a tensor by its typical component, e.g., the Maxwell stress tensor by T_{ik} and this is what we refer to as *tensor notation*. In tensor notation we have the following representations:

Vector notation	Cartesian tensor notation
$\mathbf{A}$	A_i
$\mathbf{A} \cdot \mathbf{B}$	$A_i B_i$
$\mathbf{A} \times \mathbf{B}$	$\varepsilon_{ijk} A_j B_k$
$\nabla \psi$	$\dfrac{\partial \psi}{\partial r_i}$
$\nabla \cdot \mathbf{A}$	$\dfrac{\partial A_i}{\partial r_i}$

$$\nabla \times \mathbf{A} \qquad\qquad \varepsilon_{ijk}\frac{\partial A_k}{\partial r_j}$$

$$\nabla^2\psi \qquad\qquad \frac{\partial^2\psi}{\partial r_i\,\partial r_i}$$

where

$$\varepsilon_{ijk} = \begin{cases} 1 & \text{if } i, j, k \text{ are a cyclic permutation of 1, 2, 3} \\ -1 & \text{if } i, j, k \text{ are a non-cyclic permutation of 1, 2, 3} \\ 0 & \text{if any two subscripts are the same} \end{cases}$$

is sometimes called the *permutation tensor*. The following properties of ε_{ijk} are useful for manipulating tensor formulae:

$$\varepsilon_{ijk}\varepsilon_{ilm} = -\varepsilon_{ijk}\varepsilon_{lim} = \varepsilon_{ijk}\varepsilon_{lmi} = \delta_{il}\delta_{km} - \delta_{jm}\delta_{kl}$$

These equations are easily verified using the definition of ε_{ijk}.

Another representation sometimes used for second-order tensors is the *dyadic* form, for example

$$\mathbf{P} = \mathbf{e}_i P_{ij} \mathbf{e}_j$$

where $\mathbf{e}_i$ and $\mathbf{e}_j$ are the unit vectors along the Ox_i and Ox_j axes. Similarly the unit tensor in dyadic notation is

$$\mathbf{I} = \mathbf{e}_i \delta_{ij} \mathbf{e}_j$$

Vector operations on dyadics then proceed in an obvious way, e.g.,

$$\mathbf{A}.\mathbf{B} = A_k\mathbf{e}_k.\mathbf{e}_iB_{ij}\mathbf{e}_j = A_k\delta_{ki}B_{ij}\mathbf{e}_j = A_iB_{ij}\mathbf{e}_j$$
$$\mathbf{B}.\mathbf{A} = \mathbf{e}_iB_{ij}\mathbf{e}_j.\mathbf{e}_kA_k = \mathbf{e}_iB_{ij}A_j = B_{ji}A_i\mathbf{e}_j$$
$$\mathbf{A}:\mathbf{B} = \mathbf{e}_iA_{ij}\mathbf{e}_j:\mathbf{e}_kB_{kl}\mathbf{e}_l = A_{ij}B_{kl}(\mathbf{e}_j.\mathbf{e}_k)(\mathbf{e}_i.\mathbf{e}_l) = A_{ij}B_{ji}$$

Thus the following are alternative representations:

Dyadic notation	Cartesian tensor notation
A	A_{ij}
AB	A_iB_j
A.B	A_iB_{ij}
B.A	$B_{ij}A_j$
A:B	$A_{ij}B_{ji}$
$\nabla\mathbf{A}$	$\dfrac{\partial A_j}{\partial r_i}$

$$\nabla\nabla\psi \qquad \frac{\partial^2\psi}{\partial r_i\,\partial r_j}$$

$$\nabla.\mathbf{A} \qquad \frac{\partial A_{ji}}{\partial r_j}$$

$$\nabla.\mathbf{AB} \qquad \frac{\partial(A_j B_i)}{\partial r_j}$$

$$\nabla\nabla:\mathbf{A} \qquad \frac{\partial^2 A_{ij}}{\partial r_i\,\partial r_j}$$

Appendix 2

Bessel functions

This appendix is merely a collection of some properties of Bessel functions which have been used throughout the text; for further information consult the book by Watson.[80] Bessel's equation is

$$\frac{d^2f}{dr^2} + \frac{1}{r}\frac{df}{dr} + \left(1 - \frac{p^2}{r^2}\right)f = 0 \tag{1}$$

and the solution of this equation for $p \geqslant 0$ which is bounded at $r = 0$ is the Bessel function of the first kind,

$$J_p(x) = \left(\frac{x}{2}\right)^p \sum_{k=0}^{\infty} (-1)^k \frac{(x/2)^{2k}}{k! \, \Gamma(p + k + 1)} \tag{2}$$

Since p^2 appears in (1) J_{-p} is also a solution of (1); naturally any linear combination of J_p and J_{-p} will also be a solution (p non-integral).

The Bessel function of the second kind—often known as the Neumann function—is defined by

$$N_p(x) = \frac{J_p(x) \cos p\pi - J_{-p}(x)}{\sin p\pi} \tag{3}$$

and the Bessel functions of the third kind—Hankel functions—are defined by

$$\left. \begin{array}{l} H_p^{(1)}(x) = J_p(x) + iN_p(x) \\ H_p^{(2)}(x) = J_p(x) - iN_p(x) \end{array} \right\} \tag{4}$$

Again any linear combination of two linearly independent Bessel functions will be a solution of Bessel's equation.

When the arguments of the functions are purely imaginary the name *modified* Bessel functions is used. For integral n they are defined by

$$I_n(x) = i^{-n} J_n(ix) \qquad K_n(x) = \frac{\pi}{2} i^{n+1} H_n^{(1)}(ix) \tag{5}$$

The following approximate forms valid for $|x| \ll 1$, are often useful:

$$J_p(x) \simeq \frac{(x/2)^p}{\Gamma(p+1)} \qquad I_n(x) \simeq \frac{(x/2)^n}{n!}$$

$$J_0(x) \simeq 1 - \frac{x^2}{4} \qquad I_0(x) \simeq 1 + \frac{x^2}{4} \right\} \qquad (6)$$

$$K_n(x) \simeq (2/x)^n \frac{(n-1)!}{2} \qquad K_0(x) \simeq \ln \frac{2}{\gamma x}$$

with $n \geqslant 1$ and $\ln \gamma = 0.577$.

The following recurrence relations are also widely used:

$$J_{p-1}(x) + J_{p+1}(x) = \frac{2p}{x} J_p(x) \right\}$$

$$J_{p-1}(x) - J_{p+1}(x) = 2 \frac{dJ_p(x)}{dx} \qquad (7)$$

[with analogous relations for $N_p(x)$ and $H_p^{(1)}(x)$, $H_p^{(2)}(x)$]. Also

$$I_{p-1}(x) - I_{p+1}(x) = \frac{2p}{x} I_p(x)$$

$$K_{p-1}(x) - K_{p+1}(x) = -\frac{2p}{x} K_p(x)$$

$$I_{p-1}(x) + I_{p+1}(x) = 2 \frac{dI_p(x)}{dx} \right\} \qquad (8)$$

$$K_{p-1}(x) + K_{p+1}(x) = -2 \frac{dK_p(x)}{dx}$$

Finally, some useful integral relationships are:

$$J_n(x) = \frac{1}{2\pi} \int_\alpha^{\alpha+2\pi} e^{i(x \sin \phi - n\phi)} \, d\phi \qquad (9)$$

where α is real; from this follows Eqn (9–42), i.e.,

$$e^{ix \sin y} = \sum_{n=-\infty}^{\infty} J_n(x) e^{iny}$$

Also $\qquad \int_0^\infty \frac{\cos px \, dx}{(x^2+q^2)^{s+1}} = \sqrt{\pi} \left(\frac{p}{2q}\right)^{s+1/2} \frac{K_{s+1/2}(pq)}{\Gamma(s+1)}$

Appendix 3

Plasma diagnostics

The various diagnostic techniques may be classified under the headings (A) probes, (B) propagation of electromagnetic waves, (C) spectroscopy, and (D) particle beams. We will only discuss categories (A) and (B)†; plasma spectroscopy—though of vital importance in plasma diagnostics—is too large a subject to be adequately summarized in the space available and the reader is referred to the standard work on the subject.[83] Plasma diagnostics using particle beams have been used much less widely; the interested reader is referred to the discussion in a general summary by Eubank.[109]

(A) Probes

The best known of these is the Langmuir probe which has long been a versatile diagnostic device, providing measurements of electron density and temperature and of ion density. The probe is a small metal electrode (usually cylindrical or spherical, though sometimes in the shape of a disk); the essence of the technique is to monitor the current to the probe when different voltages are applied. A typical probe characteristic displaying the variation of probe current with voltage is shown in Fig. A3–1.

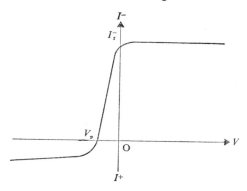

Fig. A3–1

When the applied voltage is large and negative the probe draws ion current only according to

$$i^+ = \frac{n^+ e \langle v \rangle}{4} = \frac{n^+ e}{2} \left(\frac{2\kappa T}{\pi m^+} \right)^{1/2} \tag{1}$$

† Those diagnostic methods already discussed in the text are not repeated here

However, this takes no account of the perturbation introduced by the presence of the probe in the plasma. In practice a positive sheath extends around the probe and the size of this sheath depends on the electron temperature. For $T^+ \ll T^-$ it can be shown[110] that (1) should be replaced by

$$i^+ \simeq 0 \cdot 4n^+ e \left(\frac{2\kappa T^-}{m^+} \right)^{1/2} \tag{2}$$

i.e., for a large negative potential the saturation ion current is virtually independent of ion temperature.

As V becomes less negative with respect to the plasma some of the energetic electrons from the tail of the electron distribution are collected by the probe until eventually the electron current drawn just cancels the ion current. This occurs when $V = V_p$ which is known as the *floating potential* and $V_p \simeq \kappa T^-/2$. A further increase in V leads to a rapid increase in electron current until this saturates at *space potential* V_s, on account of the space-charge limiting of electron current; from Fig. A3–1, $V_s \simeq 0$.

Consider the probe characteristic in the range $V_p < V < V_s$ in which electron current is drawn. In this region the electron sheath (cf. Section 10–11) around the probe shields it from electrons other than those energetic enough to overcome the potential barrier; denoting the potential difference across this electron sheath by V, then the threshold energy for electrons is eV. If the electrons have a Boltzmann distribution, the probability of an electron possessing energy eV is $\exp[-eV/\kappa T^-]$. Denoting the electron density at the probe surface by n_p and the ambient electron density by n_0, we have

$$n_p = n_0 \exp[-eV/\kappa T^-] \tag{3}$$

The electron current to the probe is then

$$I^- = \frac{n_0 e A}{2} \left(\frac{2\kappa T^-}{\pi m^-} \right)^{1/2} \exp[-eV/\kappa T^-] \tag{4}$$

where A is the surface area of the probe and

$$I_r^- = n_0 e A \left(\frac{\kappa T^-}{2\pi m^-} \right)^{1/2}$$

is the random electron current (cf. Fig. A3–1). From (4)

$$\ln I^- = \ln I_r^- - \frac{eV}{\kappa T^-} \tag{5}$$

i.e., the logarithm of the electron current plotted as a function of probe potential is linear and the slope of the line allows the electron temperature to be determined. Since the *measured* current to the probe has an ion component in addition to electron current, ln $(I - I^+)$ should be plotted against V. Note that there is a simple check on the assumption that the electrons have a Boltzmann distribution: if this is not the case then the above plot will not be linear.

Having determined the electron temperature one can obtain the electron density using

$$n_0 = \frac{I_r^-}{eA}\left(\frac{2\pi m^-}{\kappa T^-}\right)^{1/2}$$

though this is not a particularly reliable measurement. In addition, one can also obtain the ion density in the plasma using (2); i.e.,

$$n^+ \simeq \frac{2 \cdot 5 I^+}{eA}\left(\frac{m^+}{2\kappa T^-}\right)^{1/2}$$

The ion temperature on the other hand is not obtainable from Langmuir probe measurements.

The Langmuir probe technique is useful in plasmas of moderate density; if the density is high enough so that electron and ion mean free paths become so small as to be comparable with the dimensions of the probe, then the above discussion is no longer valid. Moreover no account has been taken of plasmas in applied magnetic fields. In that situation, if the magnetic field is strong enough, the appropriate characteristic length is the particle Larmor radius (cf. Section 3–6) and this may easily be much smaller than the dimensions of the probe. The Langmuir probe *can* provide information in the presence of a magnetic field but interpretation of the characteristics is less straightforward.

Various modifications of the Langmuir probe exist, such as the double probe technique. This avoids a serious difficulty in some discharges in which the electron current drawn to a single probe may perturb the plasma sufficiently to make the measurements meaningless.

A probe of another kind is the radio-frequency plasma probe developed by Sayers primarily for satellite measurements of ionospheric and exospheric electron density.[15] This probe consists of two disk-shaped grids mounted on insulating supports fixed to an arm projecting from the satellite. A radio-frequency voltage is applied between the grids and the resulting RF current is measured, so determining the complex impedance between the grids at the impressed frequency; from this measurement the electron density may be obtained.

(B) Electromagnetic waves

Despite the versatility of the Langmuir probe it does introduce a perturbation into the plasma. Moreover the plasma affects the probe and this may be severe enough in very hot plasmas for the probe to have a very short life. Diagnostics making use of electromagnetic waves avoid these problems. The wave frequency is chosen to be sufficiently high that ion motion and inter-particle collisions in the plasma may be ignored. Then from the theory developed in Section 7–7–1 we know that there are two types of wave, $Q_\parallel$ and $Q_\perp$ respectively. For the $Q_\perp$ wave there are two modes: one (the extra-ordinary wave) satisfies a dispersion relation involving the magnetic field $\mathbf{B}_0$, while the other (the ordinary wave) satisfies

$$n^2 = \frac{\omega^2 - \omega_p^2}{\omega^2 - \omega_p^2 \cos^2 \theta}$$

in which θ is the angle between $\mathbf{k}$ and $\mathbf{B}_0$. Since ω_p^2 is dependent solely on electron density, the ordinary wave propagating orthogonally to the magnetic field may be used for measuring the electron density in an anisotropic plasma. The plasma is inserted between microwave horns in one arm of an inter-ference bridge; if L denotes the path length in the plasma, the phase shift experienced by the wave is simply

$$\Delta\phi = \int_0^L (1 - n)\frac{2\pi}{\lambda}dx$$

$$= \int_0^L \left[1 - \left(1 - \frac{4\pi n^- e^2}{m^- \omega^2}\right)^{1/2}\right]\frac{2\pi}{\lambda}dx$$

in which n^- is the electron density.† Determination of the mean electron density requires a knowledge of the density profile $n^-(x)$. Two important conditions have to be met for this method to be viable; namely the probing frequency $\omega > \omega_p$ and $\lambda \ll L$. This requirement means that wave frequencies above ~ 10 Gc/s are needed so that the limiting electron density which may be measured is $\sim 10^9$ cm^{-3}. Until the development of monochromatic sources such as helium–neon lasers, an upper limit of $\sim 10^{14}$ cm^{-3} existed for trans-mission measurements; with such lasers the limit has been extended to $\sim 10^{18}$ cm^{-3}.

The use of waves in plasmas for diagnostic purposes is not restricted to the interferometric measurement just described. For example, the possibility exists of making use of the many resonances and cut-offs (cf. Section 7–6–1) to obtain information on ion and electron densities and the magnetic field in a plasma. This is not the place to summarize the variety of techniques avail-able but as an example consider the Alfvén wave described in Section 7–3. For a plasma of uniform density, measurement of the wave velocity allows the ion density to be determined provided the magnetic field is known.

---

† Care must be taken to avoid confusing n, the plasma refractive index, with $n^\pm$, ion and electron densities.

Appendix 4

Units

The two main systems of units for describing electromagnetic phenomena are gaussian cgs units and rationalized mks units. In this book we have used the former system throughout and the purpose of this appendix is to show how formulae and quantities may be transformed from one system to the other.

Gaussian cgs system

In the gaussian cgs system length, mass, and time are taken as fundamental units and all others are derived from these. For example, Newton's law

$$F = m\ddot{x}$$

enables the unit of force, the dyne, to be defined in terms of centimetres, grammes, and seconds: 1 dyn is that force which, acting on a mass of 1 g, produces an acceleration of 1 cm/s², i.e.,

$$1 \text{ dyn} = 1 \text{ g cm/s}^2$$

The unit of charge is derived using Coulomb's law

$$F = \frac{e_1 e_2}{r^2} \tag{1a}$$

This unit, called the electrostatic unit (esu) of charge or statcoulomb, is defined to be that charge which, placed 1 cm from a similar charge, experiences a force of 1 dyn; thus

$$1 \text{ statcoul} = 1 \text{ cm.dyn}^{1/2}$$

It then follows from

$$\phi = \frac{e}{r} \tag{2a}$$

that the unit of potential

$$1 \text{ statvolt} = 1 \text{ statcoul/cm}$$

and from

$$E = \frac{e}{r^2} \tag{3a}$$

that the unit of electric field intensity is the statvolt/cm. Current is defined as the rate of change of charge and therefore has the unit

$$1 \text{ statamp} = 1 \text{ statcoul/s}$$

A slight digression will help with the understanding of the derivation of the unit of magnetic induction, the gauss. This was defined originally in terms of the electromagnetic unit (emu) of current or abamp. The abamp is defined as that current which, flowing in a long straight wire placed 1 cm from a parallel wire carrying the same current, results in a force per unit length of 2 dyn/cm between the wires. Thus Ampère's law in emu–cgs units is

$$\frac{dF}{dl} = \frac{2I_1 I_2}{r} \quad (I_1, I_2 \text{ in abamp}) \tag{4}$$

but in gaussian cgs units is

$$\frac{dF}{dl} = \frac{2I_1 I_2}{c^2 r} \quad (I_1, I_2 \text{ in statamp}) \tag{4a}$$

where the constant of proportionality c is such that

$$I \text{ (in abamp)} = \frac{I \text{ (in statamp)}}{c}$$

From (4), 1 abamp $= 1 \text{ dyn}^{1/2}$ and hence c has the dimensions of a velocity.

The magnetic induction at a distance r from a long straight wire carrying a current I is given by

$$B = \frac{2I}{r} \quad (I \text{ in abamp})$$

in emu, and by

$$B = \frac{2I}{cr} \quad (I \text{ in statamp}) \tag{5a}$$

in gaussian cgs units. Thus

$$1 \text{ gauss} = 1 \text{ statcoul/cm}^2$$

The magnetic vector potential is given by

$$\mathbf{B} = \nabla \times \mathbf{A} \tag{6a}$$

With these derivations of units for charge, current, electric field, and magnetic induction, Maxwell's equations take the form

$$\left. \begin{array}{c} \nabla \times \mathbf{E} = -\frac{1}{c} \frac{\partial \mathbf{B}}{\partial t} \\ \nabla \times \mathbf{B} = \frac{1}{c} \frac{\partial \mathbf{E}}{\partial t} + \frac{4\pi}{c} \mathbf{j} \\ \nabla . \mathbf{E} = 4\pi q \\ \nabla . \mathbf{B} = 0 \end{array} \right\} \tag{7a}$$

where q and $\mathbf{j}$ are charge and current densities, having the units statcoul/cm^3 and statamp/cm^2 respectively. In vacuo ($q = 0 = \mathbf{j}$) the Maxwell equations may be combined to give the wave equations

$$\left.\begin{aligned}\left(\nabla^2 - \frac{1}{c^2}\frac{\partial^2}{\partial t^2}\right)\mathbf{E} = 0 \\ \left(\nabla^2 - \frac{1}{c^2}\frac{\partial^2}{\partial t^2}\right)\mathbf{B} = 0\end{aligned}\right\} \tag{8a}$$

from which c may be identified as the velocity of light, $c = 3.10^{10}$ cm/s.

The above form of the Maxwell equations is the appropriate one for microscopic phenomena. In (7a) q and $\mathbf{j}$ represent the *total* charge and current densities and, since it is convenient to discuss all charges and currents explicitly in plasma physics, this form of the Maxwell equations has been used throughout the book. (It is then customary to refer to $\mathbf{B}$ as the magnetic field.) For material media it is usual practice to distinguish between *free* charges and currents and *bound* (material) charges and currents. The latter are then represented by a polarization $\mathbf{P}$ and a magnetization $\mathbf{M}$ such that the macroscopic variables $\mathbf{D}$ and $\mathbf{H}$ are given by

$$\left.\begin{aligned}\mathbf{D} = \mathbf{E} + 4\pi\mathbf{P} \\ \mathbf{H} = \mathbf{B} - 4\pi\mathbf{M}\end{aligned}\right\} \tag{9a}$$

The Maxwell equations then take the form

$$\left.\begin{aligned}\nabla \times \mathbf{E} &= -\frac{1}{c}\frac{\partial \mathbf{B}}{\partial t} \\ \nabla \times \mathbf{H} &= \frac{1}{c}\frac{\partial \mathbf{D}}{\partial t} + \frac{4\pi}{c}\mathbf{j}_{\text{free}} \\ \nabla.\mathbf{D} &= 4\pi q_{\text{free}} \\ \nabla.\mathbf{B} &= 0\end{aligned}\right\} \tag{10a}$$

For a linear, isotropic medium

$$\mathbf{D} = \varepsilon\mathbf{E} \qquad \mathbf{B} = \mu\mathbf{H} \tag{11a}$$

where ε is the dielectric constant and μ the permeability of the medium. It is clear from (9a) that $\mathbf{D}$, $\mathbf{E}$, and $\mathbf{P}$ all have the same dimensions; hence the unit of $\mathbf{D}$ is the statvolt/cm while that of $\mathbf{P}$ is

$$1 \text{ dipole moment/cm}^3 = 1 \text{ statvolt/cm}$$

Similarly, the unit of $\mathbf{H}$,

$$1 \text{ oersted} = 1 \text{ gauss}$$

and that of $\mathbf{M}$,

$$1 \text{ magnetic moment/cm}^3 = 1 \text{ gauss}$$

From (11a) one sees that ε and μ are dimensionless. In vacuo (9a) becomes

$$\mathbf{D} = \mathbf{E} \qquad \mathbf{H} = \mathbf{B} \tag{12a}$$

Finally from the simple form of Ohm's law

$$j = \sigma E \tag{13a}$$

note that the electrical conductivity σ is measured in units of s^{-1}.

Rationalized mks system

The unit of force in the mks system, the newton, is defined as that force which, acting on a mass of 1 kg, produces an acceleration of 1 m/s^2, i.e.,

$$1 \text{ N} = 1 \text{ kg.m/s}^2$$

Next, writing Ampère's law

$$\frac{\mathrm{d}F}{\mathrm{d}l} = \frac{\mu_0}{4\pi} \cdot \frac{2I_1 I_2}{r} \tag{4b}$$

where the constant of proportionality $\mu_0/4\pi$ is 10^{-7} N/A^2 by definition and one ampère, the mks unit of current, is defined as that current which, flowing in a long straight wire placed 1 m from a parallel wire carrying the same current, results in a force per unit length of 2.10^{-7} N/m between the wires. Then B is given by

$$B = \frac{\mu_0}{4\pi} \cdot \frac{2I}{r} \tag{5b}$$

and thus the unit of magnetic induction

$$1 \text{ Weber/m}^2 = 1 \text{ N/A m}$$

The magnetic vector potential is again given by

$$B = \nabla \times A \tag{6b}$$

The unit of charge, the coulomb, is defined as that quantity of charge transported by a current of 1 A in 1 s and thus

$$1 \text{ C} = 1 \text{ A s}$$

Coulomb's law is written

$$F = \frac{1}{4\pi\varepsilon_0} \cdot \frac{e_1 e_2}{r^2} \tag{1b}$$

which defines the proportionality constant $(4\pi\varepsilon_0)^{-1}$ having the units Nm^2/C^2. Then it follows that the potential

$$\phi = \frac{1}{4\pi\varepsilon_0} \cdot \frac{e}{r} \tag{2b}$$

has the unit

$$1 \text{ volt} = 1 \text{ N.m/C}$$

and

$$E = \frac{1}{4\pi\varepsilon_0} \cdot \frac{e}{r^2} \tag{3b}$$

is measured in units of V/m.

In rationalized mks units the Maxwell equations are

$$\left.\begin{array}{c} \mathbf{V} \times \mathbf{E} = -\dfrac{\partial \mathbf{B}}{\partial t} \\[2mm] \mathbf{V} \times \mathbf{B} = \varepsilon_0\mu_0\dfrac{\partial \mathbf{E}}{\partial t} + \mu_0\mathbf{j} \\[2mm] \mathbf{V}.\mathbf{E} = \dfrac{q}{\varepsilon_0} \\[2mm] \mathbf{V}.\mathbf{B} = 0 \end{array}\right\} \tag{7b}$$

from which the vacuum wave equations are

$$\left.\begin{array}{c} \left(\nabla^2 - \varepsilon_0\mu_0\dfrac{\partial^2}{\partial t^2}\right)\mathbf{E} = 0 \\[3mm] \left(\nabla^2 - \varepsilon_0\mu_0\dfrac{\partial^2}{\partial t^2}\right)\mathbf{B} = 0 \end{array}\right\} \tag{8b}$$

and hence $c^2 = 1/\varepsilon_0\mu_0$. The equations for $\mathbf{D}$ and $\mathbf{H}$ are

$$\left.\begin{array}{c} \mathbf{D} = \varepsilon_0\mathbf{E} + \mathbf{P} \\[2mm] \mathbf{H} = \dfrac{1}{\mu_0}\mathbf{B} - \mathbf{M} \end{array}\right\} \tag{9b}$$

The alternative form of the Maxwell equations is

$$\left.\begin{array}{c} \mathbf{V} \times \mathbf{E} = -\dfrac{\partial \mathbf{B}}{\partial t} \\[2mm] \mathbf{V} \times \mathbf{H} = \dfrac{\partial \mathbf{D}}{\partial t} + \mathbf{j}_{\text{free}} \\[2mm] \mathbf{V}.\mathbf{D} = q_{\text{free}} \\[2mm] \mathbf{V}.\mathbf{B} = 0 \end{array}\right\} \tag{10b}$$

For a linear, isotropic medium

$$\mathbf{D} = \varepsilon\mathbf{E} \qquad \mathbf{B} = \mu\mathbf{H} \tag{11b}$$

From (9b) it is clear that $\mathbf{D}$ and $\mathbf{P}$ have units of C/m² while those of $\mathbf{H}$ and $\mathbf{M}$ are A/m. Hence from (11b) ε and μ have the units of ε_0 and μ_0; in fact from (9b) ε_0 and μ_0 are the vacuum values of ε and μ, i.e., in vacuo

$$\mathbf{D} = \varepsilon_0\mathbf{E} \qquad \mathbf{B} = \mu_0\mathbf{H} \tag{12b}$$

Ohm's law is written as for the gaussian system

$$\mathbf{j} = \sigma\mathbf{E} \tag{13b}$$

so that conductivity has the unit

$$1 \text{ mho m}^{-1} \equiv 1\ \Omega^{-1}\,\text{m}^{-1} = 1\ \text{A/m.V}$$

Note that the mks system has four fundamental units corresponding to length, mass, time, and charge.

Transformations

Now by comparing the pairs of equations labelled a and b one can see how to transform symbols from one system to the other. For example a comparison of (1a) with (1b) shows that e in the gaussian system must be replaced by $e(4\pi\varepsilon_0)^{-1/2}$ to obtain the equivalent mks formula. Then doing this in (2a) and comparing (2b) shows that ϕ in a gaussian formula must be replaced by $(4\pi\varepsilon_0)^{1/2}\phi$ to obtain the mks equivalent. Continuing in this way we obtain the following set of transformations (which are reversible):

Quantity	Gaussian	mks
velocity of light	c	$(\varepsilon_0\mu_0)^{-1/2}$
charge	e	$e/(4\pi\varepsilon_0)^{1/2}$
charge density	q	$q/(4\pi\varepsilon_0)^{1/2}$
current	I	$I/(4\pi\varepsilon_0)^{1/2}$
current density	$\mathbf{j}$	$\mathbf{j}/(4\pi\varepsilon_0)^{1/2}$
potential	ϕ	$(4\pi\varepsilon_0)^{1/2}\phi$
electric field	$\mathbf{E}$	$(4\pi\varepsilon_0)^{1/2}\mathbf{E}$
magnetic induction	$\mathbf{B}$	$\left(\dfrac{4\pi}{\mu_0}\right)^{1/2}\mathbf{B}$
vector potential	$\mathbf{A}$	$\left(\dfrac{4\pi}{\mu_0}\right)^{1/2}\mathbf{A}$
polarization	$\mathbf{P}$	$\mathbf{P}/(4\pi\varepsilon_0)^{1/2}$
magnetization	$\mathbf{M}$	$\left(\dfrac{\mu_0}{4\pi}\right)^{1/2}\mathbf{M}$
displacement	$\mathbf{D}$	$\left(\dfrac{4\pi}{\varepsilon_0}\right)^{1/2}\mathbf{D}$
magnetic field	$\mathbf{H}$	$(4\pi\mu_0)^{1/2}\mathbf{H}$
dielectric constant	ε	$\varepsilon/\varepsilon_0$
permeability	μ	μ/μ_0
electrical conductivity	σ	$\sigma/4\pi\varepsilon_0$

To find the transformation for a given amount of some physical quantity from one system of units to the other we again compare equations labelled a and b and also make use of the relationships between units. For example, according to (4b) a pair of long straight wires lying parallel and 1m apart give rise to a force per unit length of 2.10^{-7} N/m $= 2.10^{-7}$ kg/s^2 = 2.10^{-4} g/s^2 = 2.10^{-4} dyn/cm. If 1 A $\equiv \alpha$ statamp, the force per unit length according to (4a) is $(2\alpha^2/9 \times 10^{22})$ dyn/cm; hence $\alpha = 3 \times 10^9$. It follows immediately that 1 C $\equiv 3 \times 10^9$ statcoul and 1 C/m$^3 \equiv 3 \times 10^3$ statcoul/m^3. In this way we arrive at the following equivalences between units:

Quantity	*Gaussian*	*mks*
length	1 cm	$\equiv 10^{-2}$ m
mass	1 g	$\equiv 10^{-3}$ kg
time	1 s	$\equiv 1$ s
force	1 dyn	$\equiv 10^{-5}$ N
energy	1 erg	$\equiv 10^{-7}$ J
power	1 erg/s	$\equiv 10^{-7}$ W
charge	1 statcoul	$\equiv \frac{1}{3}.10^{-9}$ C
charge density	1 statcoul/cm³	$\equiv \frac{1}{3}.10^{-3}$ C/m³
current	1 statamp	$\equiv \frac{1}{3}.10^{-9}$ A
current density	1 statamp/cm²	$\equiv \frac{1}{3}.10^{-5}$ A/m²
potential	1 statvolt	$\equiv 300$ V
electric field	1 statvolt/cm	$\equiv 3.10^{4}$ V/m
magnetic induction	1 gauss	$\equiv 10^{-4}$ Wb/m²
polarization	1 dip.mom./cm³	$\equiv \frac{1}{3}.10^{-5}$ C/m²
magnetization	1 mag.mom./cm³	$\equiv 4\pi.10^{3}$ A/m
displacement	1 statvolt/cm	$\equiv \dfrac{1}{12\pi}.10^{-5}$ C/m²
magnetic field	1 oersted	$\equiv \dfrac{1}{4\pi}.10^{3}$ A/m
electrical conductivity	1 s^{-1}	$\equiv \frac{1}{9}.10^{-9}$ mho/m

Appendix 5

Some useful physical constants

Velocity of light	c	2.998×10^{10} cm s^{-1}
Electron charge	e	4.803×10^{-10} esu
Electron mass	m^-	9.109×10^{-28} g
Proton mass	m^+	1.673×10^{-24} g
Proton/electron mass ratio	m^+/m^-	1,837
Classical electron radius	r_e	2.818×10^{-13} cm
Thomson cross-section	σ_T	6.652×10^{-25} cm^2
Planck's constant	$\hbar$	1.055×10^{-27} erg s
Bohr radius	a_0	5.292×10^{-9} cm
Angström unit	Å	10^{-8} cm
Electron volt	eV	1.602×10^{-12} erg
Ionization potential of hydrogen		13.60 eV
Boltzmann's constant	κ	1.381×10^{-16} erg °K^{-1}
Electron plasma frequency	ω_{pe}	$5.64(n^-)^{1/2} \times 10^4$ c/s (Hz)
Electron Larmor frequency	Ω_e	$1.76B \times 10^7$ c/s (B in gauss)
Debye length	λ_D	$7.44\left(\dfrac{T^-}{n^-}\right)^{1/2} \times 10^2$ cm (T^- in eV)

Bibliography

Chapter 2

An introductory book which contains some discussion of orbit theory not included in Chapter 2 is

Schmidt, G., *Physics of high temperature plasmas*, Academic Press, New York, 1966.

Some books base all or most of their discussion of plasma dynamics on orbit theory by making suitable extrapolations of the single particle description. These include

Chandrasekhar, S., *Plasma physics*, Phoenix Books, University of Chicago Press, Chicago, 1962.
Longmire, C. L., *Elementary plasma physics*, John Wiley (Interscience), New York, 1963.

More advanced treatments of orbit theory are contained in

Northrop, T. G., *The adiabatic motion of charged particles*, John Wiley (Interscience), New York, 1963.
Chamberlain, J. W., *Motion of charged particles in the earth's magnetic field*, Gordon and Breach, New York, 1964.

Chapter 3

A comprehensive discussion of the derivation of the macroscopic equations is the subject of

Burgers, J. M., 'Statistical plasma mechanics' in *Symposium of plasma dynamics*, F. H. Clauser (ed.), Addison-Wesley, Reading, Mass., 1960.

The conditions of applicability of the hydromagnetic equations are discussed at some length in

Shkarofsky, I. P., T. W. Johnston, and M. P. Bachynski, *The particle kinetics of plasmas*, Addison-Wesley, Reading, Mass., 1966.

Rigorous discussions of the truncation procedure are to be found in ref. 4 and in

Thompson, W. B., 'Kinetic theory of plasma' in *Proceedings of the International School of Physics 'Enrico Fermi'*, Course XXV, M. N. Rosenbluth (ed.), Academic Press, New York, 1964.

Chapter 4

Hydromagnetics is treated in some of the texts included in the references (cf. 6, 13, 29, 35, and 44) to a varying degree; some emphasize different aspects of the subject from others. Introductory accounts are given by

Jeffrey, A., *Magnetohydrodynamics*, Oliver and Boyd, Edinburgh, 1966.
Kendall, P. C. and C. Plumpton, *Magnetohydrodynamics with hydrodynamics*, Pergamon Press, Oxford, 1964.

A rather more comprehensive treatment is that of

Roberts, P. H., *Introduction to magnetohydrodynamics*, Longmans Green, London, 1967.

In particular, a discussion of hydromagnetic dynamo theory is included in this book.

The treatment of hydromagnetic stability in Sections 4–5 to 4–6 is very brief; more detail may be found in the account given by

Thompson, W. B., *An introduction to plasma physics*, Pergamon Press, Oxford, 1962.

as well as in ref. 6.

A book dealing with hydromagnetics in a rather specialized sense in that it discusses mainly applications of hydromagnetics to solar and ionospheric physics is

Dungey, J. W., *Cosmic electrodynamics*, Cambridge University Press, London, 1958.

Another book, this time directed towards engineers and applied physicists, is

Shercliff, J. A., *A textbook of magnetohydrodynamics*, Pergamon Press, Oxford, 1965.

Finally one might mention two review articles; one is that of Cowling[35] and the other is

Elsasser, W. M., *Amer. J. Phys.*, **23**, 590, 1955; *ibid.*, **24**, 85, 1956.

Chapter 5

A more comprehensive treatment of hydromagnetic flows is to be found in

Shercliff, J. A., *A textbook of magnetohydrodynamics*, Pergamon Press, Oxford, 1965.

Yoler, Y. A., 'A review of magnetohydrodynamics' in *Plasma physics*, J. E. Drummond (ed.), McGraw-Hill, New York, 1961.

and in ref. 36, which also discusses engineering applications at much greater length.

Chapter 6

The jump conditions across an arbitrary shock are derived in

Jeffrey, A., *Magnetohydrodynamics*, Oliver and Boyd, Edinburgh, 1966.

Ref. 45 contains a discussion of hydromagnetic shock waves which is based on the theory of characteristics but which emphasizes the physical description of the subject.

Theoretical and experimental studies on MHD shock waves are the subjects of many of the articles in

Landshoff, R. K. M. (ed.), *Magnetohydrodynamics*, Stanford University Press, Stanford, Calif., 1957.

A comprehensive treatment of the subject is to be found in

Anderson, J. E., *Magnetohydrodynamic shock waves*, MIT Press, Cambridge, Mass., 1963.

Chapters 7, 8

The reader interested in going beyond the elementary theory of wave propagation in plasmas presented in these chapters cannot do better than read the standard work by Stix.[54] A full discussion of the CMA diagram, including its application to warm plasma waves, is given by Allis, Buchsbaum, and Bers.[64] Another book devoted entirely to waves in plasmas is

Denisse, J. F. and J. L. Delcroix, *Plasma waves*, John Wiley (Interscience), New York, 1963.

This account is less comprehensive, however, than those in refs 54 and 64.

Chapter 9

The only book specifically devoted to radiation from plasmas (in the special sense in which this has been discussed in Chapter 9) is that by Bekefi.[88] A readable review article is

Bekefi, G. and S. C. Brown, *Amer. J. Phys.*, **29**, 404, 1961.

Chapter 10

A somewhat different analysis of the linearized Vlasov equation is given in

Gartenhaus, S., *Elements of plasma physics*, Holt, Rinehart and Winston, New York, 1964.

A concise description of Chandrasekhar's derivation of relaxation times is contained in

Spitzer, L., *Physics of fully ionized gases*, John Wiley (Interscience), New York, 1956.

Ref. 103 on transport phenomena contains a brief historical review of the subject. A more comprehensive and advanced treatment of kinetic theory is given in ref. 4.

References

1. Boley, F. I., *Plasmas—laboratory and cosmic*, Van Nostrand, Princeton, N.J., 1966.
2. Budden, K. G., *Radio waves in the ionosphere*, Cambridge University Press, London, 1961.
3. Suits, C. G., *The collected works of Irving Langmuir*, vol. 5, *Plasma and oscillations*, Pergamon Press, New York, 1961.
4. Montgomery, D. C. and D. A. Tidman, *Plasma kinetic theory*, McGraw-Hill, New York, 1964.
5. Bishop, A. S., *Project Sherwood*, Addison-Wesley, Reading, Mass., 1958.
6. Rose, D. J. and M. Clark, *Plasmas and controlled fusion*, MIT Press, Cambridge, Mass., 1961.
7. Gardner, J. W., *Electricity without dynamos*, Penguin Books, Harmondsworth, Middlesex, 1963.
8. Knechtli, R. C., G. R. Brewer, and M. R. Currie, Chapter 12, *Plasma physics in theory and application*, W. Kunkel (ed.), McGraw-Hill, New York, 1966.
9. ter Haar, D., *Elements of statistical mechanics*, Holt, Rinehart and Winston, New York, 1954.
10. Chapman, S. and T. G. Cowling, *The mathematical theory of non-uniform gases*, Cambridge University Press, London, 1952.
11. Allis, W. P., 'Motion of ions and electrons' in *Handbuch der Physik*, vol. 21, Springer-Verlag, Heidelberg, 1956.
12. Sommerfeld, A., *Thermodynamics and statistical mechanics*, Academic Press, New York, 1956.
13. Longmire, C. L., *Elementary plasma physics*, John Wiley (Interscience), New York, 1963.
14. Sweetman, D. R. *et al.*, *Phys. Fluids*, **7**, 988, 1964.
15. Massey, H. S. W., *Space physics*, Cambridge University Press, London, 1964.
16. Landau, L. D. and E. M. Lifshitz, *The classical theory of fields*, Pergamon Press, Oxford, 1962.
17. Kibble, T. W. B., *Phys. Rev.*, **138**, B740, 1964.
18. Kibble, T. W. B., *Phys. Rev. Lett.*, **16**, 1054, 1233, 1966.
19. Kuiper, G. P., *The Sun*, University of Chicago Press, Chicago, 1953.
20. Landau, L. D. and E. M. Lifshitz, *Fluid mechanics*, Pergamon Press, Oxford, 1959.
21. Hirshfelder, J. O., C. F. Curtiss, and R. B. Bird, *Molecular theory of gases and liquids*. John Wiley, New York, 1954.
22. Goldstein, S., *Lectures on fluid mechanics*, John Wiley (Interscience), London, 1960.
23. Jackson, J. D., *Classical electrodynamics*, John Wiley, New York, 1962.
24. Woltjer, L., *Proc. Nat. Acad. Sci.*, **44**, 489, 833, 1958; Chandrasekhar, S. and L. Woltjer, *ibid.*, 285.
25. Glasstone, S. and R. H. Lovberg, *Controlled thermonuclear reactions*, Van Nostrand, Princeton, N.J., 1960.
26. Chandrasekhar, S., *Hydrodynamic and hydromagnetic instability*, Clarendon Press, Oxford, 1961.
27. Kruskal, M. and M. Schwarzschild, *Proc. Roy. Soc.*, **223A**, 348, 1954.

28. Rosenbluth, M., N. Krall, and N. Rostoker, *Nucl. Fusion Suppl.*, Part 1, 143, 1962.
29. Alfvén, H. and C.-G. Fälthammar, *Cosmical electrodynamics*, Clarendon Press, Oxford, 1963.
30. Weatherburn, C. E., *Advanced vector analysis*, George Bell, London, 1954.
31. Alfvén, H., *Arkiv. f. Mat.*, **29B**, No. 2, 1942.
32. Lundquist, S., *Phys. Rev.*, **83**, 307, 1951.
33. Lehnert, B., *Phys. Rev.*, **94**, 815, 1954.
34. Walén, C., *Arkiv. f. Mat.*, **30A**, No. 15, 1944; *ibid.*, **33A**, No. 18, 1946.
35. Cowling, T. G., *Magnetohydrodynamics*, John Wiley (Interscience), New York, 1957; also 'Magnetohydrodynamics' in *Reports on Progress in Physics*, **25**, 244, 1962.
36. Sutton, G. W. and A. Sherman, *Engineering magnetohydrodynamics*, McGraw-Hill, New York, 1965.
37. Grad, H., 'Mathematical problems in magneto-fluid dynamics' in *Proceedings of the International Congress of Mathematicians*, 1962, Almqvist and Wiksells, Uppsala, 1963, pp. 560–583.
38. Marshall, J., 'Hydromagnetic plasma gun' in *Plasma acceleration*, S. W. Kash (ed.), Stanford University Press, Stanford, Calif., 1960, pp. 60–72.
39. Lighthill, M. J., 'Viscosity effects in sound waves of finite amplitude' in *Surveys in mechanics*. G. K. Batchelor and R. M. Davies (eds.), Cambridge University Press, London, 1956, pp. 250–351.
40. Landau, L. D. and E. M. Lifshitz, *Electrodynamics of continuous media*, Pergamon Press, Oxford, 1960.
41. Ericson, W. B. and J. Bazar, *Phys. Fluids*, **3**, 631, 1960.
42. Bazar, J. and W. B. Ericson, *Astrophys. J.*, **129**, 758, 1958.
43. Marshall, W., *Proc. Roy. Soc.*, **233**, 367, 1956.
44. Ferraro, V. C. A. and C. Plumpton, *An introduction to magneto-fluid mechanics*, Clarendon Press, Oxford, 1966.
45. Kantrowitz, A. and H. E. Petschek, Chapter 6, *Plasma physics in theory and application*, W. Kunkel (ed.), McGraw-Hill, New York, 1966.
46. Ashby, D. E. T., K. V. Roberts, and S. J. Roberts, *Plasma Physics* (*J. Nucl. Energy* Part C), **3**, 162, 1961.
47. Roberts, K. V., *Plasma Physics* (*J. Nucl. Energy*, Part C), **5**, 365, 1963.
48. Sagdeev, R. Z., 'Cooperative phenomena and shock waves in collisionless plasmas' in *Reviews of plasma physics*, vol. 4, M. A. Leontovich (ed.), Consultants Bureau, New York, 1966, pp. 23–92.
49. Wright, J. K., *Shock tubes*, Methuen, London, 1961.
50. Kolb, A. C., 'Magnetically driven shock waves' in *Magnetohydrodynamics*, R. K. M. Landshoff (ed.), Stanford University Press, Stanford, Calif., 1957.
51. Kolb, A. C., *Proceedings 2nd International Conference on peaceful uses of atomic energy*, Geneva, **31**, 332, 1958.
52. Paul, J. W. M. *et al.*, *Nature*, **208**, 133, 1965; *ibid.*, **216**, 363, 1967.
53. Ness, N. F., C. S. Clearce, and J. B. Seek, *J. Geophys. Res.*, **69**, 3531, 1964.
54. Stix, T. H., *The theory of plasma waves*, McGraw-Hill, New York, 1962.
55. Wilcox, J. M., F. I. Boley, and A. W. De Silva, *Phys. Fluids*, **3**, 15, 1960; Wilcox, J. M., A. W. De Silva, and W. S. Cooper, *ibid.*, **4**, 1506, 1961.
56. Jephcott, D. F. and P. M. Stocker, *J. Fluid Mech.*, **13**, 587, 1962.
57. Stix, T. H. and R. W. Palladino, *Proceedings 2nd International Conference on peaceful uses of atomic energy*, Geneva, **31**, 282, 1958.
58. Boley, F. I. *et al.*, *Phys. Fluids*, **6**, 925, 1963.
59. Hooke, W. M. *et al.*, *Nucl. Fusion Suppl.*, Part 3, 1083, 1962.

60. Hooke, W. M. *et al.*, *Phys. Fluids*, **8**, 1146, 1965.
61. Stix, T. H. and R. W. Palladino, *Phys. Fluids*, **1**, 446, 1958.
62. Jephcott, D. F. and A. Malein, *Proc. Roy. Soc.*, **278A**, 243, 1964.
63. Swanson, D. G., R. W. Gould, and R. H. Hertel, *Phys. Fluids*, **7**, 269, 1964.
64. Allis, W. P., S. J. Buchsbaum, and A. Bers, *Waves in anisotropic plasmas*, MIT Press, Cambridge, Mass., 1963.
65. Ratcliffe, J. A., *Magneto-ionic theory*, Cambridge University Press, London, 1959.
66. Lehane, J. and P. C. Thonemann, *Proc. Phys. Soc. (Lond.)*, **85**, 301, 1965.
67. Bowers, R., *Scientific American*, **209**, No. 5, 1963.
68. Libchaber, A. and R. Veilex, *Phys. Rev.*, **127**, 774, 1962.
69. Harding, G. N. and P. C. Thonemann, *Proc. Phys. Soc. (Lond.)*, **85**, 317, 1965.
70. Little, P. F., *Proceedings 5th International Conference on ionization phenomena in gases*, North-Holland, Amsterdam, 1962.
71. Hatta, Y. and N. Sato, *ibid.*, 478.
72. Gould, R. W. and B. D. Fried, *Phys. Fluids*, **4**, 139, 1961.
73. Wong, A. Y., N. D'Angelo, and R. W. Motley, *Phys. Rev. Lett.*, **9**, 415, 1962.
74. Wong, A. Y., R. W. Motley, and N. D'Angelo, *Phys. Rev.*, **133**, A436, 1964.
75. Malmberg, J. H. and C. B. Wharton, *Phys. Rev. Lett.*, **13**, 184, 1964.
76. Malmberg, J. H. and C. B. Wharton, *Phys. Rev. Lett.*, **17**, 175, 1966.
77. Stringer, T. E., *Plasma Physics (J. Nucl. Energy*, Part C), **5**, 89, 1963.
78. Feynman, R. P., R. B. Leighton, and M. Sands, *The Feynman lectures on physics*, vol. II, Addison-Wesley, Reading, Mass., 1963.
79. Copson, E. T., *Theory of functions of a complex variable*, Oxford University Press, London, 1935.
80. Watson, G. N., *A treatise on the theory of Bessel functions*, Cambridge University Press, London, 1944.
81. Lichtenberg, A. J., S. Sesnic, and A. W. Trivelpiece, *Phys. Rev. Lett.*, **13**, 387, 1964.
82. Shklovsky, I. S., *Cosmic radio waves*, Harvard University Press, Cambridge, Mass., 1960.
83. Griem, H. R., *Plasma spectroscopy*, McGraw-Hill, New York, 1964.
84. Ben-Yosef, N. and A. S. Kaufman, *Electronics Letters*, **2**, 175, 1966.
85. Boyd, T. J. M., *Phys. Fluids*, **7**, 59, 1964.
86. De Silva, A. W., D. E. Evans, and M. J. Forrest, *Nature*, **203**, 1321, 1964.
87. Evans, D. E., M. J. Forrest, and J. Katzenstein, *Nature*, **211**, 23, 1966; *ibid.*, **212**, 21, 1966.
88. Bekefi, G., *Radiation processes in plasmas*, John Wiley, New York, 1966.
89. Bogolyubov, N. N., 'Problems of a dynamical theory in statistical physics', in *Studies in statistical mechanics*, vol. 1, J. de Boer and G. E. Uhlenbeck (eds), North-Holland, Amsterdam, 1962.
90. Born, M. and H. S. Green, *A general kinetic theory of liquids*, Cambridge University Press, London, 1949.
91. Green, H. S. *The molecular theory of fluids*, John Wiley (Interscience), New York, 1952.
92. Kirkwood, J. G., *J. Chem. Phys.*, **14**, 180, 1946; *ibid.*, **15**, 72, 1947.
93. Yvon, J., *La théorie des fluides et l'équation d'état: actualités scientifiques et industrielles*, Hermann, Paris, 1935.
94. Chandrasekhar, S., *Principles of stellar dynamics*, University of Chicago Press, Chicago, 1942.
95. Landau, L. D., *J. Phys. (U.S.S.R.)*, **10**, 25, 1946.
96. Jackson, J. D., *J. Nucl. Energy* Part C, **1**, 171, 1960.

97. Penrose, O., *Phys. Fluids*, **3**, 258, 1960.
98. Grad, H., 'Principles of kinetic theory of gases' in *Handbuch der Physik*, vol. 12, Springer-Verlag, Heidelberg, 1958.
99. Goldstein, H., *Classical mechanics*, Addison-Wesley, Reading, Mass., 1953.
100. Gasiorowicz, S., M. Neuman, and R. J. Riddell, *Phys. Rev.*, **101**, 922, 1956.
101. Rosenbluth, M., W. M. Macdonald, and D. L. Judd, *Phys. Rev.*, **107**, 1, 1957.
102. Kaufman, A. N., 'Plasma transport theory' in *The theory of neutral and ionized gases*, C. De Witt and J. F. Detœuf (eds.), John Wiley, New York, 1960.
103. Robinson, B. B. and I. B. Bernstein, *Ann. Phys.* (N.Y.), **18**, 110, 1962.
104. Balescu, R., *Phys. Fluids*, **3**, 52, 1960.
105. Lenard, A., *Ann. Phys.* (N.Y.), **3**, 390, 1960.
106. Guernsey, R. L., *Phys. Fluids*, **5**, 322, 1962.
107. Muskhelishvili, N. I., *Singular integral equations*, Noordhoff, Groningen, Netherlands, 1953.
108. Landau, L., *Physik. Z. Sowjetunion*, **10**, 154, 1936.
109. Eubank, H. P., *Physics today*, **19**, No. 7, 91, 1966.
110. Guthrie, A. and R. K. Wakerling (eds), *Characteristics of electrical discharges in magnetic fields*, McGraw-Hill, New York, 1949.

Index of symbols

Latin alphabet

Symbol	Description	Introduced on page
$f(\mathbf{r}, \mathbf{v}, t)$	distribution function	2
$f(1)$	$f(\mathbf{r}_1, \mathbf{v}_1, t)$	265
f^+	ion distribution function	38
f^-	electron distribution function	38
$f_i(\mathbf{r}, \mathbf{v}, t)$	distribution function for particles of type i	3
f_M	Maxwellian distribution function	46
$f_N(\mathbf{r}_1, \mathbf{v}_1, \ldots, \mathbf{r}_n, \mathbf{v}_n, t)$	generic N-particle distribution function	263
$f_s(\mathbf{r}_1, \mathbf{v}_1, \ldots, \mathbf{r}_s, \mathbf{v}_s, t)$	s-particle distribution function	264
$G(\mathbf{v}_1)$	$\int d\mathbf{v}_2 f(\mathbf{v}_2) \mid \mathbf{v}_1 - \mathbf{v}_2 \mid$	288
$G(\mathbf{r}, t; \mathbf{r}', t')$	Green function	212
$\mathbf{g}$	acceleration due to gravity	34
$\mathbf{g}$	relative velocity, $\mathbf{v}_1 - \mathbf{v}_2$	286
$\mathbf{g}'$	relative velocity after collision, $\mathbf{v}_1' - \mathbf{v}_2'$	287
$\underline{g}$	$1 - \hat{\mathbf{n}}.\boldsymbol{\beta}$	215
g_{ff}	Gaunt factor	231
$\mathbf{H}$	macroscopic magnetic field	4
H	rate of energy transfer from electrons to protons	40
H	Boltzmann's H-function	281
H	Hamiltonian	299
$H(\mathbf{v}_1)$	$\left(\dfrac{m_1 + m_2}{m_2}\right) \displaystyle\int \dfrac{d\mathbf{v}_2 f(\mathbf{v}_2)}{\mid \mathbf{v}_1 - \mathbf{v}_2 \mid}$	290
$H_l^{(1)}(x)$, $H_l^{(2)}(x)$	Hankel functions	315
$\hbar$	Planck's constant	230
$\mathbf{I}$	unit dyadic	288
I	internal energy	105
I_0	modified Bessel function	76
$I_\omega(\mathbf{s})$	radiation intensity	248
J	longitudinal invariant	24
$J_l(x)$	Bessel function	221
$\mathbf{j}$	current density	4
j_0	mean current density	96
$\mathbf{K}$	rate of momentum transfer from electrons to protons	40
K	thermal conductivity	295
$K_l(x)$	modified Bessel function	232
$\mathbf{k}$	wave propagation vector	140
$\mathbf{k}_0$	incident wave vector	240
k	axial mode number	75
k_D	λ_D^{-1}	195
k_i	imaginary part of wave number	200
k_r	real part of wave number	200
k_s	wave number of scattered radiation	240
L	$1 - \displaystyle\sum_s \dfrac{\omega_{ps}^2}{\omega^2}\left(\dfrac{\omega}{\omega - \Omega_s}\right)$	163
L	left circular polarization	168
L_N	Liouville operator for N-particle system	263
$\mathbf{M}$	magnetization	323

Symbol	Description	*Introduced on page*
$\mathbf{v}$	velocity	2
$\mathbf{v}_A$	Alfvén velocity	65
$\mathbf{v}_C$	curvature drift velocity	17
$\mathbf{v}_E$	$\mathbf{E} \times \mathbf{B}$ drift velocity	14
$\mathbf{v}_F$	constant force drift velocity	18
$\mathbf{v}_G$	grad B drift velocity	16
$\mathbf{v}_g$	guiding centre velocity	14
$\mathbf{v}_P$	drift velocity in plane wave	28
$v_{\parallel}$	velocity component parallel to magnetic field	11
$v_{\perp}$	velocity component perpendicular to magnetic field	5
v_g	group velocity	144
v_p	phase velocity	144
v_r	ray velocity	160
v_{th}	particle mean thermal velocity	7
W	particle kinetic energy	11
$W_{\parallel}$	kinetic energy parallel to magnetic field	11
$W_{\perp}$	kinetic energy perpendicular to magnetic field	11
$\mathbf{w}$	random velocity	38
X	extraordinary mode	168
Z	atomic number	230

Greek alphabet

α	$(\omega_{pe}^2 + \omega_{pi}^2)^{1/2}/\omega$	168
α	scattering parameter	243
α_ω	coefficient of absorption	249
β	$\mathbf{v}/c$	27
β	ratio of fluid to magnetic pressure	67
β	$\mid \Omega_i \Omega_e \mid^{1/2}/\omega$	168
β_+	Ω_i/ω	171
β_-	$\mid \Omega_e \mid/\omega$	171
$\beta_{\parallel}$	$v_{\parallel}/c$	221
$\beta_{\perp}$	$v_{\perp}/c$	221
$\mathbf{\Gamma}$	particle flux	295
$\mathbf{\Gamma}^+$	ion flux	296
$\mathbf{\Gamma}^-$	electron flux	296
$\Gamma(n)$	Gamma function	315
Γ_i	$(-4\pi e_i^4/m_i^2) \ln [\sin (\tfrac{1}{2}\theta_{\min})]$	287
γ	ratio of specific heats	34
γ	$(1 - \beta^2)^{-1/2}$	227
γ	Landau damping decrement	273
δ	shock thickness	125
δ	damping distance	201
$\delta(\mathbf{r})$	Dirac delta function	7
δ_{ij}	Kronecker delta	65
$\boldsymbol{\epsilon}$	dyadic form of the dielectric tensor	162
ε	dielectric constant	141
ε	azimuthal angle	277
ε_{ij}	dielectric tensor	162

Symbol	Description	*Introduced on page*
ε_{ijk}	permutation tensor	312
η	magnetic viscosity	58
η	magnetic diffusivity	59
η	coefficient of viscosity	93
η_ω	emission coefficient	249
θ	scattering angle	4
θ	pitch angle	20
$\theta(t)$	step function	302
θ_{min}	$2b_0/\lambda_D$	285
κ	Boltzmann's constant	7
Λ	$4\pi n\lambda_D^3$	8
λ	wavelength	144
λ_c	mean free path	5
λ_D	Debye length	7
μ	magnetic moment	19
μ	magnetic permeability	141
μ	mobility	294
ν	kinematic viscosity	59
ν_c	collision frequency	5
$\xi(\mathbf{r}_0, t)$	displacement of fluid element	73
ρ	mass density	34
ρ^*	$\rho - \lambda^2/4\pi$	103
σ	electrical conductivity	35
$\sigma(\theta, \mid \mathbf{v}_1 - \mathbf{v}_2 \mid)$	differential scattering cross-section	4
σ_{jk}	electrical conductivity tensor	161
σ_T	Thomson scattering cross-section	240
τ	$t - x/c$	27
τ	optical depth	250
τ_c	collision time	5
τ_D	deflection time	291
τ_D^{++}	deflection time for ions scattered by ions	307
τ_D^{+-}	deflection time for ions scattered by electrons	307
τ_D^{-+}	deflection time for electrons scattered by ions	307
τ_D^{--}	deflection time for electrons scattered by electrons	307
τ_E	energy exchange time	292
τ_E^{++}	energy exchange between ions	292
τ_E^{+-}, τ_E^{-+}	energy exchange between ions and electrons	292
τ_E^{--}	energy exchange between electrons	292
τ_s	slowing down time	290
Φ	magnetic flux	60
$\Phi(x)$	error function	290
ϕ	electrical potential	7
ϕ_{ij}	inter-particle potential	265
$\psi(x)$	$[\Phi(x) - x\Phi'(x)]/2x^2$	290
Ω	solid angle	4
Ω	Larmor frequency	5
Ω_0	non-relativistic limit of cyclotron frequency	221
Ω_e	electron Larmor frequency	9
Ω_i	ion Larmor frequency	9

Index